DEEPWATER FOUNDATIONS AND PIPELINE GEOMECHANICS

WILLIAM O. McCARRON

EDITOR-IN-CHIEF

J.ROSS PUBLISHING

Copyright © 2011 by J. Ross Publishing

ISBN-13: 978-1-60427-009-9

Printed and bound in the U.S.A. Printed on acid-free paper.

10 9 8 7 6 5 4 3 2 1

Library of Congress Cataloging-in-Publication Data

Deepwater foundations and pipeline geomechanics / edited by William O. McCarron.
   p. cm.
  Includes bibliographical references and index.
  ISBN 978-1-60427-009-9 (hardcover : alk. paper)
  1. Underwater pipelines. 2. Offshore structures—Foundations. 3. Marine geotechnique.
I. McCarron, William O., 1955–
  TC1800.D44 2011
  621.8'672—dc23

2011015501

Cover art courtesy of FMC Technologies, Inc.

Phone: (954) 727-9333
Fax: (561) 892-0700
Web: www.jrosspub.com

To Ben Murphy, for offering a few kind words

Contents

Preface

Deepwater oil and gas production has experienced phenomenal growth since the mid-1990s. Today, oil and gas production from Gulf of Mexico water depths greater than 1000 meters contributes more than 80% and 45%, respectively, of the total Gulf of Mexico production. In contrast, the associated platforms account for less than 1% of the total number in the Gulf of Mexico. To support production, there are more than 33,000 miles of pipelines on the seafloor in the Gulf of Mexico. The economic investments in new deepwater production and transportation facilities are enormous. Engineering and construction schedules for the largest projects can reach five years, longer if new transportation infrastructure is required. This handbook primarily considers deepwater geotechnical issues, but we also include other frontier-area problems as they relate to Arctic coastal processes affecting pipeline design.

The definition of *deepwater* is not precise, having evolved as design, fabrication, and installation capabilities improve. There are water depth intervals in which both floating and conventional (fixed to seafloor) platforms coexist. Platforms designed to be *compliant* to the wave, wind, and current environmental loads can reasonably be described as including deepwater technology.

The move from shallow water began in 1979 when Shell Oil installed the first fixed platform (Cognac platform, winner of the 1980 American Society of Civil Engineers Outstanding Civil Engineering Achievement Award) off the continental shelf in about 312 meters of water. Prior to this time, Gulf of Mexico production originated from thousands of conventional platforms in water depths ranging from 10 to 180 meters. But the economic and design challenges of deeper water required that the industry take a more innovative approach and develop new types of structures compliant to the environmental loading. These include the tension leg platform (TLP), the compliant tower (guyed tower), and the SPAR (large diameter deep draft floating caisson).

In 1984, Exxon installed the first compliant tower (Lena Tower) in the Mississippi Canyon in about 305 meters of water. The Lena Tower was designed to move in response to wave forces rather than resist them rigidly. Conoco (U.K.) installed the first TLP in the North Sea in 1984 for their Hutton field in about 148 meters of water. Conoco also installed the first TLP in the Gulf of Mexico in 1989 for their Jolliet development in 539 meters of water.

Oil and gas development in the Gulf of Mexico dramatically expanded into much deeper water in 1994 when Shell Oil and their partners installed the Auger tension leg platform in 872 meters of water. The Auger platform received the Outstanding Civil Engineering Achievement Award from the American Society of Civil Engineers in 1995. Over the next seven years, Shell and partners also installed the Mars in 1996 (896 meter water depth), the Ram/Powell in 1997 (980 meter water depth), the Ursa in 1999 (1159 meter water depth) and the Brutus in 2001 (1000 meter water depth), all TLPs in the Gulf of Mexico. In 2004, Shell and BP received the Offshore Technology Conference Distinguished Achievement Award for the Na Kika Gulf of Mexico project installed in 1920 meters of water.

The late 1990s also witnessed the installation of several SPAR floating systems and mini TLPs, as well as remote subsea developments tied back to surface facilities. The first SPAR platform in the Gulf of Mexico was installed by Chevron in September 1998 for their Genesis development. It was the world's first drilling/production SPAR. This huge platform consisted of a 37 m diameter by 215 m long cylinder supporting a typical fixed rig platform. The upper portion of the cylinder included an air-filled compartment that provided the buoyancy while the seawater-filled lower tanks provided weight and stability. The mooring lines were attached to the bottom of the cylinder and tethered the structure to the foundation piles on the seafloor.

Today, there are more than 45 platforms gathering oil and gas in Gulf of Mexico water depths greater than 300 meters; roughly half of these are in water depths greater than 1000 meters. The world's deepest platform is a SPAR that Shell installed for their Perdido Development in 2438 meters of water. This Gulf of Mexico platform and supporting infrastructure was installed in September 2008 at a total cost of three billion dollars.

There are now several large pipeline systems in very deep water. The Mardi Gras system, for which BP is the operator, has 780 km of pipelines from 404.4 mm (16 inches) to 762 mm (30 inches) diameter, in water depths up to 2200 m. It has a total capacity of 1 million barrels of oil per day and 1.5 billion cubic feet of gas per day, representing a gross capital expenditure of one billion dollars.

The technical challenges associated with deepwater developments have led to significant innovations. Until the 1980s, the most common design concerns for offshore foundation and pipeline engineering were associated with extreme storm loadings, earthquakes, mudflows, fatigue, and installation activities. Engineers today face additional concerns, including: submarine slope failures, thermal buckling of pipelines, catenary riser interaction with the seafloor, vortex induced vibration of flowlines, shallow water flows encountered during drilling operations, and thermal interaction of pipelines with permafrost. This handbook describes recent advances in geophysical data acquisition and evaluation as they relate to offshore developments, and also includes foundation and pipeline design considerations. The presentation is focused on deepwater geotechnics as well as subsea and Arctic pipeline design considerations, but engineers will find much of it applicable to other situations.

About the Author

Bill McCarron has more than twenty-five years of experience in the offshore industry. He was employed by Amoco Production for fourteen years, where he served in research and project engineering roles for Arctic and deepwater developments. He was the project engineer responsible for design and delivery of the tendon, foundation, and production riser components for the Gulf of Mexico Marlin tension leg platform and participated in its installation. Since then he has been involved in the development of design criteria for several subsea projects.

Dr. McCarron received his PhD from Purdue University, where he studied constitutive modeling of soils and soil-structure interaction, and is a registered professional engineer. Since founding his own consulting firm in 2001, he has specialized in the application of numerical methods to geomechanics problems, including: foundation capacity, soil-pipe interaction for thermal induced lateral buckling of flowlines, upheaval buckling of pipelines, hydraulic fracturing, pile setup, reservoir geomechanics, borehole collapse, and sand production. In addition, Dr. McCarron has managed and interpreted the results of laboratory test programs.

Contributors

Ken Been
Golder Associates

Edward C. Clukey
Geotechnical Advisor, BP America Houston

Earl H. Doyle
Geotechnical Consultant

Robert B. Gilbert
Brunswick-Abernathy Professor, The University of Texas at Austin

Vernon Kasch
Senior Geotechnical Engineer, Geoscience Earth & Marine Services

James D. Murff
Geotechnical Consultant

Andrew C. Palmer
National University of Singapore

Mark F. Randolph
Professor of Civil Engineering, Centre for Offshore Foundation Systems, University of Western Australia, Perth

Jill A. Rivette
Geotechnical Engineer, Geoscience Earth & Marine Services

Alan G Young
Vice President, Geoscience Earth & Marine Services

Civil and Environmental Engineering Series

Wai-Fah Chen, Editor-in-Chief

Semi-rigid Connections Handbook
by Wai-Fah Chen, Norimitsu Kishi, and Masato Komuro

Elastic Beam Calculations Handbook
by Jih-Jiang Chyu

Deepwater Foundations and Pipeline Geomechanics
by William O. McCarron

Sulfur Concrete for the Construction Industry: A Sustainable Development Approach
by Abdel-Mohsen Onsy Mohamed and Maisa El Gamal

Free value-added materials available from
the Download Resource Center at www.jrosspub.com

At J. Ross Publishing we are committed to providing today's professional with practical, hands-on tools that enhance the learning experience and give readers an opportunity to apply what they have learned. That is why we offer free ancillary materials available for download on this book and all participating Web Added Value™ publications. These online resources may include interactive versions of material that appears in the book or supplemental templates, worksheets, models, plans, case studies, proposals, spreadsheets and assessment tools, among other things. Whenever you see the WAV™ symbol in any of our publications, it means bonus materials accompany the book and are available from the Web Added Value Download Resource Center at www.jrosspub.com.

Downloads for *Deepwater Foundations and Pipeline Geomechanics* include spreadsheets to calculate soil-pipe interaction responses, spreadsheets demonstrating reliability calculations, color versions of geophysical profiles, and example Abaqus analysis input files for analyzing lateral buckling of subsea flowlines.

1

Deepwater Foundations and Pipeline Geomechanics

William McCarron, PhD, PE
ASGM Engineering

1.1 Introduction

Geotechnical, foundation, and pipeline engineers have faced substantial oil and gas development challenges in the past three decades. This book, which is intended to be a reference for practicing engineers, concentrates on recent geomechanics developments in deepwater foundation design, site investigation, and pipeline engineering. Practicing engineers responsible for planning site investigations or selecting foundation alternatives will find the book particularly useful.

The spatially distributed nature of pipelines and deepwater infrastructure has resulted in greater use of geophysical tools to define and assess the associated geotechnical and geologic hazards. Chapter 2 discusses integrated geoscience investigations of offshore locations, which are important to the economic acquisition and interpretation of geotechnical data. Chapters 3 and 4 discuss foundation alternatives for floating systems and seafloor supported equipment.

The 1980s saw extensive pipeline design activity related to cold regions. The fifth chapter of this book highlights several hazards to Arctic pipelines. Today, Arctic regions, including the northern Caspian Sea and the Sea of Okhotsk adjacent to Russia's Sakhalin Island, offer numerous cold-region pipeline design challenges.

The complexity of soil response is such that design needs to be supported by experimental data, which is more readily achieved through model testing than from prototype measurements. Laboratory investigations of foundation responses in centrifuge devices are discussed in Chapter 6. To evaluate hazards from a wide variety of sources, probabilistic methods are increasingly being used to differentiate between the reliability of alternative system configurations and make quantitative decisions regarding design options. Probabilistic risk analysis is gaining wide acceptance in geotechnics. Its application to deepwater foundation evaluation is illustrated through several examples in Chapter 7.

Offshore pipeline engineers face challenging design requirements for high-pressure, high-temperature conditions resulting in significant pipeline thermal expansion and lateral or upheaval buckling. In deepwater environments, novel foundation anchor systems have been developed to facilitate the mooring of floating production and exploration platforms. Complex foundation loadings and the necessity to accurately evaluate soil-structure interaction naturally led to numerical investigations to access important responses. Chapters 8 through 11 discuss numerical modeling methodologies for foundations and soil-pipe interaction phenomena for subsea flowlines.

1.2 Integrated Geophysical Investigations

The history of integrated geosciences studies over the last 20 years clearly demonstrates their benefits throughout the life of the project. As projects have moved into deeper water, the economic benefits have become more pronounced. The high cost of deepwater offshore facilities and their distribution over large areas requires economic acquisition and interpretation of seabed and subsurface geotechnical data. The offshore industry is a leader in the use of integrated geoscience teams to acquire and interpret such information. Deepwater platforms often include tiebacks to subsea production facilities. The Na Kika platform in the Gulf of Mexico produces from six separate fields as far away as 12 miles. Figure 1.1 shows an example of a deepwater field architecture, including subsea equipment tied back to a floating production center. The production wells are drilled by a separate floating drillship. Produced fluids are collected at manifolds and transported to the production facility via intra field subsea flowlines.

Figure 1.1   Deepwater field architecture (courtesy of FMC Technologies, Inc).

Data collection in deep water is expensive and generally requires coverage of a very large seafloor area as compared to most shallow water developments. An integrated study provides an important opportunity to characterize the range of geologic/geotechnical conditions in a cost effective manner. By conducting an integrated study in a logical sequence of phases, the geophysical and geotechnical programs can be carefully planned to acquire relevant data needed to define the potential subsurface variability throughout the project area. Thus, an integrated geosciences study helps achieve a reliable understanding of shallow subsurface soil conditions by providing:

1. A realistic geohazard risk assessment
2. Reliable site selection for all facilities
3. Successful foundation design and installation
4. Safe operations of planned seafloor supported structures

The high cost to construct a deepwater development means that full use of all data is essential, so an experienced team of geosciences professionals is needed to properly interpret and integrate the different data sets. Technological evolution and improvements in geophysical and geotechnical equipment have enhanced the quality of the data and increased data acquisition productivity. A diversity of skills is required of the team members to confirm that the past, current, and future geologic processes are fully understood and predictable to allow the operator to make sound risk-based decisions. Case studies for the Mad Dog and Atlantis Developments are included in the text to demonstrate a number of geo-constraints that were considered as part of the site favorability criteria to locate the final production field architecture.

1.3 Foundation Concept Selection and Design

Temporary mobile exploration and permanent production facilities make use of a wide range of foundation concepts, including driven piles, suction caissons, mudmats, drag embedded anchors, and plate anchors. Each concept will generally have a preferred or optimum application associated with its holding capacity, design requirements, installation methods, cost, and fabrication schedule. These issues are addressed in two chapters in this text.

Deepwater developments typically include:

1. A major floating structure such as a large diameter deep-draft floating caisson (SPAR) or tension leg platform (TLP)
2. An extensive number of seafloor structures such as pipeline end terminals, pipeline end manifolds, riser bases, manifolds, and holdback piles
3. A wide variety of anchor and foundation types
4. A large suite of mooring lines, pipelines, flow lines, umbilicals, risers, and steel catenary risers

To select appropriate foundation types and locations, the seafloor and subsurface conditions must be interpreted and mapped over the entire area that will encompass the field architecture. In addition, the depth of foundation interest below the seafloor must be investigated with borings, in situ testing, and/or cores in order to analyze/design all foundation types that may be considered.

A number of international specifications and guidelines have been developed by the American Petroleum Institute and the International Standards Organization that provide extensive details covering the analytical procedures appropriate for the different design applications of these offshore foundation types. Our intent is to present an overview of deepwater foundation practice and current design methods including their limitations. We will also highlight key technical improvements in the analytical methods and important innovations influencing installation.

The philosophy behind deepwater foundation design is similar to that for other offshore foundations in that the foundation must be appropriately sized to resist the applied loading without experiencing excessive deformation. For most mooring applications, the vessel is not sensitive to the small deformations of an anchor and hence *excessive deformation* typically translates to ultimate capacity. For foundations supporting seafloor facilities, storm loading is not a factor because of the large water depths. For these foundations, installation and operating loads such as those arising from hardware and mat weights, pipeline pull-in loads, and thermal expansion loads, all of which may generate moments, usually control the design. This range of design considerations complicates the problem, as the installation and operating loads are themselves very uncertain. This text discusses the required design parameters for deepwater foundations and the corresponding data required from site investigations. We further discuss the generally accepted design procedures as well as installation methods for a variety of foundation types.

1.4 Pipeline Geomechanics

In the mid-1980s, petroleum exploration in the Chukchi Sea and Beaufort Sea offshore Alaska and Canada led to the investigation of Arctic pipeline hazards associated with ice scour and coastal processes. Offshore pipelines in ice environments are buried to mitigate the effects of these hazards, and the engineering challenge is to optimize the burial depth—deeper is safer, but also more costly. We include detailed descriptions of Arctic pipeline hazards as well as methods to evaluate them. These hazards include moving ice, which gouges the seabed; stationary ice rubble piles or ridges, which form pits on the seabed; and spring flooding over coastal ice that can cause hydrodynamic scouring of the seabed.

Petroleum production and transportation in deepwater have led to advances in pipeline design techniques. Pipelines have been designed for water depths as deep as 3000 m. Deepwater reservoir fluids are often hot compared to seawater temperatures, resulting in thermal expansion of pipelines, which may result in lateral buckling of subsea pipelines. The design of systems experiencing this phenomenon requires investigation of soil-pipe interaction during the pipe movements. These studies were particularly intensive in the years from 2003 to 2010, eventually leading to development of the guidelines for analysis and design of high-pressure high-temperature pipelines (SAFEBUCK 2008). For surface-laid flowlines, the design process is typically to manage the occurrence of thermal buckles through the introduction of buckles at specific intervals. The amplitude of the lateral buckle grows in proportion to both the thermal strain and distance between lateral buckles. The flowline design must then account for the developing bending moments. Cooperative industry projects such as HOTPIPE (Vitali et al. 1999) led to improved pipeline strength design equations, including the combined effects of internal pressure, external pressure, bending moments, and axial load. At the ends of a flowline, the equipment interfacing with wellheads, manifolds, or foundations must be designed to accommodate the loads and displacements induced by thermal expansion.

This text presents alternative methods to calibrate empirical soil-pipe interaction models used to analyze buried or unburied flowlines. In the case of flowlines positioned on the seafloor surface, over the last decade there has been a general consensus developed on the key parameters controlling the evolution of soil resistance to lateral pipe movement. The lateral soil pipe resistance is dependent on its initial embedment, the soil strength at the pipe invert, and the flowline submerged weight. The models are simple in nature, allowing them to be calibrated against limit solutions, test data, or responses observed in detailed numerical simulations of soil-pipe interaction.

1.5 Foundation Analysis and Centrifuge Model Testing

As for conventional onshore foundations, the design of offshore foundations may include the use of plastic limit analysis, numerical modeling, and physical models. Each of these are discussed in detail in this text, practical applications are illustrated, and results are presented.

In the 1970s, nonlinear analyses were a novel undertaking for engineers, and continuum finite elements for two- or three-dimensional applications were just being introduced in graduate level courses. Undergraduate exposure to structural analysis mainly centered on hand methods, such as moment distribution or energy-based theories, to investigate the response of relatively simple structures. Many truss structures were analyzed using graphical methods based on joint force equilibrium. But in the early 1960s, researchers at the University of Texas began developing numeric techniques based on finite difference equations to analyze beams on elastic foundations and laterally loaded piles. These methods were extended to nonlinear analyses in 1962 (Ingram 1962). A couple years later, computer programs to analyze lateral piles were available in the offshore industry (Matlock 1963; Matlock and Haliburton 1964) in stand-alone versions not directly including interaction with the supported structure response. Indeed, in the early 1980s, the interaction between nonlinear pile response and offshore structural analysis was only approximately included through a semi-automated iterative solution. This was more a reflection of lack of demand than capability, since fully nonlinear finite element programs such as MARC and Abaqus were available in the 1970s. The technology has moved on, as more and more engineers wanted to carry out large-deformation analysis, which included substantial changes of geometry, and those analyses are now being applied to seabed indentation by catenary risers, to ploughs and other soil-moving equipment, to plate anchors, and to suction piles.

In the late 1960s and 1970s, there were strong development efforts aimed at constitutive modeling of soils, resulting in the critical state soil mechanics models and the so-called capped-plasticity models, which included a compactive yield surface. The inclusion of the compactive cap allowed realistic representation of volumetric plastic strains and the undrained response of lightly overconsolidated soils. It was not until the 1980s that widespread investigation of foundation responses including plasticity effects was within the reach of practicing engineers, often through the economics of time-sharing on large mainframe computers. Even into the 1990s, many oil companies shipped their most resource-demanding geophysical processing work, using finite difference calculations, to time-share super computers for processing. The present state of computer hardware and software is such that typical foundations may be analyzed in an economic manner with realistic constitutive models. The smallest engineering firms can purchase computers capable of solving finite element problems with millions of degrees of freedom. Today, lack of sufficient information on material mechanical properties or in situ conditions can be the primary analysis obstacle.

Over the course of the past 20 years, centrifuge testing has evolved from primarily a research tool to a tool capable of resolving critical design issues for foundations. This transition has become no more obvious than in the offshore industry where, because of the size of the foundations, it is often difficult to perform reasonably scaled 1-g model tests to verify designs. The offshore industry and centrifuge testing have evolved, with much of the attention now directed at deepwater applications. As the industry expanded into deep water, new types of foundations were required. One of the most significant changes once water depths reached about 1300 to 1400 m was the switch from driven piles to suction caissons for both exploration and production facilities. Centrifuge testing, carried out at various testing facilities throughout the world, provided critical and timely information to verify design approaches for suction caissons. In addition to suction caissons, centrifuge testing has helped establish the viability of other types of alternate anchorage systems such as suction embedded plate anchors and torpedo anchors. The technique has also been successfully used to determine the response of conductors, flowlines, and pipelines when subjected to complex loadings from fatigue to thermal expansion.

Modeling of soil, rock, and concrete, collectively referred to as geomaterials, has an extensive history. We shall present some of the significant historical accomplishments related to the modeling of geomaterials but at a relatively high level. Plasticity-based models are the most popular and available geomaterial models. While there are many functional forms describing their strength characteristics, most realistic models fall within a *critical state* or capped-plasticity family. In this text, we will confine the discussion to plasticity-based models. We demonstrate the use of geomaterial constitutive models to analyze simple and complex boundary value problems under drained, undrained, and coupled consolidation conditions. The ability of classical plasticity models implemented in finite element programs to replicate the results of limit analysis solutions for bearing capacity problems is illustrated.

1.6 Probabilistic Risk Analysis

Reliability is an important consideration in planning, designing, constructing, and operating off-shore facilities. The consequences of a failure for these facilities can be severe and the costs to increase the reliability can be large. An assessment and analysis of the reliability helps to balance the risks of failure with the costs of conservatism. Risk analysis has been used as a decision tool for offshore installations for more than 30 years. Designs of Gulf of Mexico platforms before the 1980s were based on a sparse hurricane wave heights database, and thus many platform decks were positioned too low. When operators recognized this, risk analyses were employed to priori-tize platform reinforcement, wave loading mitigation (through the removal of equipment), as well as maintenance and inspection. The risk analyses considered platform strength and redundancy, probability of hurricane impact, and economic consequences of damage.

Risk analysis has been applied to a variety of different types of foundations for structures: driven piles for steel jackets, shallow foundations for gravity-based structures, shallow founda-tions for jack-up drill rigs, tension piles for TLPs, and suction caisson anchors for floating produc-tion systems. With facilities moving into deeper water, there has been greater emphasis recently on the reliability of seafloor facilities such as pipelines and flowlines. Another common applica-tion for offshore reliability analyses is for natural hazards such as earthquakes, submarine slopes, and ice loads.

The use of probabilistic methods to evaluate the probability and consequence of foundation failure or the effects of spatial variation of material properties is still an evolving field. We can expect significant advances in the techniques as they are applied to geotechnical applications. In this text, we illustrate the use of probabilistic risk analysis for design and decision making with several foundation design examples.

1.7 References

Ingram, W. B., 1962. Solution of generalized beam-columns on nonlinear foundations, MS thesis, University of Texas.

Matlock, H., 1963. Application of numerical methods to some problems in offshore operations. In *Proceed-ings 1st Conference on Drilling and Rock Mechanics*, University of Texas.

Matlock, H. and Haliburton, T. A., 1964. BMCOL 28—A program for finite element solution of beam-columns on nonlinear supports, Report to Shell Group.

SAFEBUCK, 2008, SAFEBUCK JIP—Safe design of pipelines with lateral buckling, Design Guideline, Phase II, Report BR02050, Atkins Boreas.

Vitali, L. R., Bruschi, R., Mork, K., Levold, E., and Verley, R., 1999. HOTPIPE Project: Capacity of pipes subject to internal pressure, axial force and bending moment, International Society of Offshore and Polar Engineers (ISOPE) Conference, 2:22-33, Brest, France.

2

Deepwater Integrated Geosciences Studies

Alan G Young, PE
Vice President, Geoscience Earth & Marine Services

Vernon Kasch, PE
Senior Geotechnical Engineer, Geoscience Earth & Marine Services

2.1 Introduction

The worldwide demand for offshore hydrocarbon production and the diminishing production onshore have dramatically increased the number of offshore developments either installed or planned in water depths greater than 500 m (deepwater). The challenge for the hydrocarbon industry is that deepwater geologic settings around the world exhibit a wide range of seafloor and subsurface conditions. The geologic conditions in some offshore environments are quite uniform and present few seafloor and subsurface risks for a deepwater development while other areas are often quite irregular and exhibit extensive variability in spatial soil conditions. The Sigsbee Escarpment in the deepwater Gulf of Mexico (see Figure 2.1) exhibits such conditions.

A cursory review of Figure 2.1 indicates a number of geologic constraints need to be addressed in order for subsea facilities to be safely sited, foundations properly designed, and pipelines appropriately routed. Consequently, the design of production facilities in demanding deepwater geologic environments, such as the Sigsbee Escarpment, call for far greater data collection and analytical efforts than those typically needed for shallow water platform installations. A critical component of an integrated geosciences study is to define the scope of field investigations necessary to efficiently collect all geosciences data needed to make an accurate assessment of all risks posed by seafloor and subsurface conditions.

Data collection in deepwater is expensive and generally requires coverage of a very large seafloor area as compared to most shallow water developments. An integrated study provides an important opportunity to characterize the range of geologic/geotechnical conditions in a cost effective manner. By conducting an integrated study in a logical sequence of phases, the geophysical and geotechnical programs can be carefully planned to acquire relevant data needed to define the potential subsurface variability throughout the project area. Thus, an integrated geosciences study helps achieve a reliable understanding of shallow subsurface soil conditions by providing: (1) a realistic geohazard risk assessment, (2) reliable site selection for all facilities, (3) successful foundation design and installation, and (4) safe operations of planned seafloor supported structures.

The high cost to construct a deepwater development means that full use of all data is essential, so an experienced team of geosciences professionals is needed to properly interpret and integrate

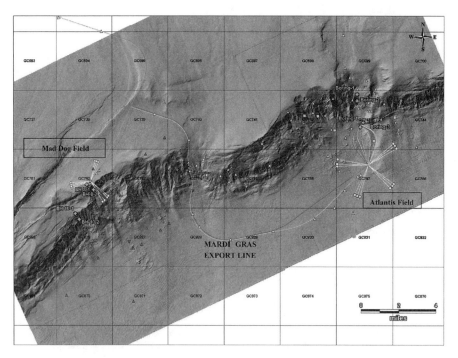

Figure 2.1   Overview of Sigsbee Escarpment near Mad Dog and Atlantis Developments © 2003 OTC (Jeanjean et al. 2003): Reproduced with permission of owner. Further reproduction prohibited without permission.

the different data sets. All professionals must be comfortable working as part of an integrated team. A diversity of skills is required of the team members to confirm that the past, current, and future geologic processes are fully understood and predictable to allow the operator to make sound risk-based decisions.

2.1.1 Chapter Overview

The main purpose of the chapter is to describe the different phases, data requirements, and work processes for conducting an integrated geosciences study. The coauthors present the results of two case studies as described by Jeanjean et al. (2005), involving the Mad Dog and Atlantis developments located on the Sigsbee Escarpment in the Green Canyon area in the deepwater Gulf of Mexico. These case studies illustrate the process for conducting an integrated study and show how the geophysical and geotechnical data are used to develop a working geologic/geotechnical model.

The coauthors discuss the evolution of the deepwater geophysical and geotechnical acquisition systems and explain the current state of practice. Technological evolution and improvements in geophysical and geotechnical equipment have enhanced the quality of the data and increased data acquisition productivity as described by Kolk and Campbell (1997) and Kolk and Wegerif (2005). The advantages of each system over another when applied to various soil types and geologic settings also are discussed.

We describe how the integrated geosciences team must be vigilant to avoid the pitfalls of allowing data overload to mask their understanding of the potential range in geologic/geotechnical conditions existing within the project area. Examples from the case studies illustrate how the geophysical data is interpreted and correlated with the geotechnical data. The impact of data variability and uncertainty is discussed in terms of the philosophy that should be taken when selecting a design soil strength profile. Finally, we describe the importance of data quality and coverage in terms of achieving the desired reliability expected by the owner and certification authorities.

The chapter further describes how the geosciences team defines the geologic risks in terms of potential impacts posed to the planned facilities. The coauthors illustrate how each site must be individually evaluated to avoid foundation problems during the installation and operating life of the facility.

The chapter discusses the individual roles of each project professional, and the critical interactions that take place during the different phases of an integrated study. The importance of using a Geographic Information System (GIS) to manage the large volume of different data sets is explained. By using this system, the different layers of data are always readily available to key professionals and are permanently archived throughout the duration of the project to serve any future needs or modifications to the field architecture.

2.1.2 Technological Improvements

Over the last thirty years, technological improvements have dramatically changed the way geophysical and geotechnical site investigations are conducted. The first offshore geotechnical borings in the late 1940s, as described by McClelland (1972), were drilled from small platforms or anchored barges, as illustrated in Figure 2.2. Geophysical surveys were conducted as part of the

Figure 2.2   Drilling geotechnical boring from an anchored barge. Figure provided courtesy of Alan G Young.

geohazard study with a small vessel using analog systems. The output of the geophysical survey was a series of paper seismic records that were interpreted by a team of geologists to develop geohazard maps. The geohazard study was submitted to the regulatory agency to obtain a drilling permit and generally was not seen by the geotechnical engineer.

The geotechnical investigation generally consisted of a single boring at a proposed platform site to satisfy the regulatory requirements for foundation design. In some cases, a soil boring was not drilled, and design was based on a boring from a nearby block (several miles away). Refusal during pile installation occasionally occurred due to unforeseen sand layers. Derrick barges frequently had jetting equipment onboard for removing the soil plug allowing the pile to be driven to the planned depth.

These two independent efforts (i.e., collection of geophysical data and geotechnical soil borings) were conducted to satisfy different requirements for the exploration and production groups within the oil company. The concept of integrating the geophysical and geotechnical data was not likely to occur and only occasionally done if potential problems occurred while placing a seafloor supported drilling rig or installing the piles for production facilities. The quality of the geotechnical and geophysical data was sometimes poor and typically, there was not a good opportunity to correlate the two data sets.

Today, high quality, digital, geophysical data is routinely acquired very efficiently in deepwater with an Autonomous Underwater Vehicle (AUV) (see Figure 2.3) as described by Bingham et al. (2002). The AUV is equipped with a large suite of geophysical equipment to acquire high-resolution data such as swath bathymetry, sub-bottom profiles, and side-scan sonar imagery. These digital data are then processed and interpreted to develop a clear 3-D picture of the seafloor and subsurface geologic conditions.

Specialized geotechnical drillships are now available, as shown in Figure 2.4, with the capability to drill soil borings to penetrations of 600 m in water depths up to 3000 m as described by Ehlers and Lobley (2007). New sampling and testing methods have evolved that dramatically improve sample quality and reliability of in situ testing data. Seafloor platforms and down-hole in situ testing tools are now available to perform a reliable cone penetrometer test (CPT) and in situ vane shear test (VST) in a variety of soil types. Continuous CPT data aids the geotechnical engineer in selecting the appropriate stratigraphic boundaries encountered at each site. The in situ

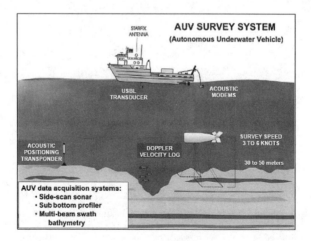

Figure 2.3   Schematic of deepwater Autonomous Vehicle (AUV) System Operations © 2005 (Kolk and Wegerif, 2005): Reprinted with permission from Dr. Susan Gourvenec, Professor at University of Western Australia, Center for Offshore Foundation Systems, Perth.

Figure 2.4   Specialized Deepwater Geotechnical Drillship © (Courtesy of Fugro website © Fugro N.V.)

data supplements the laboratory soils data for use in foundation design and also confirms that the laboratory testing was performed on high quality samples.

The chapter will provide a historical overview of the most significant technological improvements in data acquisition equipment made over the last 30 years. These improvements have been critical for obtaining high quality data essential for a deepwater integrated study. We will describe the current state of practice and some new developments in equipment design that will further enhance data acquisition (Spencer, 2008).

2.2 Objective of an Integrated Site Investigation

Typical deepwater projects involving floating production units, mooring systems, and seafloor infrastructure may cover an area of 15 to 20 sq km, as shown in Figure 2.5. Within an area, this large, spatial variation in geologic features and lateral variation in soil stratigraphy and properties may be very significant in many geologic settings. Thus, the scope of both the geophysical and geotechnical investigations needs to be carefully planned and extensive enough to fully define the foundation design conditions for each planned facility.

Over 20 years ago, Campbell et al. (1988) described the importance of conducting an integrated study for a deepwater project. The primary reason for conducting an integrated site investigation is to define the seafloor and subsurface conditions that will influence the layout of the field architecture and impact the design of seafloor pipelines and foundations supporting seafloor facilities. The principal objective is to use the data sets in a fashion where they are mutually supportive in developing the geologic model that defines potential geo-constraints and geotechnical engineering properties throughout the project area. The objective is most effectively achieved when the integrated study is conducted in a sequence of phases as described by Campbell et al. (2008).

To achieve this objective the integrated team of geologists, geophysicists, geotechnical engineers, facility designers, and other supporting geoscience professionals will closely interact

Figure 2.5 Typical seafloor infrastructure of deepwater development. Figure provided courtesy of Alan G. Young.

throughout each phase. The joint role of the geologist and geotechnical engineer begins at the start of the planning study and continues through all subsequent phases. Clear communication of all potential geologic and geotechnical constraints is essential to confirm that the final field architecture places all components of the seafloor infrastructure at optimum sites. The geologist and geotechnical engineer will work closely together in selecting the most appropriate foundation types for the soil conditions existing throughout the development.

One of the early efforts to approach an offshore site investigation as an integrated study took place after the failure of three platforms in South Pass, Block 70, due to Hurricane Camille in 1969 as described by Sterling and Strohbeck (1973). Both geophysical and geotechnical data were acquired to evaluate the thickness of the mudflow events that contributed to the failure of the three fixed platforms.

The oil industry's first truly deepwater integrated study involving a comprehensive inter-disciplinary geophysical, geologic, and geotechnical assessment was conducted in the early 1980s by Exxon for their Zinc Development (Templeton et. al. 1985). Their study produced a revolutionary new interpretation of the complex geologic region known as the Mississippi Canyon in the Gulf of Mexico. Exxon took advantage of the full benefits of an integrated geosciences study by using long term field and laboratory data acquisition programs and numerical analyses to assess the effects of seafloor creep on the installation of the first guyed tower facility in the Gulf of Mexico.

2.3 Spatial Variability and Data Uncertainty

The uncertainty of the data obtained during a site investigation is both epistemic and aleatory, and deepwater site investigations should be designed to investigate both types of uncertainty in engineering properties. Epistemic uncertainties are defined as uncertainty due to lack of knowledge that can be reduced with additional data (Christian 2003). Aleatory uncertainties are due to the random nature of the phenomenon, and more information will not reduce this uncertainty.

Geologic processes may have deposited sediments in both predictable circumstances and under chaotic conditions. Thus, interpretation of the engineering properties throughout the site is probably the most difficult task to be accomplished, bringing into focus the need for the site investigations to identity the spatial trends in the development area. Unfortunately, our lack of knowledge of the spatial distribution of specific soil properties deposited in chaotic (i.e., high energy) conditions creates the epistemic uncertainty. Random errors in testing or sample disturbance can create much of the aleatory uncertainty, although some spatial variability in identified sediment trends are often considered as aleatory uncertainty.

Dr. Ralph Peck (1962), an early pioneer in geotechnical engineering, has described the importance of fully understanding the geologic model if one hopes to properly assess soil properties used in the foundation design. Some of his quotes that best describe this importance are presented below.

"Subsurface engineering is an art; soil mechanics is an engineering science. This distinction, often expressed but seldom fully appreciated, must be understood if we are to achieve progress and proficiency in both fields of endeavor."

"Soil mechanics, in the first place, provides qualitative and quantitative data concerning the stress-strain-time characteristics of earth materials. This knowledge gives us a feeling for the behavior of soil, under idealized conditions, which may guide us in anticipating behavior under the more complex conditions in the field. Similarly, the theories of soil mechanics provide insight into behavior under simple, ideal conditions."

"Geology, the third ingredient, is as basic to subsurface engineering as is soil mechanics. Possibly its most significant role is to make us aware of the departures from reality inherent in our simplifying assumptions. Whereas the theories and computational procedures of soil mechanics would be impracticable without the simplifying assumptions regarding the subsurface materials, nature is not simple. The geology of a site must be understood before any reasonable assessment can be made of errors involved in our calculations or predictions. Indeed, in some instances the geologic structure or the results of geologic processes may completely override all considerations of soil mechanics."

"Whether we realize it or not, every interpretation of the results of a test boring and every interpolation between two borings is an exercise in geology."

Dr. Peck described the building blocks of an integrated study. Dr. Peck also emphasized that we must understand the natural processes that created the soil deposits if we are going to appreciate their inherent variability. Thus, we must approach all geotechnical engineering problems from a geological point of view. The quotes by Dr. Peck indicate that the disciplines of geology and geotechnical engineering are mutually dependent upon each other to achieve a reliable and competent job.

The coauthors want to define the importance of each discipline's role in developing a proper geologic model. As stated by one of the earliest pioneers in geotechnical engineering, geology plays an essential and significant role in the design process and should guide all data acquisition activities. Even though broad ranges of geologic processes exist in deepwater, major improvements in the quality and coverage of the data acquired have evolved since Dr. Peck emphasized the critical role of geology.

The volume and quality of subsurface data that can be acquired today in the offshore environment allows the team to develop a much clearer picture of the potential geologic and sediment variability throughout the project area. The credibility of the integrated assessment depends upon the resolution and quality of the geophysical and geotechnical data as described by Moore et al. (2007). Figure 2.6 shows the range of equipment and sampling resolution that can be acquired with current AUV systems and geotechnical equipment. The quality of the ultra high-resolution data obtained with the AUV systems, such as the multi-beam swath bathymetry, side-scan sonar, and sub-bottom profiler, provides an opportunity to understand the subsurface conditions far better than ever attained with onshore or near shore projects.

Moore et al. (2007) has compared the excellent data quality of the AUV with seabed images acquired with 3-D seismic data used for hydrocarbon exploration. Examples of seabed images obtained with the different geophysical systems are shown in Figure 2.7 illustrating the excellent quality of the AUV data. These improvements in the data provide two important benefits when used appropriately: (1) the cost of the site investigations is reduced, and (2) the entire foundation

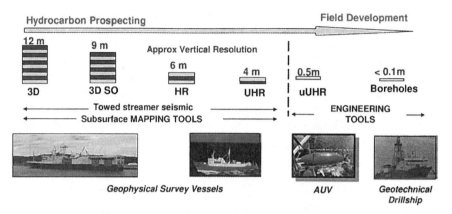

- **3D** 　 Conventional Exploration 3D data (c. 5000m penetration below seabed)
- **3D SO** 　 Reprocessed Short Offset Exploration 3D data, (c. 3000m below seabed)
- **HR** 　 High Resolution Exploration Seismic (c. 1500m below seabed)
- **UHR** 　 2D Ultra-High Resolution seismic (c. 750m below seabed)
- **AUV** 　 2D Ultra-Ultra-High Resolution seismic (c. 60m below seabed)

Figure 2.6 Data acquisition technologies and vertical resolution (Moore et. al. 2007): Reprinted with permission from Mr. Gary Nicol, Consultant, Dr. Roger Moore with Halcrow Group Ltd., Nat Usher, and Trevor Evans with BP.

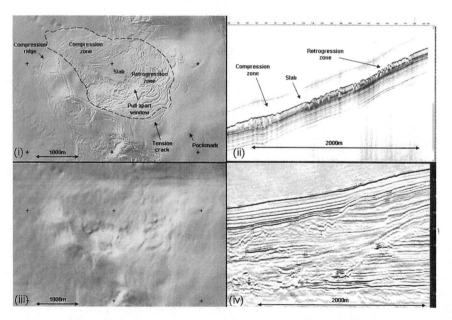

Figure 2.7 Contrasts in resolution between 3-D imagery for landslides on the West Nile Delta (Moore et. al. 2007): Reprinted with permission from Dr. Roger Moore with Halcrow Group Ltd., Nat Usher, and Trevor Evans with BP.

design process is conducted with less uncertainty and fewer risks. The appropriate equipment and the state of practice employing this equipment to achieve these benefits are described below.

2.4 Planning Study with 3-D Seismic Data

The first step of an integrated study utilizes available data to plan the field architecture for a major offshore development. The desktop study relies upon a good understanding of the seafloor conditions, shallow stratigraphy, and soil conditions, and how these conditions will impact the selection of pipeline routes and placement of seafloor-supported infrastructure. A methodology described by Doyle et al. (1996) provided a breakthrough in identifying near-surface geohazards by using conventional 3-D seismic data collected for exploration purposes. The method allows drilling hazards to be evaluated using 3-D seismic data instead of performing high-resolution deepwater surveys at this early stage in a project. The resolution of the 3-D data is frequently sufficient to evaluate seafloor topography, shallow geologic conditions, and potential constraints for evaluation of preliminary production options. Additional processing techniques, such as spectral whitening and selecting short offset volumes, may further enhance the use of the collected data. The methodology has also proven quite valuable for shallow engineering evaluations as required for a desk study.

The 3-D seismic data sets can provide valuable information regarding the seafloor and subsurface geologic conditions. The difficult task is to interpret and map all the data in a consistent manner to develop an accurate three-dimensional picture of the subsurface geologic structure, as illustrated in Figure 2.8. There are wide varieties of computer programs that can be used to interpret and construct geologic maps in 3-D. These programs can be used to construct a digital elevation model (DEM). As described by Bolstad (2002), a DEM is a grid of elevation values with corresponding (x, y) coordinates.

Groshong (2006) describes how the DEM can be contoured easily by triangulation to produce a visualization of the topographic surface. The closely gridded data can be used to produce a three-dimensional image of identifiable seismic horizons. Structural and stratigraphic relationships can be correlated and evaluated using the vertical cross-sectional profiles. Enhanced seafloor renderings, as shown in Figure 2.8, can be derived from the 3-D seismic data. The enhanced seafloor

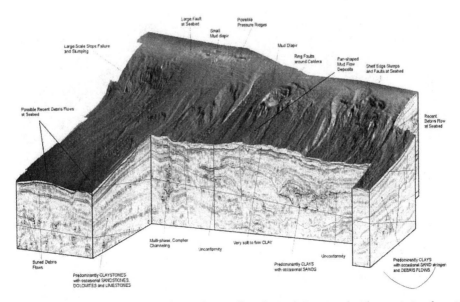

Figure 2.8 Enhanced 3-D seafloor rendering (Horsnell et al. 2009): Reprinted with permission from Mr. Mike R. Horsnell, Robert L. Little, and Kerry J. Campbell with Fugro Geoconsultants.

renderings may be used to make assessments of near-seafloor geological features and provide insight toward understanding the geologic processes existing within the area of interest. During the planning study, the 3-D seismic data may be used to identify near seafloor geologic conditions before conventional side-scan sonar or sub-bottom profiler data has been acquired.

The 3-D seismic amplitudes associated with the seafloor in the 3-D seismic data can be used to infer the physical properties of the shallow sediments and identify geologic processes as described by Roberts et al. (1996) and Brand et al. (2003). The amplitude of a seismic wavelet can be overlain onto a rendering, as illustrated in Figure 2.9. By associating amplitudes with seafloor morphology the rendering will show, for example, areas where strong soils, authigenic rock outcrops, active fluid expulsion, and potential chemosynthetic communities may exist and pose a constraint to future production operations.

The 3-D seismic data generally provides appropriate resolution to cover the needs of a desktop planning study. This data should be supplemented with any available geotechnical data, such as shallow gravity or piston cores, to gain insight relative to the soil properties of the near seafloor sediments. The available data is used to develop an initial geologic model that allows the design team to plan the preliminary layout of different prospective field architectures. As the integrated

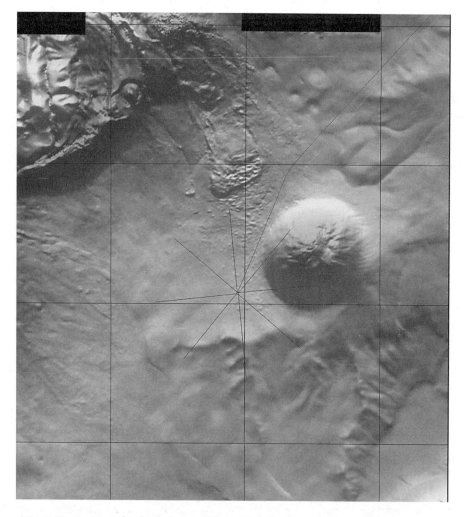

Figure 2.9   3-D seismic amplitude rendering of Auger Site. Figure provided courtesy of Mr. Earl Doyle.

study progresses and better data become available, the geologic model is refined, and then adjustments in the layout of planned facilities can be made to accommodate identified geologic/geotechnical constraints.

The best approach for conducting an integrated study is to optimize the amount of data to be acquired during geophysical and geotechnical field programs. Thus, the initial planning study plays a very important role in setting the scopes of geophysical survey and the geotechnical investigation. The following sections will provide an overview of the state of practice for conducting a shallow geophysical survey and performing a geotechnical investigation.

2.5 Geophysical Site Survey

The high-resolution geophysical site survey should be conducted first since it provides an opportunity to achieve maximum data coverage most economically over the entire project area to refine the geologic model.

2.5.1 Data Acquisition

A geophysical investigation (survey) generally employs a suite of geophysical systems that are simultaneously transported over the seafloor as a towed fish behind a vessel (see Figure 2.10) or as part of an AUV (see Figure 2.3). The high-resolution systems are more accurately classified as multisensory acoustic systems since each system has a sound generating device, sound receiver, and data recording equipment. These systems are generally low energy, high frequency devices designed to provide high-quality resolution data instead of greater seafloor penetration, as described by Campbell and Hough (1986).

The different acoustic systems provide a suite of data sets, as illustrated in Figures 2.11-2.13 that are useful for defining different geologic conditions, specifically:

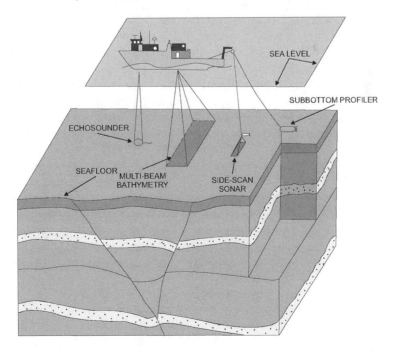

Figure 2.10 Acoustic systems used during geophysical survey. Figure provided courtesy of Mr. Alan G Young.

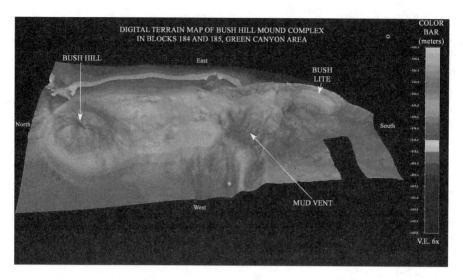

Figure 2.11   Multi-beam swath bathymetry of seafloor typography. Figure provided courtesy of Mr. Earl Doyle.

Multi-beam system provides a swath of water depth (bathymetry) data that defines the seafloor topography and slope gradients (see Figure 2.11).

Side-scan sonar defines seafloor irregularities, outcrops, gas expulsion features, man-made features, and a general texture of the seafloor materials (see Figure 2.12).

Sub-bottom profiler penetrates the seafloor to provide seismic cross-sections showing subsurface conditions useful for mapping shallow soil stratigraphy and enhancing the structural interpretation of the near seafloor geologic conditions (see Figure 2.13).

This high-resolution data, when combined with data for the shallow geotechnical cores, provides a wealth of knowledge relative to the geologic conditions of the area including stratigraphy, existence of unconformities, lithology, structural deformation, sedimentary processes, faulting, boundaries of gas charged sediments, and geochemistry. This data also has broad application to the engineering design studies by providing information relative to the spatial variation in foundation conditions.

The acoustic-systems all work on the same geophysical principle that the transmitted energy is reflected at different sediment interfaces. The contrast in acoustic properties at the interface depends upon the impedance of the two materials on each side. Impedance is a function of the density, strength, and elastic properties of each material. The very high-frequency systems, such as the depth recorder or (multi-beam swath bathymetry system) and the side-scan sonar, reflect their energy from the first acoustic interface, typically interpreted as the seafloor.

The lower energy systems, such as the sub-bottom profiler, penetrate the seafloor with a portion of the transmitted energy being reflected at each interface. At each interface, the amount of energy that continues to travel down is reduced by the amount that is reflected. The limits of depth penetration depend upon the number and reflectivity of each interface penetrated.

A number of different sub-bottom profiling systems have evolved, as shown in Table 2.1, that have various operating features. The different operating features of each system are shown in Table 2.1 in terms of frequency, resolution, and final penetration depth.

Each profiler system has key attributes, operating features for various terrain types, limitations, and safety considerations as listed in Table 2.2.

The geologists and geophysicists need to understand these operating features, attributes, and limitations in order to select the best equipment for different sediment types and geologic

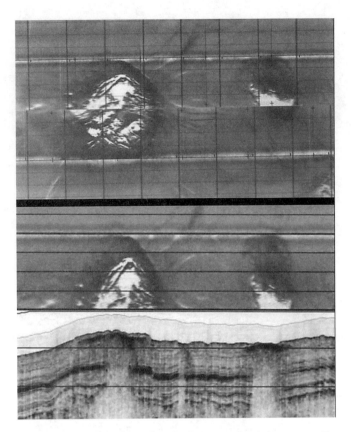

Figure 2.12   EDO deep-tow side-scan imagery of the seafloor with sub-bottom profiler section. Figure provided courtesy of Mr. Earl Doyle.

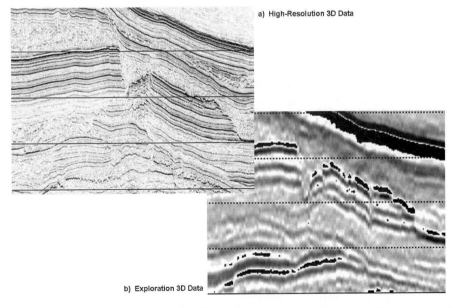

Figure 2.13   High-resolution sub-bottom profiler seismic cross-section. Figure provided courtesy of Mr. Earl Doyle.

Table 2.1 Operating features of various sub-bottom profiler systems (ConocoPhillips 2009)

System	Frequency	Source	Receiver	Resolution	Penetration	Configuration
Chirp	0.5-15 kHz swept	Transducer	Transducer	5-10 cm	10 m	Hull mounted array or tow body
Pinger	2-14 kHz tuned	Transducer	Transducer	15 cm	10-25 m	Hull mounted array or tow body
Boomer	750 Hz-5 kHz	Electromechanical plate	Integral or towed hydrophone	25 cm	50 m	Surface towed and various depth tow bodies
Sparker	50 Hz-4 kHz	Electrical discharge	Towed or (rarely) integral hydrophone	2-5 m	100-200 m	Surface towed frame or tow body

Table 2.2 Key attributes, suitable terrain types, limitations, and safety considerations of sub-bottom profiler systems (ConocoPhillips 2009)

System	Key attributes	Suitable type of terrain and uses	Limitations	Safety considerations
Chirp	Highest frequency of any profiler, very high resolution Can be used in extremely shallow water	Profiling in soft clays and silts, deltas Narrow beam Frail tow body	Low output power Very limited in hard or coarse seabed	Hull mounted array or light easily deployed tow body Constant voltage supply required
Pinger	Good resolution Robust tow body	Profiling in soft clays and silts Multiple transducers produce narrow beam Depth of pipeline burial surveys; and very deep water	Low output power Limited use in coarse seabed	Readily hull mounted Heavy tow body Constant voltage supply required
Boomer	Good combination of resolution and penetration Several configurations	All continental shelf areas	Surface tow models very limited by sea state	Surface tow and various depth tow bodies—all heavy High voltages required
Sparker	Simple source construction	Deep penetration in marine areas Works reasonably in hard/coarse materials	Poor resolution Generally crude pulse signature	Heavy surface towed frame or tow body Very high voltages involved

conditions. Proper selection of the appropriate equipment is one of the key decisions required to acquire high quality data. ISSMGE (2005) and Oil & Gas Producers (OGP) (2011) provide guidance for shallow geophysical investigations for offshore and near-shore developments that describes many of the important operational considerations also applicable in deepwater.

The following section will describe the technological improvements in various geophysical systems that evolved to fulfill the needs of deepwater developments. The section will also present illustrations showing the graphic data and examples of the features that can be interpreted and mapped as part of defining the 3-D geologic/geotechnical model, the backbone of the integrated study.

2.5.2 Deepwater Geophysical Tools

The suite of geophysical tools used in shallow water is generally deployed as a group of tow fishes pulled behind the vessel as it transits a planned grid of survey lines. The move into deepwater in the 1980s led to the development of a new suite of equipment that could obtain the type of high quality geophysical data required for an integrated geosciences study.

Two deep-tow systems were initially developed to serve the needs of the offshore oil and gas industry. Shell Oil worked with Edo-Western to develop a system that was operated by C&C Technologies. The AMS 120 was developed and operated by Williamson and Associates. The paper by Prior and Doyle (1984) and web site of Williamson and Associates, Inc. (2010) describe many of the operational details of these systems.

The early deepwater surveys were conducted in two phases using a suite of different geophysical equipment as described by Prior et al. (1988). The first phase survey was conducted with surface-tow equipment such as the narrow-beam echo-sounder (Edo-Western 4077) for bathymetric profiles, a sparker for subsurface stratigraphy to 400 to 500 m, and a mini-sparker for subsurface stratigraphy to a depth of 100 to 200 m. The second survey required a special deep-tow fish that included a side-scan sonar (100 kHz) and a sub-bottom profiler (3.5 to 7.0 kHz) that were part of the Edo-Western 4075 system.

The operations of the deep-tow fish, as illustrated in Figure 2.14, was designed to resist large hydrostatic pressures (up to 2350 m) and remain about 30 m above the seafloor in order to acquire high quality data. The extreme water depths required a cable (umbilical) with lengths up to 10,000 m and a drag chain to keep the positively buoyant tow fish near the seafloor. A large shipboard winch was mounted on the vessel deck to accommodate the long tow cable. A large handling system was required to place and remove the deep-tow fish from the water. Survey positioning was achieved with a ship mounted GPS and a Sonardyne system requiring a grid of seafloor transponders.

Survey speeds with the deep-tow systems were in the 2.0 to 2.5 knot range and extensive time (four to six hours) was consumed in making the turn at the end of each track line. The inefficiencies of performing deepwater geophysical surveys with deep-tow systems prompted BP to pursue an improved system. BP contracted with C&C Technologies in 2000 to develop an AUV called HUGIN 3000 in conjunction with Kongsberg Simrad, and initial project work began in early 2001.

The AUV operations with the HUGIN 3000 requires only one support vessel. The vessel has complete support equipment including launch-and-recovery hardware (see Figure 2.15), navigation and positioning, data collection and processing, and system maintenance. The HUGIN 3000 includes a suite of geophysical tools: (1) Simrad EM2000 Multi-beam Bathymetry and Imagery, (2) Edgetech Chirp Side-Scan Sonar, (3) Edgetech Chirp Sub-bottom profiler, and (4) an optional Magnetometer.

The AUV has become the workhorse of deepwater geophysical surveys due to its many commercial benefits. Today, there are at least three companies (C&C Technologies, DOF Subsea, and

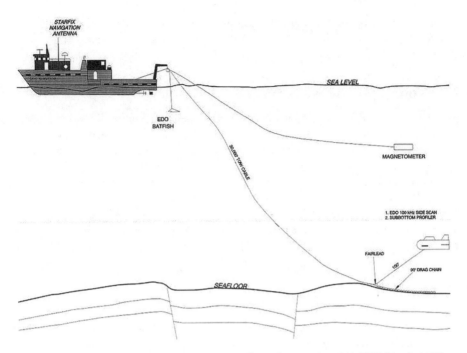

Figure 2.14 Deep-tow system used for deepwater geophysical surveys © 1998 OTC (Doyle 1998): Reproduced with permission of owner. Further reproduction prohibited without permission.

Figure 2.15 HUGIN 3000 launch and recovery system. Figure provided courtesy of Mr. Charlie Spann with C&C Technologies.

Fugro) capable of performing a deepwater geophysical survey with an AUV. The following section will describe the type of data that is acquired with each seismic tool commonly used during a deepwater survey.

2.5.3 Deepwater Geophysical Survey

Acoustics is the preferred way to efficiently and accurately investigate the water column and seafloor conditions, according to Blondel (2009). A number of different instruments have evolved that rely upon acoustic methods as previously illustrated in Figure 2.10. Acoustic waves generated with these instruments are capable of traveling long distances without attenuating in deepwater. The acoustic methods are generally divided into three broad categories of sonar mapping systems: (1) single-beam echo-sounder, (2) multi-beam echo-sounder, and (3) side-scan sonar. The acoustic echoes generated by these systems provide detailed information about the distance traveled (water depth) and how the beam was reflected (seabed type or obstacles). The data from each type of sonar (originally an acronym for SOund Navigation And Ranging) provides details on the water depth (bathymetry) and imagery (seafloor types).

2.5.3.1 Single-Beam Echo-Sounder
Water depth is usually computed with an echo-sounder by measuring the two-way travel time for an acoustic pulse to travel from the transducer on the ship to the seafloor and return. The echo-sounder is a single beam, low-frequency (20 kHz) system that collects the data directly beneath the vessel in a single line along the vessel route.

2.5.3.2 Multi-beam Echo-sounder
The move into deepwater established the need for swath bathymetry. The multi-beam echo-sounder that included a series of transducers became accessible in the 1980s. The transducers were arranged at different angles to produce an array of acoustic beams on either side of the vessel track. The transducer array was designed to obtain a swath of water depth measurements along the vessel path that are repeated at very close time intervals (1 to 2 seconds). The interferometric unit measures the pattern of wave interference by superposition and corrects for vessel or tow-fish motions, such as heave, pitch, roll, yaw, etc., to produce a mosaic of adjusted water depth measurements that reflect the true seafloor typography, as previously illustrated in Figure 2.11.

Some swath bathymetry systems are permanently mounted to the hull of the vessel, pole mounted, or towed as a fish behind a vessel. To obtain more accurate measurements of water depth in deepwater, the swath system is integrated into an AUV and deployed closer to the seafloor to increase the density of the data. The swath bathymetry system can also be designed to record the reflected energy (backscatter) to produce an excellent seafloor model integrated with the swath bathymetry. The combination of data provides details about the seafloor topography and texture that help define past and current seafloor processes, as illustrated in Figure 2.16.

2.5.3.3 Side-Scan Sonar
The side-scan sonar should be flown close to the seafloor to obtain high quality two-dimensional imagery of the seafloor, as shown in Figure 2.12, that is closely analogous to low-oblique aerial photos. In earlier years, the side-scan sonar was deployed as a deep-tow fish, but problems with short base line surveying necessitated the use of a second vessel (chase boat). Positioning the deep-tow fish at the proper height above the seafloor was also problematic to obtain high quality data. Both these obstacles prompted the mounting of side-scan sonar on an AUV. Deployment with the AUV provides a number of benefits, including the flexibility of the field operations and the efficiency of performing the survey.

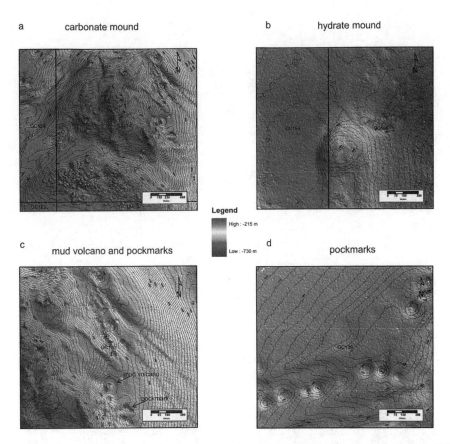

Figure 2.16 Backscatter data draped over seafloor bathymetry © 2010 OTC (Zang and McConnell 2010): Reproduced with permission of owner. Further reproduction prohibited without permission.

The side-scan sonar image is generated as two arrays of transducers emitting seismic pulses in a range from 100 to 500 kHz down to the seafloor that are then reflected back to the receivers, as illustrated in Figure 2.17. The fan-shaped beams are emitted perpendicularly along each side of the vessel transit to provide swath imagery on both sides. The intensity and distribution of the reflectivity of the seismic pulse from the seafloor provides an indication of the composition and texture of the seafloor. The side-scan sonar image also shows objects resting on the seafloor and details of topographic irregularities.

Some of the newer side-scan sonar systems are capable of draping the bathymetry over the imagery to produce high-resolution 3-D views of the seafloor, as shown in Figure 2.18, that are similar in quality to satellite images.

2.5.3.4 Sub-Bottom Profiler

The sub-bottom profiler is another important geophysical tool that has the capability of providing high quality shallow profiles of the sediments underlying the seafloor. This seismic reflection system is high resolution and achieves typical seafloor penetrations of 50 to 80 m. Typical resolution is about 0.1 to 0.3 m depending upon the frequency of the system and the composition of the sediments. Seismic profiling has been used to study deep geologic structures for many years. The sub-bottom profiles provide continuous images along the path of the vessel transit showing subsurface conditions as previously illustrated in Figure 2.13. The sub-bottom profiles clearly display

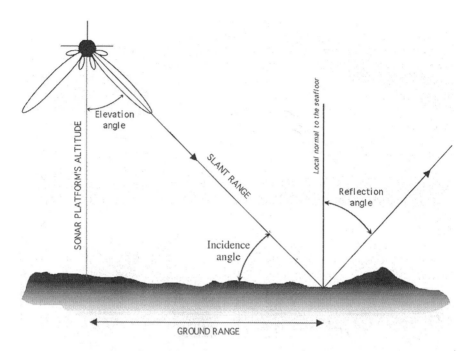

Figure 2.17   Illustration showing side-scan system operation (Blondel, P. 2009): Reprinted with permission from Dr. Philippe Blondel, head of Remote Sensing Laboratory at University of Bath, UK.

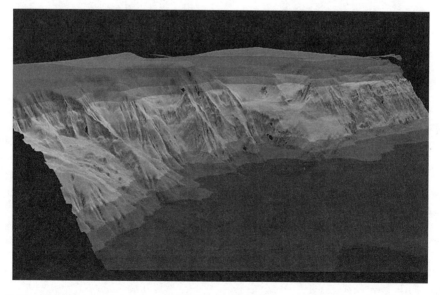

Figure 2.18   Side-scan sonar image draped with bathymetry data. Figure provided courtesy of Mr. Earl Doyle.

the sediment stratigraphy, geologic structure, and various geologic features and processes, such as faults, gas charged sediments, unconformities, etc.

Developments over the last two decades have advanced the science from recognition of stratigraphic horizons to the ability to recognize the shape of stratigraphic sequences, to interpret their depositional history, and to distinguish unconformities and reconstruct the transgressionalregressional history of an area (Prothero and Schwab, 1996). The sub-bottom data is displayed graphically on a vertical scale showing the time taken for the transmitted energy to travel from the source to the interface and back to the receiver. The velocity of sound in the sediment can be used to compute the depth of the interface. However, the velocity depends upon the acoustic properties that typically change with depth as the sediment density increases. An accurate measurement of each interface depth requires borehole velocity data to reflect the change in velocity with depth. The calibration of the velocity through the sediment profile is critical and generally requires a reliable measurement of the sediment interfaces by logging the hole after completing the soil boring. The logging data can then be used to develop a seismic velocity profile that can be correlated with the depths of the seismic horizons interpreted from the seismic cross-section.

Seismic profiles may appear relatively easy to read, but their interpretation is complicated by the fact that the vertical scale is two-way travel time, not actual layer thickness. Lithology and seismic velocity variations also confound the issue. Additionally, reflector horizons may result from abrupt density changes rather than stratigraphic boundaries. Likewise, boundaries may appear weak if there is little density contrast between layers. Consequently, experienced geologists must exercise care to distinguish subsurface depositional sequences, unconformities, and approximate depths (Prothero and Schwab, 1996).

2.6 Geotechnical Site Investigation

The three-dimensional geologic model developed during the initial planning study defines the geologic conditions, subsurface stratigraphy, and spatial soil variability throughout the project area. This information is very valuable in terms of understanding the subsurface conditions and selecting the sampling and in situ testing locations and procedures most appropriate for the interpreted sediment types. The geologic model also allows one to select sites where the most uniform sediment stratigraphy exists; thus, potential uncertainty in soil properties can be minimized within the foundation footprint of the facility and apparent anomalies can be avoided. All this information plays a critical role in establishing the final scope of the geotechnical investigation.

2.6.1 Deepwater Data Acquisition

There is a wide range of geotechnical equipment available to conduct the deepwater geotechnical program. The most appropriate sampling and in situ testing methods depend upon a number of site and project specific factors. Water depth, sediment type, potential foundation depth, and soil properties all influence selection of the most appropriate tools for the conditions expected at each facility site.

Most equipment available for conducting the geotechnical investigation is divided into two broad categories: (1) seabed systems that acquire samples or in situ data to depths of about 30 to 35 m and (2) down-hole systems that are deployed from a drilling vessel through the bore of the drill-pipe used to advance the soil boring. In recent years, new seabed systems have evolved that perform both seabed and down-hole sampling and in situ testing with a robotic drilling rig placed on the seafloor. The innovative systems have the capability to operate to depths of 100 m below the seafloor.

Typical down-hole systems deployed from a deepwater drillship have the capability to sample or to perform in situ tests to depths of 600 m below the seafloor. The seabed systems have some advantages for shallow site investigations since they can be deployed from a number of different types of vessels that are less expensive than the specialized geotechnical drillships and are more readily available.

The seabed geotechnical equipment generally includes: (1) Jumbo Piston Cores or JPC, (2) Piston Cores or PC, (3) Box Cores or BC, (4) seabed Cone Penetrometer Test (CPT) system, and (5) seabed in situ Vane Shear Test (VST) system. Continuous sampling and in situ test data can be acquired with these systems from the seafloor to depths up to 35 m, except for the seabed in situ VST System that is limited to depths of about 10 m.

The down-hole geotechnical equipment is designed to operate down the bore of the drill-pipe to obtain a sample or perform an in situ test in the undisturbed soil below the base of the open-centered drill bit, as shown in Figure 2.19. The specialized down-hole tool is designed as a self-contained system that free-falls down the drill-pipe and latches into the drill bit. Upon

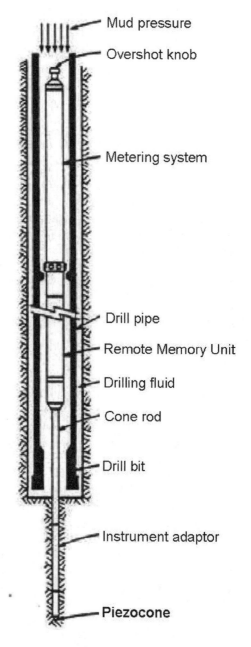

Mud pressure

Overshot knob

Metering system

Drill pipe

Remote Memory Unit

Drilling fluid

Cone rod

Drill bit

Instrument adaptor

Piezocone

Figure 2.19 Down-hole Dolphin CPT System. Figure provided courtesy of Mr. Alan G Young.

completion of the in situ test or sample acquisition, the tool is retrieved with a wire-line overshot connected to a high-speed winch (Spencer et. al., 2011).

The down-hole equipment has a disadvantage in that each down-hole operation requires the retrieval of the equipment back to the deck of the drillship. In deepwater, each sample or in situ test may take one to two hours, and a deepwater boring may require three to five days of drillship time to complete a single boring. Ehlers and Lobley (2007) state that the day rate of a deepwater geotechnical drillship was about $150,000 per day. Therefore, deepwater borings are very costly and typically kept to the minimum possible. To optimize the cost of a single boring, multiple runs with different tools are performed within the same borehole to maximize data collection and correlation of the different data sets.

The following sections will describe in more detail each of the seabed and down-hole sampling and in situ testing methods. Examples of data acquired with the different methods will be presented to illustrate how the data can be used to select the appropriate design soil parameters for deepwater foundations.

2.7 Performing the Geotechnical Site Investigation

Once the geophysical data is acquired and the geologic model refined, the final scope of the geotechnical investigation can be defined and the planning process started to select the preferred vessel and equipment. After the appropriate vessel and equipment have been selected, secured, and mobilized to the site, the site investigation can begin. A detailed field operations plan is critical to make sure that the program satisfies all technical requirements. The plan should include a description and requirements of sampling and in situ testing activities for each location. The plan should also identify which laboratory tests might be performed and a schedule when all phases of testing can be completed.

In addition, we recommend taking all or the most important portions of the geophysical data available for each proposed geotechnical site to provide subsurface information that can support the drilling operations. This information can be used to alert the driller when to expect a different soil stratum or soil type, so he can anticipate the potential changes in his drilling mud program. In addition, the geophysical data can be correlated with the initial geotechnical data to confirm that a sufficient number of undisturbed samples are being obtained within each soil layer to accurately characterize their physical and engineering properties.

The swath bathymetry data should be available to confirm that the seabed is flat enough for the seabed testing platform or seabed drilling and in situ testing reaction mass to be properly placed for the field operations. The water depth at the proposed boring site is also critical information for the vessel and drilling operations. Reliable determination of each sample depth or in situ test depth requires an accurate water depth measurement at the start. Since the length of drill pipe is probably the most direct measurement, it can be used as a yardstick to confirm the water depth measurement if currents are not excessive. Bottom sensors mounted in the center bore of the bit are also important when measuring the water depth in very soft soils.

2.7.1 Sampling Procedures

The procedure for sampling sediments below the seafloor is generally divided into two broad categories: (1) seabed piston samplers and (2) down-hole samplers. The seabed samplers generally are deployed from a vessel using an A-frame to launch the equipment off the vessel deck, as shown in Figure 2.20. Seabed piston samplers can acquire continuous cores with lengths up to 30 m in a very timely manner with a single deployment to the seafloor as described by Moore

Figure 2.20   Vessel using an A-frame to deploy Jumbo Piston Core (JPC). Figure provided courtesy of Dr. Bernie Bernard with TDI-Brooks International.

and Heath (1978). Conversely, the down-hole sampler requires a geotechnical drillship and drill-pipe to advance the boring down to the desired sampling depth numerous times for each boring. Down-hole samplers required the vessel to remain relatively stationary within a small distance (less than 30 to 40 m) for periods from 3 to 5 days. This type of position keeping is accomplished by the use of a dynamic positioning (DP) system that uses DGPS, Ultra Short Base Line (USBL) or Long Base Line (LBL) transducers. In addition, the down-hole sampler requires a motion compensation system to avoid problems associated with vessel heave and usually a seabed reaction mass with a hydraulic clamping device to activate the drill string compensator.

2.7.1.1 Seabed Samplers

This sampling method provides many benefits since it can be deployed from a smaller, more economical vessel, and the actual time to acquire the sample is extremely quick (typically 30 to 40 minutes). The three seabed samplers most commonly used on a deepwater geotechnical investigation are the PC, JPC, and BC samplers. Another type of seabed sampler named the Deepwater Sampler (DWS) as described by Lunne et al. (2008) has been developed for obtaining extremely high-quality continuous push samplers.

The BC is used to obtain a large cube of high quality sample, as illustrated in Figure 2.21. The sampling box generally has equal dimensions of either 0.5 or 1.0 m. Due to the large area of the sampling box, a number of special tests can be run in situ within the sample. A continuous T-bar test can be performed throughout the depth of the core to define the shear strength profile, as described by Randolph et al. (2004). In addition, a series of miniature vane tests can also be performed throughout the depth of the large box sample. Both sets of test data are very useful for pipeline design since the upper 0.5 m below the seafloor is the zone that actually influences the pipeline performance. Data acquired in this fashion is the most representative of the near seafloor conditions and provides a vivid understanding of this important interface.

Figure 2.21   T-bar testing in Box Core (BC). Figure provided courtesy of Dr. Bernie Bernard with TDI-Brooks International.

The PC and JPC have a number of benefits that make them excellent tools to use during the initial portion of an integrated study. When the project schedule needs to be expedited, the seabed sampling activity may be moved forward to coincide with the geophysical acquisition phase. The lower daily cost of the vessel generally allows more samples to be obtained before the final boring and in situ testing locations are known. Thus, there is an opportunity to explore more areas with different geologic and soil conditions. One can correlate these additional cores with the sub-bottom geophysical data to improve our confidence level in understanding the potential range of soil and geologic conditions.

Typically, the PC consists of a 3- to 9-meter-long continuous sampling tube with a diameter of about 76 mm. Another more robust type of seabed sampler is the JPC as described by Young et al. (2000). The JPC is a larger version PC, and it has a sampler barrel ranging from 20 to 30 m in length with an inside diameter of 100 mm.

This seabed sampler, as shown in Figure 2.20, is deployed with a high-speed winch and cable, generally using an A-frame positioned off the stern of the vessel. The lower end of the PC or JPC is lowered to about 3 m above the seafloor and then triggered, allowing the sampler barrel to free-fall and to rapidly penetrate into the seafloor, as shown in Figure 2.22. The sampler barrel is then pulled from the seafloor using the high-speed wire-line winch.

Both the PC and JPC have a plastic or PVC liner that is extracted from the sampler barrel after recovery and cut in sections for transportation and storage. The liner must be continuous or joined within the sampler to form a seamless joint if individual sections are used. The PC and JPC have a piston within the sampler barrel that remains in contact with the seafloor and creates a small suction within the sample liner as the sampler barrel penetrates into the sediments. The piston greatly improves the sample recovery and the quality of the continuous sample.

The STACOR sampler is another free-fall seabed sampler that can obtain 30 m samples, 100 mm in diameter. The STACOR (see Figure 2.23) as described by Borel et al. (2002, includes

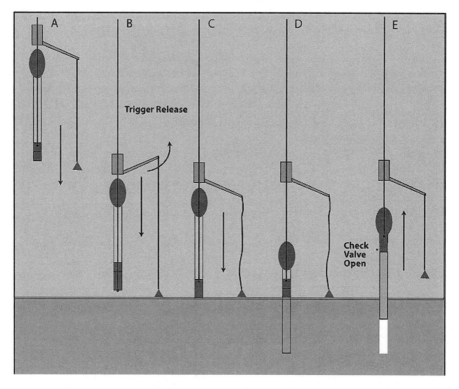

Figure 2.22 JPC sampling operation. Figure provided courtesy of Dr. Bernie Bernard with TDI-Brooks International.

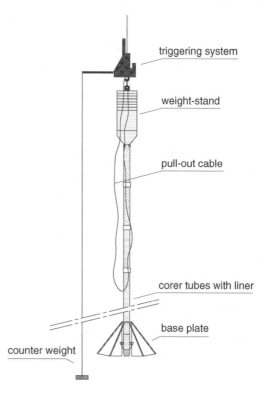

Figure 2.23 STACOR sampler © 2005 (Kolk and Wegerif, 2005): Reprinted with permission from Dr. Susan Gourvenec, Professor at University of Western Australia, Center for Offshore Foundation Systems, Perth.

a base reference plate system that keeps the piston relatively stationary at the sediment surface as the corer penetrates. Core recovery is generally in the order of 95 to 100% due to the stationary position of the piston.

2.7.1.2 Down-hole Samplers

The down-hole sampling operation is more time consuming and results in a costly operation since the entire operation must be conducted with a deepwater geotechnical drillship. However, the method is necessary when the foundation depth of interest extends to depths of 40 m or more, such as would be required for foundations employing deep driven piles.

There are a number of different methods to obtain down-hole samples, as illustrated in Figure 2.24. In the early years, a wire-line hammer tool was used during offshore site investigations as described by McClelland (1972) to obtain samples off a floating vessel. This sampler caused a significant degree of sample disturbance due to the action of driving a small dia. (57 mm) sampling tube as described by Emrich (1970).

To overcome the problems with sample quality, sampling systems were significantly improved in the North Sea in the late 1970s, resulting in a 76 mm dia. sampler that could push the sampling tube into the soil in one fast, continuous motion. Fugro developed the first down-hole hydraulic push sampler and CPT system called Wison, as described by Zuidberg and Windle (1979). The system featured a down-hole hydraulic electrical hose connected to a drill-string hydraulic packer

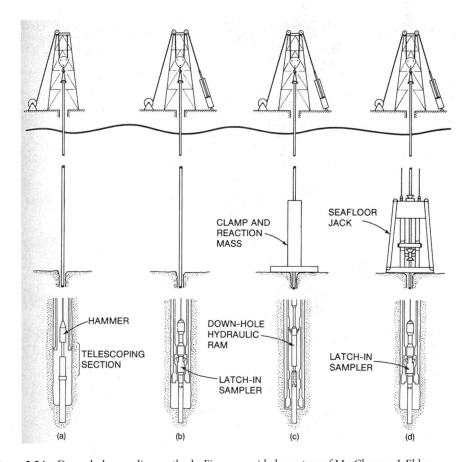

Figure 2.24 Down-hole sampling methods. Figure provided courtesy of Mr. Clarence J. Ehlers.

system located just above the drill bit. The packer system when inflated provided the necessary reaction force but many times failed to deflate or the packer element was torn, resulting in loss time due to recovery and replacement. This system was later replaced with a seafloor reaction mass that could clamp onto the drill string and transfer its weight to the drill string in order to compensate against vessel motions.

McClelland Engineers developed another system called Stingray, described by Semple and Johnston (1979), that featured a seafloor jacking system, as illustrated in Figure 2.25. The system included two pairs of hydraulic cylinders. One pair clamped the drill pipe while the second pair was used to thrust the drill string down the borehole with the sampler fixed within the drill bit as the sampling tube was pushed into the soil. The seafloor reaction mass had a submerged weight of about 9 kN that could resist a thrust during sampling up to about 90% of this weight. The seafloor reaction mass was lowered with a heavy lift winch that also had to be heave compensated.

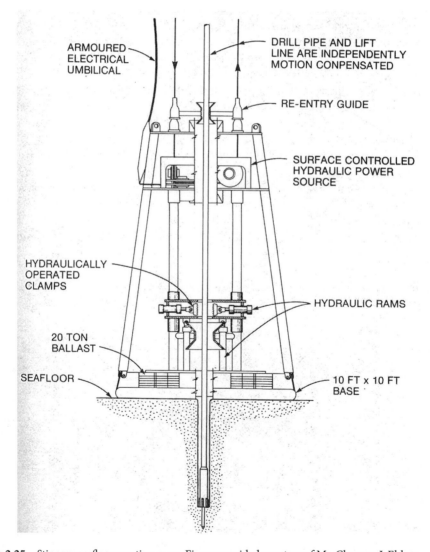

Figure 2.25   Stingray seafloor reaction mass. Figure provided courtesy of Mr. Clarence J. Ehlers.

This new development in the offshore geotechnical market further improved operations since the drill-string could be recovered, if required, via the use of a bit guide system that was raised and lowered with a device that tracked the two heavy lift lines that were suspended from the seabed frame. This same type of reaction mass system is still used for down-hole sampling and in situ testing, although the vertical thrusting mechanism is typically not used today.

The Dolphin System as described by Peterson et al. (1986) was developed by McClelland Engineers in 1982 and is still today routinely used by Fugro-McClelland. This tool is allowed to free fall down the pipe as a complete sampling or in situ testing unit. The tool could actually be configured into three types of in situ testing or sampling devices. These included the piston sampler, CPT, or VST. The thin-walled sampling tube at the end of the sampler is thrust into the soil beneath the bottom of the borehole using the mud pressure produced in the sealed drill string. Once the sampler is pushed into the soil, the drill-string is detached at the drill floor. An overshot tool is lowered down the pipe on a wire-line with a high-speed winch that is used to retrieve the sampler back to the vessel deck.

To avoid problems with vessel heave, a motion compensation system is required that applies a constant tension load to the drill-string while a portion of its weight is applied to the reaction mass. In heavy seas, the motion compensation system requires a constant tensioning device typically consisting of a gas-over-oil cylinder that expands and contracts with the ship heave, allowing the tool to push with a constant rate of penetration.

Typically, deepwater borings extend to depths ranging from 45 to 150 m, although boring depths up to 600 m are needed for certain geologic settings or for obtaining exploration top-hole information. The sampling interval routinely has been continuous or semi-continuous in the upper 15 m and at 3 m intervals below to the terminal depth of the boring. In more recent years, the sampling intervals have been more random to match the need for data in each soil stratum. Sample locations have been selected to provide data near the interface of each soil unit and at appropriate intervals in each stratum in order to achieve optimum coverage needed to interpret a design strength profile. In addition, in situ testing with the CPT has become the predominant data acquisition tool, so fewer samples are required to supplement the in situ data with laboratory data.

Deepwater borings are drilled using the open-hole technique, meaning that the drilling mud is discharged through an open-center bit during the drilling process. The drilling mud cleans the borehole during circulation by removing the cuttings produced by the rotating bit. The cuttings are moved up the borehole annulus and discharged on the seafloor. A properly planned drilling mud program is required: (1) to avoid hydraulic fracturing of the soil formation, (2) to properly clear the borehole of cuttings, and (3) to assure high quality soil samples are acquired.

There have been a large number of different down-hole samplers developed over the years as described by Schjetne and Brylawski (1979), McClelland and Ehlers (1986), and Lunne et al. (2001). The preferred method, according to Terzaghi and Peck (1948), is to acquire a sample with a rapid continuous push. A study by Young et al. (1983) highlighted the deleterious effects of the hammer sampling method and proposed the use of a new push sampler using a 76-mm diameter thin-walled sampling tube. The new sampler, as illustrated in Figure 2.26, was pushed into the soil using the weight of the drill string. In many offshore regions where only clay soils exist, this sampling procedure has become the preferred method since it can be performed using a drilling rig mounted on a vessel of opportunity and does not require a heave compensation system.

The paper by Young et al. (1983) also described a number of other field operational practices that need to be carefully monitored to obtain high quality data and avoid sample disturbance. Some of the most important considerations that must be addressed:

1. Avoid excessive vessel heave and movement of the drill string
2. Exercise care during the drilling operations

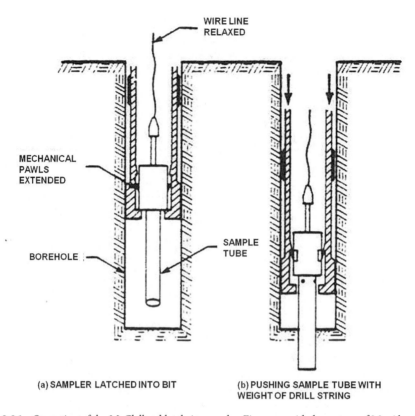

Figure 2.26 Operation of the McClelland latch-in sampler. Figure provided courtesy of Mr. Alan G Young.

3. Recognize the effects of stress relief during sample recovery
4. Use proper sample extrusion process
5. Carefully conduct sample handling, packaging, and transportation procedures
6. Properly store the samples
7. Conduct laboratory testing using proper procedures
8. Consider the influence of geologic and physio-chemical conditions upon measured soil properties

These important considerations have become part of the standard practice for conducting an offshore geotechnical investigation. The considerations are very critical to completing a successful geotechnical investigation and need to be carefully controlled by the geotechnical engineer and drilling crew. The coauthors recommend that the drilling crew be properly trained to assure that high quality samples are obtained to avoid disturbed measurements of soil properties.

A large number of samples are typically extruded and tested offshore. The coauthors believe it is beneficial to collect a number of samples and retain them in the sample tubes during a deepwater investigation. These Save tubes are usually collected where advanced testing is to be performed at a later date in an onshore laboratory. The Save tubes can also be X-rayed or CT-scanned upon arrival to the laboratory to select the undisturbed portion for testing.

To reduce the time to drill a deepwater boring, a new series of down-hole samplers have been introduced for geotechnical investigations. Many of the sampler concepts were adopted from the Ocean Drilling Program (ODP). These new geotechnical samplers allow a continuous sample with lengths ranging from 4.5 to 9.5 m, as reported by Kolk and Wegerif (2005). Based on ODP technology, Fugro and ConocoPhillips introduced a new Hydraulic Piston Corer (HPC), as shown in Figure 2.27, that is similar to the Advanced Piston Sampler (APS). The system has the capability to shoot a 66-mm diameter sampler into the soil at a very high speed using stored energy in the drill-string available from pressuring up the drilling fluid inside the pipe. A comparison of the sample test data reported by Jeanjean (2005) with adjacent borehole data from push samples indicates the samples have similar densities. There has been limited comparison of reported strength data, but the limited data indicates the smaller diameter and high-speed insertion force may cause a high degree of sample disturbance.

2.7.2 Deepwater In Situ Testing Methods

There are a wide variety of in situ testing methods available to use during an offshore geotechnical investigation, but most deepwater investigations rely primarily upon the CPT, VST, and piezo-probe. These tools have been dramatically improved and now provide reliable operation in this deep and rugged operating environment. In addition, there have been many innovations to reduce the sampling and testing time associated with operating through 1 to 2 km of drill-string. The following sections will describe the most pertinent details involved with using these tools in deepwater.

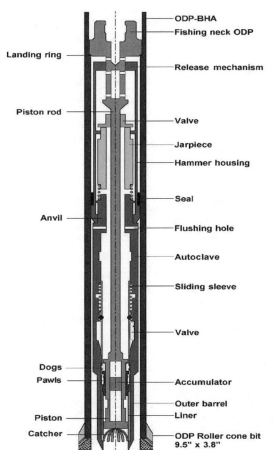

Figure 2.27 Fugro hydraulic piston corer (HPC) © 2005 (Kolk and Wegerif, 2005): Reprinted with permission from Dr. Susan Gourvenec, Professor at University of Western Australia, Center for Offshore Foundation Systems, Perth.

2.7.3 Cone Penetrometer Testing

The CPT has played a key role in marine geotechnical site investigations for over thirty years. The equipment used to deploy the CPT has improved dramatically over this time, allowing high quality CPT data to be acquired in a wide variety of soil types in water depths approaching 3500 m.

A number of key references, such as Lunne et al. (1997); Robertson (2009); ASTM (1995); Schnaid (2009); and ISO (2006), describe: (1) procedures, (2) equipment specifications, (3) intended use, (4) performance requirements, and (5) data interpretation methods available for the CPT. The chapter cannot repeat all the available details; however, this section will highlight the important role and critical operational factors when using the CPT during a deepwater integrated study. Our chapter will briefly describe key portions of the current CPT state of practice for interpreting physical and engineering properties of the sediments and understanding their spatial variability.

Briaud and Meyer (1983) presented a historical summary of the various types of CPT equipment and described the pertinent details of each of their operations. The CPT is used on a deepwater site investigation typically in two different modes depending upon the depth of sediment interest below the seafloor.

2.7.3.1 Seabed CPT Equipment

For depths up to about 30 to 40 m, the CPT operation typically involves a seafloor platform that houses mechanical thrusting equipment and provides the required reaction mass. Fugro developed the Seacalf system in 1972, as illustrated in Figure 2.28 (Zuidberg 1975). The system has the capability to provide a vertical cone thrust of 5 kN. A CPT probe attached to the end of a continuous cone rod is thrust into the seafloor at a continuous rate of 2 cm per sec. A 10 or 15

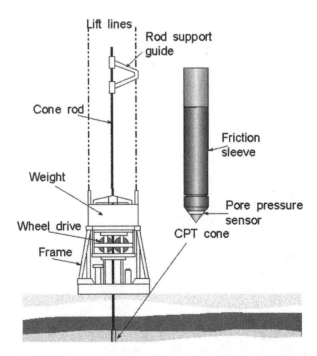

Figure 2.28   Seacalf CPT system (ISSMGE 2005): Reprinted with permission from Ed Danson with Swan Consultants Ltd.

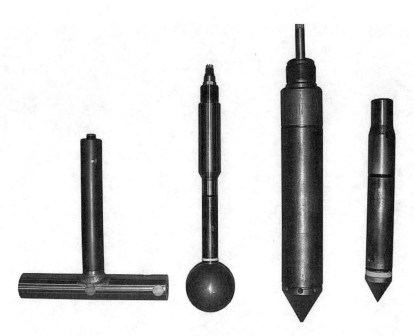

Figure 2.29   T-Bar, ball penetrometer, and 33 cm² and 15 cm² piezocone penetrometers (Kolk and Wegerif, 2005): Reprinted with permission from Dr. Susan Gourvenec, Professor at University of Western Australia, Center for Offshore Foundation Systems, Perth.

cm² cone, as shown in Figure 2.29, is pushed up to maximum depths of about 35 m in normally consolidated soils.

A number of other seabed CPT systems were developed over the years by several researchers. A small version of a seabed CPT system was first developed in 1987 as described by Young et al. (1988) that features a 5 cm² cone that proved very useful for pipeline or cable route surveys. This system had the capability to coil the push rod during CPT retrieval and then straighten the rod as the CPT was being thrust into the seabed. Many such devices are still in use today. An upgraded version of this system, as shown in Figure 2.30, has the capability to perform a continuous CPT push to a 10 m depth in water depths up to 3000 m. Gregg Drilling recently demonstrated the successful use of this system (Boggess and Robertson 2010). They performed about 120 mini-cone tests offshore in the Gulf of Mexico in water as deep as 1800 m to investigate seafloor variability for a large number of mudmat sites.

A new CPT Stinger system was developed in 2010 by TDI-Brooks International that can perform CPT testing in deepwater to seafloor penetrations of about 35 m off a standard oilfield supply vessel equipped with launch and recovery system for JPC operations. The system, as shown in Figure 2.31, simply replaces the standard JPC core liner with a module that includes the CPT attached to the thrusting rod, power, control, and logging system for the CPT data. The submerged weight of the complete system plus the axial soil resistance on the JPC barrel provides the required weight to resist the vertical thrust developed as the CPT and push rod penetrate the sediments (Young et al. 2011).

2.7.3.2 Down-hole CPT Equipment

Downhole CPT equipment that evolved over the last thirty years included the Wison, Stingray, and Dolphin systems previously described in the soil sampling section. All these systems require

Figure 2.30 Launching Gregg Drilling mini-cone system off vessel. Figure provided courtesy of Dr. Peter Robertson with Gregg Drilling.

a seafloor reaction mass to resist the force required to push the cone into the undisturbed soil beneath the bottom of the borehole. After each CPT push, the entire system must be retrieved from the drill-string while the rig drills down to the next testing depth. Unlike the seabed CPT equipment, the data from a single CPT push is limited to about 3 m. To acquire continuous data in a borehole, a series of tests must be performed at repeated 3 m intervals or less.

2.7.3.3 Operational and Technological Improvements

An overview of the most important operational procedures is briefly discussed to give the reader a general understanding of their importance. The book by Lunne et al. (1997) includes a list of standards that describe the proper CPT procedures. The International Society of Soil Mechanics and Foundation Engineering (ISSMFE 1989; ISSMGE 1999) issued their standards that describe the equipment, procedures, terminology, and minimum requirements for properly conducting a CPT program.

Based on the coauthors experience, the key operational procedures that need to be carefully monitored during the field program include: (1) cone calibration, (2) measurement of verticality, (3) rate of penetration, (4) cone depth reference, (5) piezocone saturation, (6) sensitivity of cone compared to soil strength, and (7) resolution of measuring and recording systems. The CPT equipment is typically very reliable and provides high quality data, but close monitoring of all these procedures is essential to confirm the systems have been maintained and are operating properly.

A number of technological deficiencies emerged as use of CPT has shifted into deeper water. The book by Lunne et al. (1997) summarizes many of the more important improvements with the CPT equipment, procedures, corrections, and new standards.

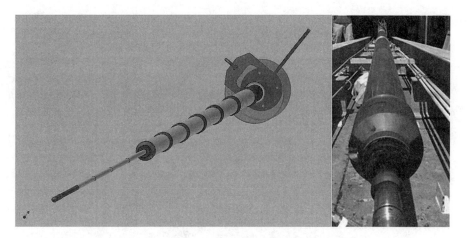

Figure 2.31 TDI-Brooks' *CPT-Stinger System*. Figure provided courtesy of Dr. Bernie Bernard with TDI-Brooks International.

The measurement of in situ characteristics of the near surface, soft sediments requires careful attention to details in cone design and procedures. A very large hydrostatic pressure exists as compared to the low shear strength of the near seafloor soils in deepwater. A new compensated cone as described by Boggess and Robertson compensates for the large hydrostatic pressure by filling the inside with oil and connecting the inner oil pressure with the ambient seawater pressure outside the cone. The load cell within the compensated cone can be designed with a lower capacity. Since the compensated cone does not require measuring the high water pressures, it provides more accuracy for measuring the resistance of weak soils commonly encountered in deepwater areas.

There have been a number of recent new systems and developments that improve the capability of performing seabed or down-hole CPT testing and/or sampling. Benthic Geotech developed the Portable Remotely Operated Drill (PROD) as described by Carter et al. (1999). The seabed system, as shown in Figure 2.32, has the capability to take piston samples and to perform CPTs in the same borehole in water depths up to 2000 m. More recent developments with deepwater seabed systems include telemetry and robotic components, such as the Seabed CPT System (Boggess and Robertson, 2010) developed by Gregg Drilling & Testing Inc. (see Figure 2.33).

The Rovdrill, as shown in Figure 2.34, is another new geotechnical sampling and in situ testing system capable of operating in 3000 m of water. This system can operate off any type of DP vessel and uses an ROV for electrical and hydraulic power, telemetry, and high definition video and operator interface to the operator control room onboard the vessel. The system as described by Spencer (2007, 2008, 2011) has the capability to drill, to obtain continuous cores and samples, and/or to perform in situ testing to depths of 100 m. These technological improvements continue to provide the opportunity to improve the quality of samples and data and to reduce vessel time. Working from the seafloor eliminates delays associated with using a wire-line to sample or to perform in situ tests at depths of 1 to 2 km below sea level.

Figure 2.32   Portable remotely operated drill (PROD) system © 2005 (Kolk and Wegerif, 2005): Reprinted with permission from Dr. Susan Gourvenec, Professor at University of Western Australia, Center for Offshore Foundation Systems, Perth.

Figure 2.33   Gregg Drilling—
Seabed CPT System. Figure
provided courtesy of Dr. Peter
Robertson with Gregg Drilling.

Figure 2.34 Seafloor Geoservices'—*Rovdrill* drilling, sampling, and in situ testing system. Figure provided courtesy of Mr. Allan Spencer with Seafloor Geoservices Inc.

2.7.4 Use of CPT Data

The standard cone penetrometer, as shown in Figure 2.29, includes instrumentation to measure the resistance on the cone tip and along the sleeve continuously as the cone is thrust into the soil. The CPT data has one important benefit over other in situ tools since it provides continuous profiles, as shown in Figure 2.35, of corrected point resistance (q_t) and sleeve friction (f_s). These two measurements also provide a continuous plot of friction ratio (f_r) defined as the f_s divided by q_t. The friction ratio is often used as described by Robertson (2008) to classify soil types.

The continuous CPT data provides an improved understanding of the variation of the soil properties throughout each sediment layer. The continuous measurements can also be used to derive various soil parameters, as shown in Table 2.3.

The table presents a reliability rating for each soil parameter in terms of the confidence level for deriving the parameters based on the level of experience with the particular sediment type (clay or sand). A number of different references (Young et al. 1983; Lunne et al. 1997; Robertson 2009; ATSM 1995; Schanid 2009) provide significant details on the state of practice and describe the procedures for estimating these parameters.

The addition of other sensors within the CPT tool allows other soil properties to be measured, such as the dynamic pore pressure and the shear wave velocity. These two properties are especially important when inter-layered sediments of granular and cohesive soils may exist. The dynamic pore pressure provides a more vivid indication of changes in soil stratigraphy than the cone friction ratio.

The Piezocone (PCPT) generally includes an external filter connected to a pore pressure transducer mounted at one of three locations: (1) tip of the cone, (2) in the gap between the cone tip and sleeve, and (3) behind the sleeve. The pore pressure transducer, as shown in Figure 2.36, is shorter than the length of the cone sleeve, so the friction ratio represents an average measurement over a sleeve length of about 130 mm. The pore pressure transducer is typically only about 6 mm in length. Thus, the dynamic pore pressure provides a more immediate response and less averaging of the changes in the soil profile as the cone is pushed into different soil stratum.

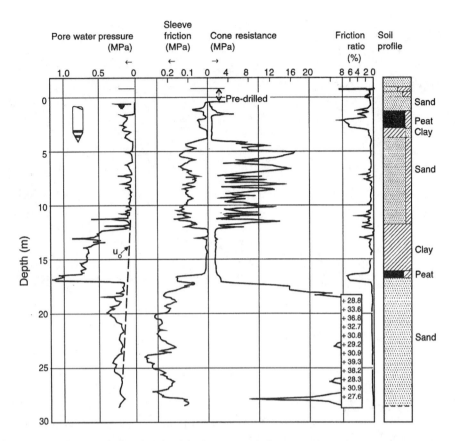

Figure 2.35 Continuous profile of PCPT data showing q_t, fs, friction ratio, and pore pressure measurements (Lunne et al. 1997): Reprinted with permission from Mr. Tom Lunne with NGI, Dr. Peter Robertson with Gregg Drilling, and Mr. John Powell with Geolabs Ltd.

Table 2.3 Perceived applicability of CPT for deriving soil parameters (Lunne et al., 1997)

Soil Type	D_r	Ψ	K_0	OCR	S_t	s_u	ϕ'	E,G*	M	G_0*	k	c_v
Sand	2-3	2-3	5	5			2-3	2-3	2-3	2-3	3	3-4
Clay			2	1	2	1-2	4	2-4	2-3	2-4	2-3	2-3

1 = high, 2 = high to moderate, 3 = moderate, 4 = moderate to low, 5 = low reliability, Blank = no applicability, *improved with SCPT

Where:

D_r	Relative density	ϕ'	Fiction angle
Ψ	State Parameter	K_0	In situ stress ratio
E, G	Young's and Shear moduli	G_0	Small strain shear moduli
OCR	Over consolidation ratio	M	Compressibility
s_u	Undrained shear strength	S_t	Sensitivity
c_v	Coefficient of consolidation	k	Permeability

In the last decade, the T-Bar Test (TBT), as described by Stewart and Randolph (1994) and Randolph et al. (1998), and the Ball Penetrometer Test (BPT) (Watson et al. 2008; Randolph et al. 2000), as illustrated in Figure 2.29, were introduced as an alternate to the CPT. The TBT is suitable only for seabed systems, but the BPT can be used with either the seabed or down-

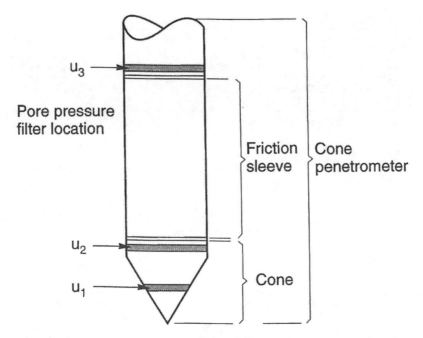

Figure 2.36 Cone penetrometer measurements including different pore pressure transducer locations (Lunne et al. 1997): Reprinted with permission from Mr. Tom Lunne with NGI, Dr. Peter Robertson with Gregg Drilling, and Mr. John Powell with Geolabs Ltd.

hole systems, as long as the diameter of the ball cone is less than 100 mm. Both have been used instead of the CPT to assess the undrained shear strength of clay soils since theoretical studies suggest a more unique correlation exists for strength interpretation as proposed by Randolph (2004).

The limited space available for this chapter does not allow a detailed description of the many soil properties that can be interpreted from the CPT data. We will provide a brief description of the key parameters in clay soils that include: (1) soil classification, (2) undrained shear strength, (3) normalized undrained shear strength (S_u/σ'_{vo}), and (4) over-consolidation ratio (OCR). To describe the methods available to interpret these important soil parameters, Table 2.4 presents the key terms, definitions, and equations most commonly used by geotechnical engineers in their published references.

The CPT data also provides a key benefit for granular soils since it can be used to evaluate the state of in situ conditions in terms of relative density and whether the site may be susceptible to liquefaction in seismic areas. Use of the CPT to interpret the relative density of granular soils has provided a dramatic improvement in foundation design over relying upon blow counts obtained with a wire-line percussion sampler.

A brief description of the procedures for interpreting the various parameters is described below.

2.7.4.1 Soil Classification

The initial work to use the CPT to classify the soil penetrated was achieved by Begemann (1953) when he added the adhesion jacket (sleeve) to the Dutch static cone penetrometer. He was also the first to propose the friction ratio, f_r, (sleeve friction/cone resistance) to be used to identify soil stratigraphy in terms of soil type (Begemann, 1965). Douglas and Olson (1981) improved the

Table 2.4 Key CPT terms and equations required to interpret soil parameters

Key CPT terms, parameters, and equations			
Terms	Definitions	Equations	Definition of other terms
Q_c	Total force acting on cone tip		
q_c	Measured cone tip resistance	$q_c = Q_c/A_c$	A_c = area of projected cone tip
q_t	Corrected cone resistance	$q_t = q_c + (1 - a)u_2$	u_2 = pore pressure measured behind cone tip see Figure 2.36
a	Cone area ratio	$a = A_n/A_c$	A_n = cross-sectional area of cone shaft or load cell A_c = projected area of cone
Q_t	Normalized cone resistance	$Q_t = (q_t - \sigma_{vo})/\sigma'_{vo}$	σ_{vo} = total overburden stress σ'_{vo} = effective overburden stress
q_{net}	Net cone resistance	$q_{net} = q_t - \sigma_{v0}$	
N_{kt}	Empirical cone factor	$N_{kt} = q_{net}/S_u$	S_u = undrained shear strength based on various lab tests

practice of using the CPT to classify soil by recognizing that sandy soils tend to produce high cone resistance and low friction ratio. On the contrary, clay soils produce low cone resistance and high friction ratio. The procedure was based on a classification chart that compared the behavior of the cone resistance with the friction ratio.

The sleeve friction is not always accurately measured, since cones of different designs will produce different measurements of sleeve friction. To overcome these problems, the CPT was modified to allow dynamic pore pressure to be measured continuously during cone penetration. This type of cone is generally called a Piezocone (PCPT or CPTU). To overcome the problem of using only sleeve friction, classification charts as described by Baligh et al. (1980) were based on the q_t and pore pressure parameter, B_q. Robertson (1990) later recommended the two charts, as shown in Figure 2.37, that require all three measurements from the PCPT data.

The Robertson (1990) method was modified to include an expanded database, so the method is considered global in nature and can be used as a guide for defining soil behavior type (SBT). Robertson recommends that the CPT classification method based on SBT should be used with caution when comparing with traditional classification methods based on grain size distribution and plasticity. Some soil types will fall within different zones on the two charts, so judgment and additional laboratory classification testing will be needed to finalize the classification for some soils.

2.7.4.2 Undrained Shear Strength

A number of different approaches exist for interpreting the undrained shear strength, S_u, from CPT or PCPT data as described by Lunne et al. (1997). There is no unique measure of S_u since the undrained response to loading in fine-grained soils is a function of soil anisotropy, strain rate, stress history, and loading mechanism. The different theoretical approaches for using CPT data relate the cone tip resistance to different theoretical cone factors (N_k or N_{kt}), depending upon whether the measured cone tip resistance (q_c) or net cone tip resistance (q_{net}) is used. In cohesive

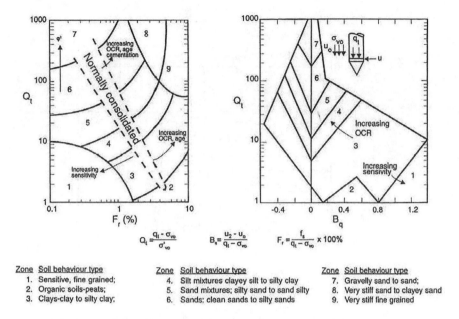

$$Q_t = \frac{q_t - \sigma_{vo}}{\sigma'_{vo}} \qquad B_s = \frac{u_2 - u_o}{q_t - \sigma_{vo}} \qquad F_r = \frac{f_s}{q_t - \sigma_{vo}} \times 100\%$$

Zone	Soil behaviour type	Zone	Soil behaviour type	Zone	Soil behaviour type
1.	Sensitive, fine grained;	4.	Silt mixtures clayey silt to silty clay	7.	Gravelly sand to sand;
2.	Organic soils-peats;	5.	Sand mixtures; silty sand to sand silty	8.	Very stiff sand to clayey sand
3.	Clays-clay to silty clay;	6.	Sands; clean sands to silty sands	9.	Very stiff fine grained

Figure 2.37 Robertson soil classification chart (Lunne et al. 1997): Reprinted with permission from Mr. Tom Lunne with NGI, Dr. Peter Robertson with Gregg Drilling, and Mr. John Powell with Geolabs Ltd.

soils, the recommended practice is to use the relationship between undrained shear strength, S_u, and net cone tip resistance, q_{net}, expressed as follows:

$$q_{net} = N_{kt} S_u \qquad (2.1)$$

where N_{kt} is the cone factor that is analogous to a bearing capacity factor, N_c.

The values of N_k or N_{kt} are not unique for various soil types and typically are empirically derived. A broad range of reported values have been used to compute the undrained shear strength based on a variety of reference strengths, such as UU-triaxial compression, in situ vane, or average values based on simple shear, triaxial compression, and extension. The coauthors recommend using a consistent reference such as the static direct simple shear (DSS) when correlating with q_{net} since the hydrostatic component in deepwater can have a major influence on the measured value.

Kjekstad et al. (1978) initially recommended an average value of N_k equal to 17 based on UU-triaxial compression. Lunne and Kleven (1981) later suggested that N_k typically range from 11 to 19 depending upon the plasticity index of the clay. They reported an average value of 15.

Aas et al. (1986) later presented a correlation between N_{kt} and plasticity index, as shown in Figure 2.38, based on an average undrained strength from the triaxial compression and extension and direct simple shear. The actual value of N_{kt} depends upon many variables, such as state of consolidation and type of laboratory test. The authors believe that N_{kt} should be used instead of N_k in the offshore environment. Our experience with deepwater normally consolidated clays in the Gulf of Mexico indicates an appropriate value of N_{kt} equal to 17.5 for UU-triaxial compression and direct simple shear tests.

There have many other studies correlating N_k and N_{kt} with undrained shear strength. These values range from about 15 to 20 as indicated by the papers at ISOPT-1 (Penetration Testing 1988). Reported values of N_{kt} for different offshore regions will vary due to the type of laboratory test and the degree of sample disturbance. Since the value of N_{kt} is not unique, we recommend correlating

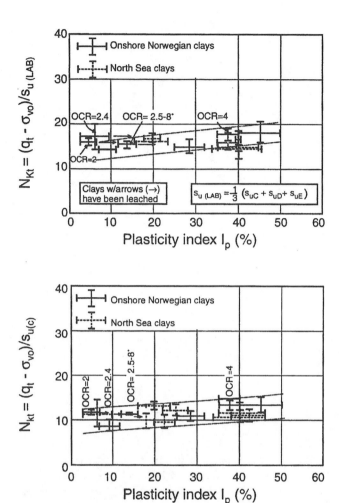

Figure 2.38 N_{kt} correlation with plasticity index (Aas et al. 1986; Lunne et al. 1997): Reprinted with permission from Mr. Tom Lunne with NGI, Dr. Peter Robertson with Gregg Drilling, and Mr. John Powell with Geolabs Ltd.

laboratory and CPT test data in other regions to make sure the proper N_{kt} value is understood and used for foundation design.

2.7.4.3 Strength Ratio (S_u/σ'_{vo})
For normally consolidated clays, the undrained strength ratio, S_u/σ'_{vo}, typically ranges from 0.22 to 0.26 based on static direct simple shear tests (Wroth, 1984; Ladd, 1991). Using an average value of N_{kt} equal to 17.5, then the strength ratio can be rewritten as follows:

$$(S_u/\sigma'_{vo})_{nc} = Q_t/N_{kt} \tag{2.2}$$

or

$$(S_u/\sigma'_{vo})_{nc} = 0.057Q_t \qquad (2.3)$$

For a normally consolidated clay, Q_t will range from 3.9 to 4.6 for $(S_u/\sigma'_{vo})_{nc} = 0.22$ to 0.26.

2.7.4.4 Over-consolidation Ratio (OCR)

The over-consolidation ratio (OCR) is one of the most important parameters measured during a laboratory testing program since it helps define the stress history of the soils. The state of consolidation is a critical parameter for understanding how the soil will respond to future foundation loading and helps define the geologic history of the sediments.

The importance of OCR has resulted in a long list of methods using three different parameters available from the CPT/PCPT data. The methods are generally divided into the three categories: (1) using the undisturbed strength S_u, (2) based on the shape of the PCPT profile, and (3) based directly on PCPT data.

The method using the undrained strength was generally adapted by Andersen et al. (1979) that relies upon the methods proposed by Schmertmann (1974, 1976). The method uses a chart that includes parameters such as S_u/σ'_{vo}, OCR, and plasticity index, as shown in Figure 2.39. The value of S_u is based on strength measurements from the CAUC triaxial tests.

The shape of the PCPT profile has also been used to estimate the preconsolidation pressure and in turn the OCR. The range of normalized behavior obtained from the PCPT is compared to normally consolidated clays to see if it is larger. This value is then plotted on a normalized soil behavior classification chart, shown in Figure 2.37 as proposed by Robertson (1990). When the q_t profile is close to the theoretical band, the clay is normally consolidated. The clay is considered over-consolidated when q_t is significantly larger than the theoretical band.

Kulhawy and Mayne (1990) recommended the method most commonly used today to estimate OCR in clay soils. Their equation for over-consolidated soils due to removal of overburden stress is defined as follows:

$$\text{OCR} = \sigma'_p/\sigma'_{vo} \qquad (2.4)$$

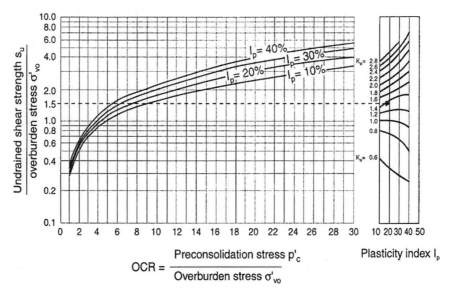

Figure 2.39 Andersen chart to interpret OCR (Lunne et al. 1997): Reprinted with permission from Mr. Tom Lunne with NGI, Dr. Peter Robertson with Gregg Drilling, and Mr. John Powell with Geolabs Ltd.

where σ'_p = preconsolidation or yield stress, and

$$(S_u/\sigma'_{vo})_{oc} = (S_u/\sigma'_{vo})_{nc}(\text{OCR})^{0.8} \qquad (2.5)$$

since $(S_u/\sigma'_{vo})_{nc}$ typically ranges from 0.22 to 0.26 for direct simple shear

$$(S_u/\sigma'_{vo})_{oc} = 0.22\,(\text{OCR})^{0.8} \text{ to } 0.26(\text{OCR})^{0.8} \qquad (2.6)$$

A simpler expression for estimating OCR in cohesive soil can be computed from the following expressions:

$$\text{OCR} = \sigma'_p/\sigma'_{vo} \text{ and } \sigma'_p/ = k(q_t - \sigma_{vo})$$

substituting:

$$\text{OCR} = k(q_t - \sigma_{vo})/\sigma'_{vo} = kQ_t \qquad (2.8)$$

where

$$Q_t = \text{normalized cone resistance, and}$$
$$k = \text{nondimensional lateral coefficient}$$

The value of k can range from 0.25 to 0.50 with the higher values proposed for aged, heavily over-consolidated clays. An average value of 0.33 is recommended unless previous experience exists with the soil deposit or laboratory measurements have been made to correlate with the site-specific value of k.

2.7.5 In Situ Vane Shear Testing

The use of the VST in the marine environment increased dramatically as equipment designs improved over the last thirty years. The VST has become an important tool to avoid the effects of sampling disturbance.

The initial use of the in situ vane shear test (VST) offshore was performed from a fixed platform or with a small seafloor-supported platform. The first use from a floating vessel was performed in 1970 with a down-hole VST device (see Figure 2.40) called the Remote Vane (Doyle et al. 1971). This testing was performed as part of a major program for Shell Oil to investigate the mudslide failures that caused the complete loss of three fixed offshore platforms located within the Mississippi Delta in about 90 m of water (Sterling and Strohbeck 1973).

The Remote Vane was deployed down the bore of ordinary oilfield drill-pipe using a multi-conductor armored wire-line. In situ tests were conducted by using the weight of the drill-pipe to insert the VST tool below the bottom of the borehole. The drill-string was then lifted free of the latch-out pushing pawls, isolating the tool during the testing operation. At this point, the vane blades were rotated at a constant rate while a torque transducer measured the maximum torque required to shear the soil.

2.7.5.1 Seabed VST

The low soil strengths at the seafloor require that the VST tool be deployed with a seabed template, as shown in Figure 2.41, to keep the tool from sinking into the soil under its own weight. This seabed VST system, called *Halibut* (Ehlers et al. 1983), was designed in 1976 to allow a single vane test without the use of a drill-pipe. The seabed platform is lowered over the side of the vessel to rest on the seafloor while the vane tool is pushed to desired test depth by clamping a fixed length of its shaft below the seabed template. There is sufficient weight in the seabed template to push the test blade to depths up to 7 m below the seafloor.

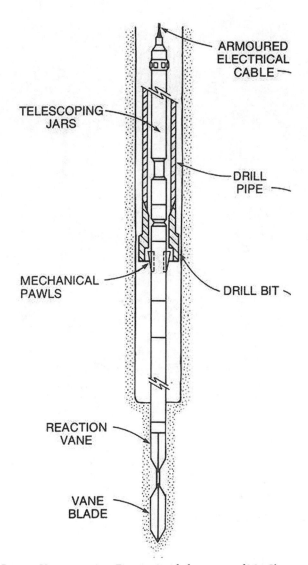

Figure 2.40 In situ Remote Vane operation. Figure provided courtesy of Mr. Clarence J. Ehlers.

The seabed system was improved later to include a seabed jacking system that can push the test blade to multiple test depths without retrieving the system back to the vessel to lengthen the shaft of the VST tool.

2.7.5.2 Down-hole VST tool

The down-hole deployment of the VST tool was greatly improved in 1985 with the introduction of the Dolphin system as described by Peterson et al. (1986). The free falling down-hole tool included a battery, electronics control, and remote memory module. The wireless tool body was embedded in the soil beneath the drill-pipe to isolate it from vertical heave by the use of a seabed frame (SBF) system. This system is still used today and has been used in water depths approaching 3000 m.

To acquire good data with the VST, the most important operational procedures should be briefly discussed. Randolph et al. (2005) suggests that the strain rate is a major cause for the

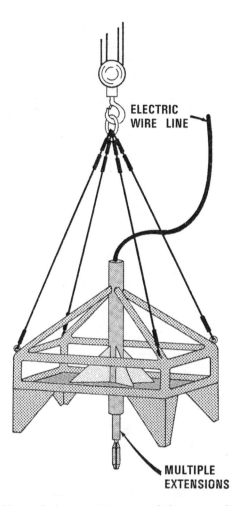

Figure 2.41   VST with Halibut seabed system. Figure provided courtesy of Mr. Alan G Young.

strengths measured with the VST being greater than values measured from laboratory tests. The coauthors agree and believe other operational factors are equally important. These other operational procedures as described by Young et al. (1988) must be followed to produce high-quality VST data. Some of the most important procedures include: (1) vane blade geometry, (2) vane rotation rate, (3) test penetration below bottom of borehole, (4) drilling mud weight and pump pressure, and (5) torque calibration.

The coauthors recommend a height-to-diameter ratio of 2 for the blade geometry and area ratios (vane blade area divided by failure plane area) in the range of 12 to 15%. In weaker clays, a vane blade with a diameter of 65 mm and height of 130 mm is recommended. In stronger clays, the vane blade diameter may decrease to diameters of 55 and 40 mm for strengths greater than 50 and 100 kPa, respectively.

The rotation of the vane blade should be in the range of 12 to 18 degrees per minute. The coauthors recommend a minimum vane test penetration of 1.0 m below the bottom of the borehole. Excessive fluid pressure at the bottom of the borehole is detrimental, so the drilling crew should carefully monitor mud weights and pump pressure when approaching the test depth penetration. The vane torque system should be calibrated before the tool is first used in the boring.

There are a number of key references that address these operational procedures used for sampling and the CPT and VST operations. These are presented in more detail in reference standards such as NORSOK (2004), ISO 19901 (2006), SUT (2005), ASTM (2008), and ASTM (2005).

2.7.6 Use of VST Data

Earlier studies by Ehlers et al. (1980), Chandler, R. J. (1988), and Quiros and Young (1987) confirm that laboratory measurement of S_u on samples can be significantly understated due to effects of low saturation and the high sensitivity of marine sediments. The advantages of VST data make it ideally suited for reducing effects of sample disturbance. Its advantages are clearly evident for the following conditions:

1. Improved evaluation of areal shear strength variation
2. Better definition of strength profiles in crustal and gas charged sediments
3. Clear identification of zones experiencing mass soil movements

An additional advantage is the reduction of scatter in the strength data that allows a better interpretation of the design shear strength profile as described by Young et al. (1987). The VST strengths are generally higher than any laboratory measurements on recovered samples, as shown in Figure 2.42.

2.7.6.1 Undrained Shear Strength

Although the VST data are a more direct measure of the undrained shear strength and avoid the effects of sample disturbance, most investigators believe that the values cannot be used directly for foundation design. Adjustment factors are typically applied to the VST data to confirm that VST strength data is consistent with the strength data used to develop the empirical design method.

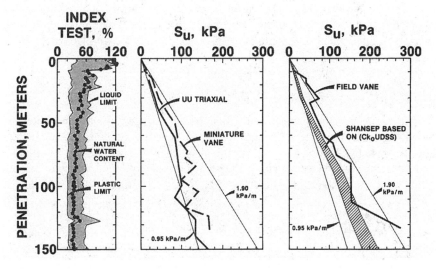

Figure 2.42 VST data compared with laboratory test data (Young et al. 1987): Reprinted with permission from *STP 1014, Vane Shear Strength Testing in Soils: Field and Laboratory Studies*, copyright ASTM International, 100 Barr Harbor Drive, West Conshohocken, PA 19428.

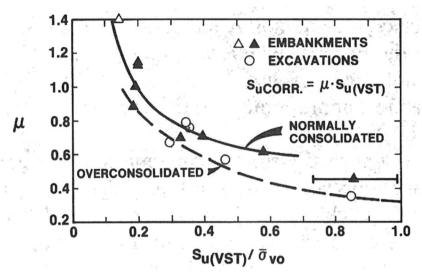

Figure 2.43   Aas VST strength adjustment factors (Young et al. 1987): Reprinted with permission from *STP 1014, Vane Shear Strength Testing in Soils: Field and Laboratory Studies*, copyright ASTM International, 100 Barr Harbor Drive, West Conshohocken, PA 19428.

Adjustment factors to use with VST data were first recommended by Bjerrum (1973) for onshore slope stability and bearing capacity analyses. His adjustment factors range from 0.60 for high plasticity soils to 1.00 for low plasticity soils. For a soil with a plasticity index of 60, his recommended adjustment factor was 0.75. Ehlers et al. (1980) later proposed an adjustment factor of 0.75 for pile design in normally consolidated soils in the Gulf of Mexico. Aas et al. (1986) then proposed the adjustment factor should be correlated with the strength ratio, S_u/σ'_{vo}, to give a more consistent interpretation for foundation design, as shown in Figure 2.43. Their adjustment factors range from 1.00 to 0.60 for a strength ratio of 0.20 to 0.60, respectively. The adjustment factor is not unique and may vary due to disturbance of the laboratory samples and variations in the type of laboratory test. The coauthors have found that values ranging from 0.75 to 0.85 are most consistent for deepwater sites around the world when correlated with UU-triaxial and direct simple shear test.

2.7.7   In Situ Piezoprobe

In situ pore pressure measurements are sometimes obtained during an offshore geotechnical investigation to understand if excess pore pressures are present that can influence slope stability or water flow associated with conductor/well casing installation. Wissa et al. (1975) initially developed a piezometer probe for use on land projects that measured the pore pressure during penetration and pauses in penetration. The offshore piezometer was first developed as described by Preslan and Babb (1979) to make pore pressure measurements by pushing the piezoprobe about 1 m below the bottom of a borehole.

The new Fugro MkII piezoprobe (see Figure 2.44) described by Kolk and Wegerif (2005) is designed to operate in water depths approaching 2000 m with seafloor penetrations up to 500 m. An example showing results of the pore pressure dissipation over time made with the MkII piezoprobe installed in combination with the Wison XP system is presented in Figure 2.45. A test typically takes about four to six hours to achieve 90% dissipation. The results have been used to measure pore pressures in excess of hydrostatic above over-pressured sand layers and measure the pore pressures within steep submarine slopes.

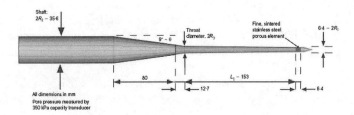

Figure 2.44 Fugro MkII piezoprobe.

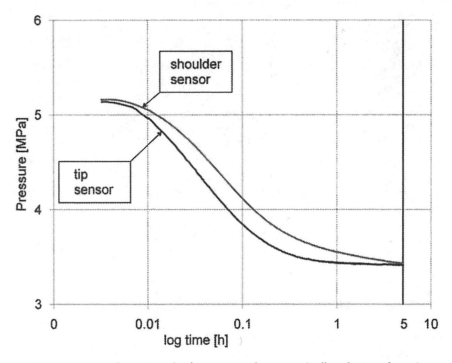

Figure 2.45 Pore pressure dissipation plot from piezoprobe © 2005 (Kolk and Wegerif, 2005): Reprinted with permission from Dr. Susan Gourvenec, Professor at University of Western Australia, Center for Offshore Foundation Systems, Perth.

To reduce the time for excess pressure generated during insertion to dissipate, the spear-like piezoprobe was designed, as shown in Figure 2.44, with a tapered diameter down to the pore pressure transducer mounted at the tip. The smallest diameter exists at the tip to minimize the area of induced pore pressure developed during insertion of the piezoprobe since dissipation time is proportional to the square of the diameter. The smaller diameter also reduces the time for pore pressure dissipation during the field test.

2.8 Integrated Case Studies

The first portion of this chapter provides an overview of geophysical and geotechnical field investigations practices that have evolved over the last few decades to accommodate the data requirements for a deepwater integrated study. There has been a significant investment in tools and procedures needed to obtain high quality data in this challenging frontier environment. We can now describe the process for integrating the different data sets and interpreting the geologic-geotechnical model needed to assess any geo-constraints and to evaluate potential seafloor and subsurface risks that could impact architecture for various production scenarios. Our approach will involve the use of case studies previously conducted in offshore areas exhibiting complex geologic conditions.

The coauthors have been involved with two major deepwater projects in the Gulf of Mexico, the Mad Dog and Atlantis Developments operated by BP, both located on the Sigsbee Escarpment. Since both developments have complex geologic conditions associated with the escarpment, they serve as excellent examples for illustrating the importance of conducting an integrated geosciences study as described by Jeanjean et al. (2005).

2.8.1 Geologic Setting and Field Architectures

The Sigsbee Escarpment is the most prominent geologic feature in the Gulf of Mexico. The Mad Dog and Atlantis Developments straddle this complex geomorphic region, as shown in Figure 2.46. Different components of the two field developments are located within three different geologic provinces: lower continental slope, Sigsbee Escarpment, and the upper continental rise. There have been an extensive number of technical papers listed in the text or tables in this chapter that describe the geologic features and geotechnical conditions throughout both developments. The reader is encouraged to consult these references to obtain more details than can be covered in this chapter.

The Mad Dog Development in Green Canyon, Block 782, as shown in Figure 2.47 features a SPAR platform located in 1348 m of water about 365 m away from the top of the escarpment and a major slump (submarine landslide named Slump 8). The mooring spread for the SPAR at Mad Dog includes eleven mooring lines connected to 18- to 25-ft diameter suction caissons.

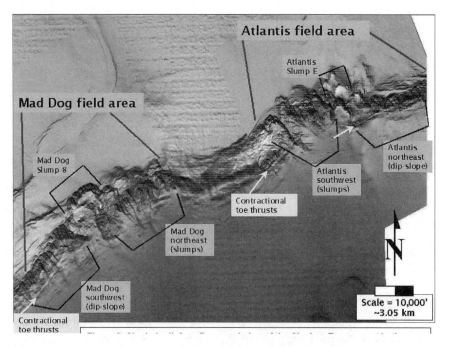

Figure 2.46   Mad Dog and Atlantis Developments on Sigsbee Escarpment © 2003 OTC (Orange et al. 2003a): Reproduced with permission of owner. Further reproduction prohibited without permission.

Figure 2.47   Mad Dog Production SPAR © (Courtesy of BP website).

The Atlantis Development includes a semi-submersible platform positioned above the production well cluster that is located about 1.8 km from the toe of the escarpment, as shown in Figure 2.48. The twelve point mooring system consists of four clusters of three suction caissons located below the escarpment. The water depth at the Atlantis DC-1 Drill Center is 2077 m.

The seabed geomorphology along the Sigsbee Escarpment is extremely complex and exhibits numerous signs of the dynamic behavior of the underlying allochthonous salt nappe. The thin-skinned sediment section overlying the series of en echelon, coalescing Pleistocene emplaced salt intrusions has been uplifted and deformed, resulting in the seaward movement of sediments and subsequent over-steepening of the slope as described by Sweirz, A. M. (1992). The upward and lateral movement of the underlying salt has deformed the sediments resulting in steep slopes, faults, and slumps, as illustrated in Figure 2.49.

Thirty-five large-scale slope failures (slumps or submarine slides) have been mapped along the escarpment, as shown in Figure 2.46, between Atlantis and Mad Dog, as described by Young et al. (2003). The largest landslide feature in the Mad Dog area is Slump 8, as illustrated in

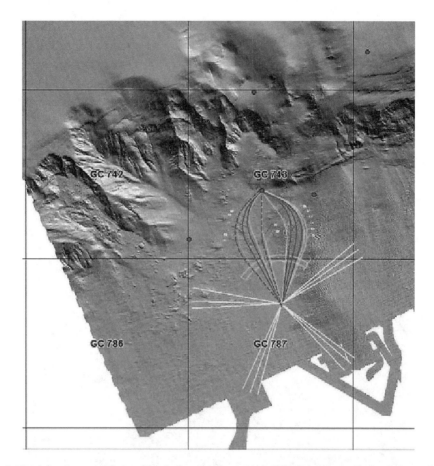

Figure 2.48   Atlantis semi-submersible platform layout © 2003 OTC (Jeanjean et al. 2003): Reproduced with permission of owner. Further reproduction prohibited without permission.

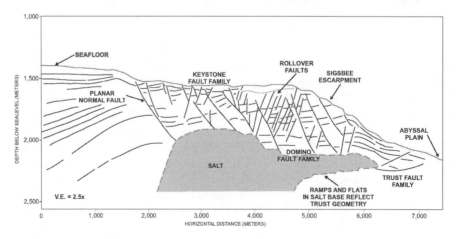

Figure 2.49   Seismic profile illustrating salt/fault interaction across Sigsbee Escarpment. Figure provided courtesy of Mr. Alan G Young.

Figure 2.50. The depleted area covers over 8.1 km² with an estimated volume of 1.02 billion m³ as estimated by Young et al. (2003). Slump A is the largest slump in the Atlantis development, and it formed in a similar fashion as Slump 8 at Mad Dog. The deformational style of each slump (landslide) is different and is a function of the location of the salt with respect to the geometry and physical properties of the overlying sediment.

These slumps produced four types and magnitudes of mass-transport events:

Slumps—landslide characterized by a large shearing and rotary movement of an independent mass of earth along a curved slip surface, as shown in Figure 2.50

Debris Flows—a moving mass of soil and mud with large clumps of intact sediment

Mudflows—a flowing mass of predominantly fine-grained sediment possessing a high degree of fluid

Turbidity Currents—a sediment-laden, bottom-flowing current that moves swiftly down the slope and spreads laterally as it churns up sediment in the surrounding water.

All of these mass transport types represent different levels of risk to various seafloor facilities. An important component of the Mad Dog and Atlantis integrated studies was the assessment that each type of landslide might pose to the field architecture planned for these two developments.

2.8.2 Mad Dog and Atlantis Geo-constraints

There were a large number of other geo-constraints that had to be addressed in the area around the two developments, as described by Jeanjean et al. (2003). The coauthors want to highlight an important distinction between the words geohazard and geologic constraint since we prefer to use the word geo-constraint. We prefer not to use the word geohazard in this chapter because we believe that the word is a misnomer and implies overstated risks to nontechnical personnel involved with the project.

The ISO 17776:2000E (2000) defines a hazard as a source of harm. The environmental loadings associated with winds, waves, and currents are not classified as hazards. These loading conditions are considered design constraints. The word hazard implies a judgment has been made in terms of

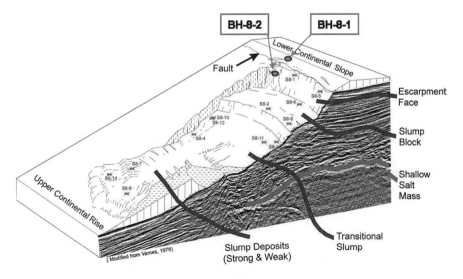

Figure 2.50   Mad Dog slump 8—3-D perspective © 2003 OTC (Young et al. 2003): Reproduced with permission of owner. Reproduction prohibited without permission.

risk before any geologic assessments have been made in combination with engineering studies to design the facility for all exiting constraints. In this context, even a flat seafloor can be a geohazard if the foundation supporting a facility was not properly designed for the soil strength present at the site. Our experience indicates that very few geologic conditions exist where a realistic and safe engineering solution cannot be developed. For this reason, we prefer to use the word geo-constraint in this chapter and propose that other industry professionals follow suit.

Table 2.5 will address each of the geo-constraints that exist along the Sigsbee Escarpment between Mad Dog and Atlantis. The table includes a checklist of data requirements and references that describes the scope of work and methods of analyses used to evaluate the design impact of each geo-constraint.

The chapter will now focus on each of these geo-constraints and describe the approach used to assess the risk to the field development architecture.

2.8.3 Refining Integrated Geologic Model

The initial geologic model developed during the planning study was used to ensure that the planned geophysical and geotechnical site investigations provided the proper level of coverage. The coverage needed to be extensive enough to adequately characterize the geologic and geotechnical conditions throughout the planned area of the field architecture. In some areas, the coverage extended beyond the planned area of the developments to confirm that the risks associated with slope instabilities upslope or downslope would not adversely influence the developments.

Once the geophysical survey was complete and the data was available, then the process of refining the geologic model began in order to allow adequate time to finalize the scope of the geotechnical investigation. The process started by loading all the geophysical data into a GIS database.

The marine geologic team then interpreted the data and mapped the geologic conditions and processes within the three-dimensional space identified as the project area. The goal was to use all the data to refine the geologic/geotechnical model by characterizing the seafloor and subsurface features, past and active processes, and the sequence and variability of subsurface conditions. The intent was to map all geologic conditions and stratigraphic horizons within the plan area of the development to a depth (about 200 to 400 m below the seafloor) greater than planned foundation penetration to investigate potential risk of slope instability. This was important to define the deeper geologic structure and processes, such as salt intrusion, that have impacted the area.

The objective of building the working geologic model was to understand the processes that influenced the area in order to:

- Understand the historic events
- Identify the stratigraphic boundaries (horizons)
- Recognize the geologic structure and deformation
- Identify geologic constraints to infrastructure layout
- Reconstruct the sequence stratigraphy and age of each horizon
- Correlate seismic stratigraphy with soil properties
- Use seismic data to select key stratigraphic targets to plan borings and cores
- Develop soil province maps defining spatial variability
- Project the historic model into the future to assess future risks

All seabed conditions, geologic processes, and subsurface conditions that would impact shallow or deep foundations and infrastructure lineaments, such as pipelines, were carefully investigated. In addition, the geosciences information gained during the integrated study provided valuable input to the exploration groups for planning the location and designing top-hole conductor installation for future exploration wells.

Table 2.5 Mad Dog and Atlantis geo-constraints—data requirements and references

Geo-constraint	Data requirements							References
	Deep cores	MSCL	Borings	Sub-bottom profiles	Swath bathymetry	Advanced testing	Age dating control	
Steep slope gradients				Maybe	Yes			
Slope reversal (irregular seafloor topography)				Maybe	Yes			
Fault displacement/offsets	Yes	Yes	Maybe	Yes	Yes		Yes	Angell et al. (2003) Orange et al. (2003) Slowey et al. (2003)
Shallow-seated seafloor instability	Yes	Yes	Maybe	Yes		Yes	Yes	Brand et al. (2003b)
Debris/turbidity flows	Yes	Yes		Yes	Yes	Yes	Yes	Niedoroda et al. (2000)
Deep-seated seafloor instability	Maybe		Yes	Yes		Yes	Yes	Niedoroda et al. (2003) Orange et al. (2003) Nowacki et al. (2003) Al-Khafaji et al. 2003
Spatial soil variability	Yes	Yes	Yes	Yes		Yes	Yes	Young et al. (2003) Brand et al. (2003b)
Current and erosion	Yes	Yes		Yes	Yes	Yes	Yes	Brand et al. (2003) Niedoroda et al.(2003a)
Gas/fluid expulsion shallow water flow			Maybe	Yes	Maybe			Eaton (1999) Pelletier et al. (1999)

2.8.3.1 Refined Geologic Model

A realistic geologic model for the Mad Dog Atlantis development required a good understanding of the impact of the salt underlying the Sigsbee Escarpment. The deformation of the supra-salt played a critical role in understanding the resulting morphology and the stratigraphic section of the sediment overriding the salt. The geologic model required a proper understanding of the deformation history associated with the underlying coalescing salt masses. The model had to define the mechanism associated with over-steepening of the escarpment slopes and the resulting gravitational instability and slumping along the face of the escarpment as described by Orange et al. (2003) and Young et al. (2003).

The paper by Orange indicates that the appearance of the escarpment, as shown in Figure 2.46, is related:

1. Variations in the geometry and structural interactions between shallow sediments and the underlying salt
2. Stratigraphic sources of anomalously high pore-fluid pressure (conduits)
3. Long-term patterns of sediment erosion and deposition

The style and timing of faulting in different areas is strongly influenced by the salt-sediment interaction. The faulting in particular influences the magnitude and recurrence interval of the mass wasting events. Thus, the geomorphic character of the escarpment is related to the style of fault-related deformation, the distribution and geometry of the salt nappe, and sedimentary loading of the basin behind the escarpment.

Placement of any planned facilities throughout the Mad Dog and Atlantis Developments required a clear understanding of their proximity to past and future deformation of the supra-salt section and the processes that contributed to evolution of the escarpment. Thus, the final geologic/geotechnical model required deciphering the factors that helped form the escarpment that included:

- Sea level change and sedimentation rate
- Pore water pressure development and variability
- Salt movement and fault response
- Rate and location of salt deformation
- Seafloor gradient changes
- Slope adjustment mechanisms

A number of key seafloor and geologic conditions/processes were identified and evaluated as listed in Table 2.6 to complete the geologic/geotechnical model. The potential degree of risk posed by each geo-constraints is also indicated in Table 2.6 for different foundation types.

The type of geophysical data required to assess the risks associated with each geo-constraint is also presented in Table 2.6. The sub-bottom profiler serves a very important role since its coverage provides a comprehensive three-dimensional grid of seismic sections covering the entire cube of geologic conditions and processes. Most of the geologic conditions that can impose a potential geo-constraint were identified and mapped in the subsurface volume. The sub-bottom profiles were also used to understand the sedimentary processes that formed each depositional layer. Each stratigraphic unit was closely examined to assess the depositional and deformation processes that formed the shallow stratigraphic section. For example, sediment layers formed by debris or turbidity flows exhibit different seismic signatures. Turbidity flows and debris flows, as illustrated in Figure 2.51, produce a very irregular seismic signature as compared to uniform layering associated with normal deepwater sedimentary processes.

Table 2.6 Potential foundation risks for various geo-constraints

Geologic process or condition (geo-constraint)	Seafloor lineaments (pipelines, mooring lines, etc.)	Shallow foundation (mudmats, suction piles, etc)	Deep foundation (driven piles, conductors, etc.)	Geophysical data required
Steep slope gradients	Medium	High	Low	Multi-beam bathymetry
Slope reversal (irregular seafloor topography)	High	High	None	Multi-beam bathymetry
Fault displacement/offsets	Low	Medium	High	Side-scan sonar & sub-bottom profiler
Shallow/deep-seated slope instability	High	High	Medium	Side-scan sonar & sub-bottom profiler
Debris/turbidity flows	High	Medium	Low	Side-scan sonar & sub-bottom profiler
Spatial soil variability	High	High	Low	Side-scan sonar & sub-bottom profiler
Currents and erosion	High	Medium	Low	Multi-beam bathymetry Side-scan sonar & sub-bottom profiler
Gas/fluid expulsion shallow water flow	Low	Medium	High	3-D seismic & 2-D High resolution

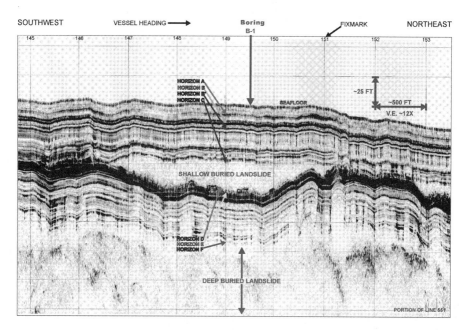

Figure 2.51   Comparison of seismic profiles for different sedimentary processes © 1998 OTC (Doyle 1998): Reproduced with permission of owner. Further reproduction prohibited without permission.

At this stage, the refined geologic model had progressed to the point where the work plan and work scope for the geotechnical investigation could be finalized. The next section will describe briefly the scope of the geotechnical investigation and the process of planning and conducting the laboratory program.

2.8.4   Geotechnical Investigation Objectives

One of the key objectives of the geotechnical investigation performed for the Mad Dog and Atlantis projects was to define the physical and engineering properties of the sediments. Another key objective of the program was to estimate the time frames when the mass movements occurred along the escarpment. This objective also required the measurement of soil properties needed for the slope stability analyses and the properties of the sediments required to perform the mass gravity flow analyses.

The geotechnical program included JPCs and PCs at 95 sites and deep soil borings at 8 locations. The paper by Young et al. (2003) describes the details of the coring program. Al-Khafaji et al. (2003) describes the details for the soil-boring program and compares the strength profiles for the soil borings drilled, as shown in Figure 2.52 and upon the Sigsbee Escarpment.

2.8.5   Laboratory Testing Program

After the samples and field data were acquired at the Mad Dog and Atlantis developments, then it was reviewed to plan a comprehensive program of laboratory testing. The laboratory program was divided into two stages. The first stage of laboratory testing was conducted on the shallow seabed samples, and the down-hole samples were tested during Stage 2.

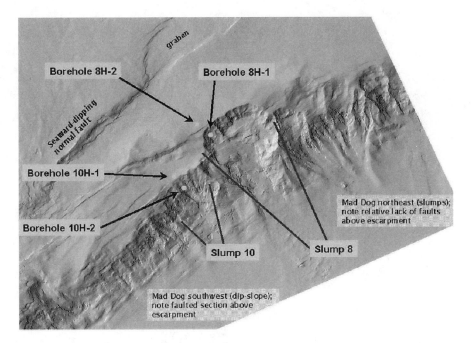

Figure 2.52 Mad Dog and Atlantis geotechnical sites 2003 OTC (Orange et al. 2003b): Reproduced with permission of owner. Further reproduction prohibited without permission.

2.8.5.1 Shallow Seabed Samples

Each core acquired with the JPC or PC was extensively tested as described by Young et al. (2003). The samples obtained with the PC and JPC are continuous which provided an opportunity for direct correlation with the continuous CPT data. The continuous samples were logged with a Multi-Sensor Core Logger (MSCL) manufactured by Geotek (Schultheiss and Weaver 1992) to provide a continuous profile of sediment properties. The MSCL is a computer-controlled, automated system that takes nonintrusive measurements of bulk density, compressional wave (P-wave) velocity, and magnetic susceptibility of a sediment core. Physical soil properties that can be computed from the MSCL measurements include: (1) moisture content, (2) bulk density, and (3) impedance. These continuous profiles of data, as shown in Figure 2.53, were correlated with the sub-bottom profile data to help identify marker horizons and confirm the variation in properties between horizons.

Another key benefit of a continuous sample was the ability to split the sample and take color photographs, as shown in Figure 2.54. Visual inspection of the cores allowed one to detect debris flow deposits from normal sedimentary deposits. The continuous visual picture of the sediment lithology allowed the selection of appropriate intervals for age dating and other testing with more confidence.

2.8.5.2 Age Dating Shallow Sediments

The principles of chronostratigraphic sedimentation played a very important role in mapping marine sediments and understanding the close connection with the changes in sea level and the cyclic change in sedimentation rates. The sedimentary layers underlying the seafloor provided a fundamental framework for understanding the geologic history and the geologic processes that influenced the development area as described by Healy-Williams (1984).

A reference JPC designated as CSS-1, illustrated in Figure 2.55, was taken near the Atlantis Development above the escarpment where normal sedimentation was anticipated to have occurred

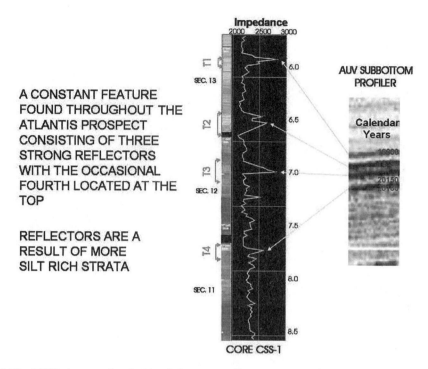

Figure 2.53 MSCL data correlated with sub-bottom profiles © 2003 OTC (Young et al. 2003): Reproduced with permission of owner. Further reproduction prohibited without permission.

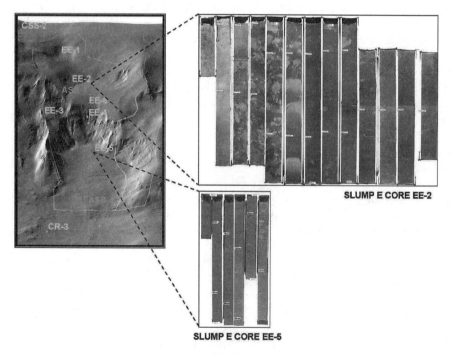

Figure 2.54 Photograph of Atlantis split core showing chaotic debris flow deposits in JPC (continuous 100-mm-diameter sample) © 2003 OTC (Young et al. 2003): Reproduced with permission of owner. Further reproduction prohibited without permission.

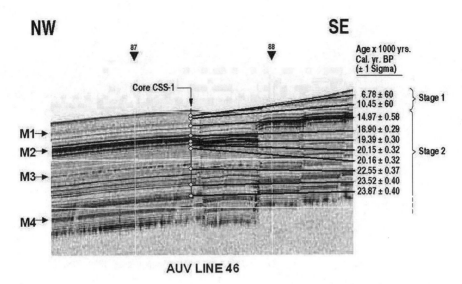

Figure 2.55 Reference core CSS-1 with age dates © 2003 OTC (Slowey et al. 2003): Reproduced with permission of owner. Further reproduction prohibited without permission.

over the last 40,000 years. Foraminifera shells were picked from this core and cleaned in an ultrasonic bath, dried, and then converted to graphite. The ratio of radioactive C_{14} to stable C_{12} was measured by an FN tandem accelerator mass spectrometer (AMS). The resultant absolute AMS-C_{14} ages were reported in conventional radiocarbon years. A correction to calendar years before 1950 (BP) was made for ocean reservoir age effect and temporal variations in the production of radiocarbon as described by Slowey et al. (2003).

There were a total of 79 samples from Mad Dog and Atlantis that were radiocarbon age dated. The CSS-1 core included 10 age dates, as shown in Figure 2.55. Radiocarbon dating is generally considered accurate in the age range from a few hundred years to about 50,000 years. Details of how these radiocarbon age dates were used to investigate the risk of various geo-constraints are described later in the chapter.

2.8.5.3 Down-hole Samples

The series of laboratory tests performed during the boring program were plotted on an individual boring log for each soil boring. The core log, as shown in Figure 2.56, includes the strata breaks and a written description of the major soil layers encountered in the boring and any secondary soil components existing throughout each soil stratum. The descriptions were based on a standard classification system such as the Unified Soil Classification System (1953). This integrated boring log also shows the results of the MSCL and a cross-section obtained with the sub-bottom profiler showing the stratigraphy where the JPC was obtained. Other important descriptors included structure, texture, inclusions, and color. For consistency, the colors were based on a standard color description chart (Munsell chart).

The laboratory tests performed during the field investigation included moisture content, miniature vane, remolded miniature vane, and UU-Triaxial tests. Additional classification tests were then assigned to improve our knowledge of the soil types encountered at each depth in the soil profile. Supplemental classification tests included: (1) moisture content, (2) Atterberg limits, (3) grain size distribution (sieves and hydrometer), (4) bulk density, and (5) specific gravity. Once these classification results were available along with the results from the site investigation,

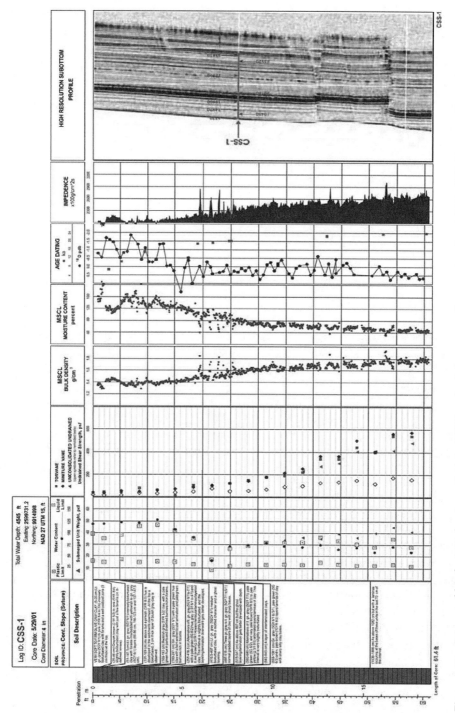

Figure 2.56 Integrated log for JPC showing soils data, MSCL and sub-bottom profiler cross-section © 2003 OTC (Young et al. 2003): Reproduced with permission of owner. Further reproduction prohibited without permission.

an advanced testing program was assigned to determine important geologic and geotechnical engineering parameters.

The framework for the advanced testing was to characterize the stress history and measure drained peak and undrained strength properties. The advanced laboratory testing program as described by Al-Khafaji et al. (2003) included: (1) controlled-rate-of-strain (CRS) one-dimensional consolidation, (2) static K_o consolidated drained direct simple shear (CK$_o$U'-DSS) at lab induced OCR ≥ 1.0, (3) static CK$_o$U'-TC (triaxial-compression) and CK$_o$U'-TE (triaxial-extension), (4) ring shear, undisturbed and remolded, (5) x-ray radiograph, (6) MSCL, (7) C_{14} radiocarbon age dating, and (8) erosion tests. The ring shear tests were performed to provide input soil parameters for the slope stability analyses to assess whether slope instability posed a potential constraint to planned facilities. Slope instability in the over consolidated marine clays is generally associated with progressive (delayed) slope failures as described by Skempton (1985) and Stark and Eid (1997).

The laboratory program was carefully monitored to confirm that tests were conducted in accordance with recognized standards (ASTM 2007). All laboratory data sheets were reviewed to confirm that individual tests were performed according to the test assignments. All test results were checked and the advanced test results were generally plotted individually while the standard testing data were presented on the core or boring logs.

2.8.5.4 Shear Strength Testing Interpretation

The final geotechnical engineering analysis required more detailed laboratory test results to supplement the in situ testing data and develop undrained shear strength profiles for analyses of deep-seated slope failures. Consequently, an extensive program of advanced testing was conducted on the samples from the Mad Dog and Atlantis cores and borings. The Stress History and Normalized Soil Engineering Properties (SHANSEP) approach was selected to develop the shear strength profiles (Ladd and Foott 1974). The SHANSEP technique also allows for extrapolation of test results to other locations within a soil deposit and can be extremely beneficial for reducing the scope of a deepwater site investigation.

The SHANSEP concept utilizes the following equation (Ladd et al. 1977) to relate in situ undrained shear strength S_u to parameters developed from results of CRS consolidation tests and K_o-consolidated undrained shear strength tests:

$$S_u = \sigma'_v \left(S_{u_{DSS}} / \sigma'_{vc} \right)_{nc} OCR^m \tag{2.9}$$

where S_u = computed in situ undrained shear strength

σ'_v = in situ effective vertical stress

$\left(S_{u_{DSS}} / \sigma'_{vc} \right)_{nc}$ = normalized shear strength for normally consolidated soil

where S_u = undrained shear strength obtained from laboratory sample consolidated to effective consolidation pressure σ'_{vc}

OCR = overconsolidation ratio

m = strength rebound exponent relating the normalized shear strength ratio to the OCR

For the Mad Dog and Atlantis samples, $\left(S_{u_{DSS}} / \sigma'_{vc} \right)_{nc}$ was based on the results of DSS testing. The consolidation test results were used to estimate the state of stress (preconsolidation pressure) and to select consolidation pressures for the DSS tests. The DSS tests were performed to determine the normalized parameters utilized by the SHANSEP method. The state of stress and normalized data then were used to develop undrained shear strength profiles (Ladd and Foott 1974). The range of parameters developed for the SHANSEP analyses is presented in Table 2.7.

Table 2.7 Best estimate of SHANSEP parameters, Jeanjean et al. (2003)

Site	Normalized shear strength ratio	m
Mad Dog	0.18 to 0.23*	0.8
Atlantis	0.25	0.67

Note: Range of values measured in different layers

The SHANSEP method assumes that the normalized shear strength ratio $(S_{u_{DSS}}/\sigma'_{vc})_{nc}$ is a constant value. However, Quiros et al. (2000) have presented DSS test results for soils from various parts of the world, including significant data from the Gulf of Mexico. The data set utilized by Quiros et al. is plotted on Figure 2.57 in the form of normalized shear strength ratio versus effective consolidation pressure. As indicated by the plot, a trend of decreasing $(S_{u_{DSS}}/\sigma'_{vc})_{nc}$ values with increasing values of vertical consolidation pressure is established. Quiros et al. utilized a least-squares regression on their plotted data to yield the following relationship:

$$(S_{u_{DSS}}/\sigma'_{vc})_{nc} = 0.294(\sigma'_{vc})^{-0.113} \tag{2.10}$$

where the equation components are as previously defined.

The Quiros et al. paper cautions that if the above laboratory correlation is employed to estimate the in situ shear strength, the results are likely to be unconservative due to a decrease in void ratio compared to the in situ void ratio. In an effort to further extend the usefulness of their DSS data set, Quiros et al. studied the possibility of relating the laboratory shear strength data with the pressure-water content ratio, the effective consolidation pressure divided by the measured final (after consolidation) water content (σ'_{vc}/W_c). Their data set plotted as DSS shear strength versus the (σ'_{vc}/W_c) ratio is shown on Figure 2.58. Their equation based on a best-fit line correlation of the data is as follows:

$$S_u = 0.258(\sigma'_{vc}/W_c)^{0.686} \tag{2.11}$$

where the equation components are as previously defined.

They state that the above equation may be useful in preliminary evaluations of shear strength where no advanced testing is available but caution that site-specific determination of the relation between undrained shear strength and the pressure-water content ratio (termed the SPW line) is advisable. They also state that the history of successful application of the stress-history approach to offshore in situ shear strength evaluation may demonstrate that sample strength in offshore environments generally are more strongly influenced by past maximum consolidation than soil structure. Their study also indicates that the database shows no reliable correlation between the normalized shear strength ratio and the soil plasticity indices.

The following sections will describe procedures most routinely used to interpret soil parameters from the laboratory and in situ testing data.

2.8.6 In Situ Data Interpretation

The CPT/PCPT data obtained during the field investigations was used according to Lunne et al. (1997) for three primary applications: (1) profiling the subsurface stratigraphy, (2) identifying the soil type, and (3) estimating geotechnical soil parameters. We did not use the data directly in the geotechnical engineering analyses as proposed by Lunne for the third application.

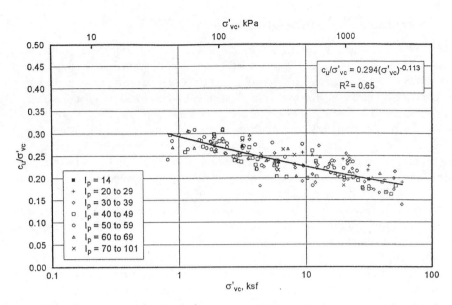

Figure 2.57 Quiros plot of $S_{u_{DSS}}/\sigma'_{vc}$ versus σ'_{vc} © 2000 OTC (Quiros et al. 2000): Reproduced with permission of owner. Further reproduction prohibited without permission.

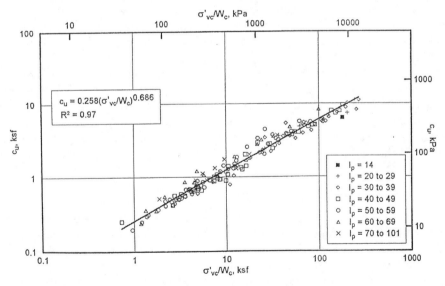

Figure 2.58 Quiros plot of $S_{u_{DSS}}$ versus σ'_{vc}/W_c © 2000 OTC (Quiros et al. 2000): Reproduced with permission of owner. Further reproduction prohibited without permission.

Continuous measurements made during a PCPT sounding, as shown in Figure 2.35, are very useful for selecting the best foundation site and interpreting the range in soil parameters appropriate for foundation design. The continuous measurements during the PCPT sounding, as shown in Figure 2.35, are the cone point resistance q_{net}, the sleeve friction f_s, and the dynamic pore pressure μ.

2.8.7 Interpreted Shear Strength Profile

Once the field and laboratory testing was available, then the geotechnical engineers interpreted the shear strength profile to be used for foundation design as described in Chapter 3 or to perform slope stability analyses. The results of in situ and laboratory tests are typically plotted on a boring log, as shown in Figure 2.56. The log generally includes plots of addition data sets, such as classification test results and MSCL measurements.

One of the most difficult tasks in the design process is to interpret a design shear strength profile applicable for various analyses to be performed for the different foundation types. There may be significant scatter in the values of S_u measured on laboratory samples as described by Quiros et al. (1983). Significant differences may also exist between the measured values of laboratory and in situ strength data. Our experience indicates that SHANSEP DSS data corrected for OCR typically compares closely with strengths interpreted from CPT data using an appropriate value of N_{kt}. These strengths also typically compare closely with VST data corrected with a proper adjustment factor. We recommend using all three data types (SHANSEP DSS, CPT, and VST) to interpret a consistent strength profile for foundation design. The site-specific SPW profile should also be plotted as an additional set of data for comparison. Table 2.8 details the field and laboratory shear strength data and procedures frequently used to interpret the final strength profile.

Soil sensitivity is a parameter that has proven to be very useful to support interpretation of the design strength profile. Samples that are disturbed during the sampling procedure or during transport from the field to the laboratory will obviously have lower values of soil sensitivity than those that are properly sampled and handled. A comparison of the measured sensitivities with trends in moisture content and plasticity indices will help reveal the samples that are highly disturbed and should be eliminated in interpreting the design strength profile. This approach will improve final selection of the design strength profile by eliminating individual strength measurements on disturbed samples.

The process of interpreting the strength profile for foundation design is complex and must involve a close review of the multiple data sets. A good understanding of the geologic setting that influenced the stress history and depositional processes also play a very important consideration. Finally, sound engineering judgment is required while using the data sets described in Table 2.8 to select the final strength profile for design representing each soil unit that will influence foundation performance.

2.8.8 Uncertainty in Data Interpretation

The geotechnical engineer must develop a good understanding of the epistemic and aleatory uncertainty of the data throughout the project area. As previously discussed, Christian believes that geotechnical engineers generally face epistemic uncertainties. The coauthors believe that the geosciences team faces both types of uncertainties during an integrated study.

The dictionary defines aleatory as a state of uncertainty that is beset with perils, full of risk, and fraught with danger. During an integrated study, both the geophysical and geotechnical data can be used to assess the epistemic and aleatory uncertainties. By comparing both the geophysical and geotechnical data at a potential foundation site, one can assess the range in soil strength profiles applicable for the foundation design. An example of the extreme variation in the aleatory uncertainty is illustrated in Figure 2.59. The measured strengths in the debris flow deposits reveal extreme differences due to depositional nature of these materials. The wide variation is real and should not be neglected in interpreting the design strength profile.

Jeanjean et al. (2003) describes the aleatory uncertainty associated with the soil conditions in Slump 8 at Mad Dog where one cluster of suction caissons was installed. Three anchor clusters moor the Mad Dog SPAR, as shown in Figure 2.60.

Table 2.8 Types of shear strength data available for interpreting design strength profile

Undrained shear strength data	Comments
Standard field and laboratory tests	
Undisturbed miniature vane	Excellent data to interpret the initial design strength profile if one eliminates measurements on disturbed samples.
Remolded miniature vane	Critical data that provides a check on the trend for interpreting design strength profile and determining where MV data were measured on disturbed samples.
Unconsolidated undrained triaxial compression (UU)	Trend of UU strength data typically closely correlates with other strength data measured on high-quality samples. The data is often used as the reference strength to correlate with CPT data.
Remolded unconsolidated undrained triaxial compression	Trend of strength data generally follows the MV_r trend.
Plot of $MV_r \times S_t$ versus depth. Sensitivity S_t = undisturbed MV strength/remolded MV strength	The plot, when based on realistic values of S_t, will produce a good correlation with the design strength profile obtained with other strength data.
UUr $\times S_t$ where S_t = undisturbed UU-triaxial shear strength/remolded UU-triaxial shear strength	Realistic values of S_t will produce good correlation to the design strength profile obtained with other strength data.
In situ field tests	
CPT	Key method for interpreting the design strength profile if a realistic value of N_{kt} is used. The S_u profile is based on the equation ($S_u = q_{net}/N_{kt}$) where N_{kt} = regional factor. (17.5 for Gulf of Mexico)
Vane shear test (VST)	S_u = VST times adjustment factor (ranging from 0.75 to 0.85).
Advanced laboratory tests	
Direct simple shear (DSS)	Best data available to interpret the design shear strength data when high quality samples are consolidated to appropriate effective overburden pressures.
SHANSEP $S_u = \sigma_v' (S_{u_{DSS}}/\sigma_{vc}')_{nc} OCR^m$	One of the best methods for interpreting the design strength profile and serves as an excellent baseline for comparison with other strength types.
SPW $S_u = 0.258 (\sigma_{vc}'/W_c)^{0.686}$	Useful correlation when site-specific SHANSEP data is not available, but final determination should be based on actual site data.

As illustrated in Figure 2.61, the CPT profiles for Anchor Cluster 2 are highly variable as compared to the CPT soundings at Anchor Clusters 1 and 3. The soils in Slump 8 were highly variable since they were deposited during a debris flow as interbedded zones of soft debris flow material, silt and sand layers, and stiff clay debris flow blocks. The soil conditions throughout the area of

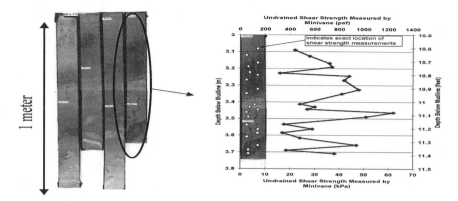

Figure 2.59 Variability in soil strength profile © 2003 OTC (Jeanjean et al. 2003): Reproduced with permission of owner. Further reproduction prohibited without permission.

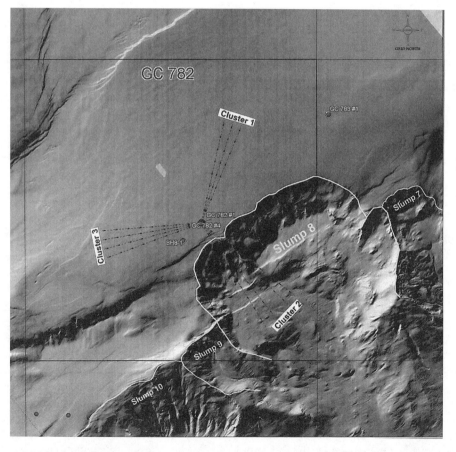

Figure 2.60 Mad Dog SPAR anchor clusters © 2006 OTC (Berger et al. 2006): Reproduced with permission of owner. Further reproduction prohibited without permission.

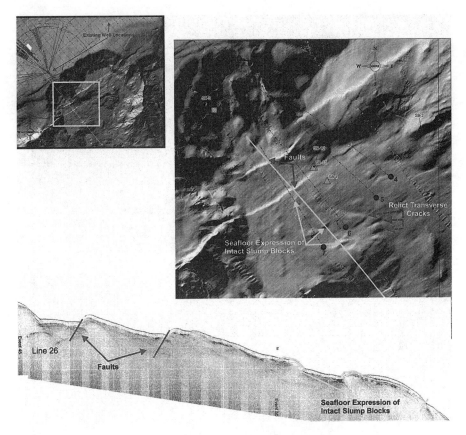

Figure 2.61 CPT profiles for Anchor Cluster 2 © 2006 OTC (Berger et al. 2006): Reproduced with permission of owner. Further reproduction prohibited without permission.

Clusters 1 and 3 were very uniform and continuous, as shown in Figure 2.62. The sub-bottom profiles and four CPTs helped define the systemic uncertainty, but the data demonstrated that the aleatory uncertainty was not as great as would be expected without the additional data. Thus, the integrated data sets allowed the geotechnical engineers to more accurately understand how different geologic processes impacted the epistemic and aleatory uncertainties for the various soil profiles.

2.8.9 Final Geologic/Geotechnical Model Integration

Once the results from field and laboratory programs were available, the geosciences team began the process of integrating the results into the final geologic/geotechnical model. The word geotechnical has been added as a descriptor to the term geologic model since it was the first opportunity to make a real comparison of the stratigraphic section and the soil properties associated with each stratum.

At this point, we want to emphasize the important role of stratigraphy in defining the geologic/geotechnical model. Weller (1947) provides an excellent definition of stratigraphy as follows:

"Stratigraphy is the great unifying agency of geology that makes possible the synthesis of a unified geological science from its component part."

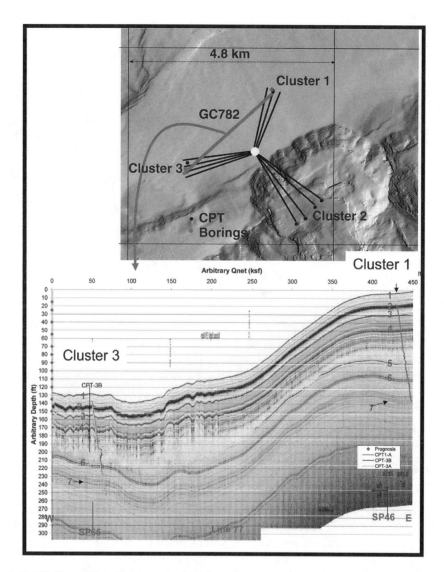

Figure 2.62 Uniform soil conditions around Anchor Clusters 1 and 3 at Mad Dog Spar © 2005 (Jeanjean et al. 2005): Reprinted with permission from Dr. Susan Gourvenec, Professor at University of Western Australia, Center for Offshore Foundation Systems, Perth.

In other words, stratigraphy is the unifying component that allows integration of the geology and geotechnical engineering data into a comprehensive geologic/geotechnical model. The activities associated with this phase of the integrated study included:

- Interpreting critical geologic and geotechnical parameters that defined the spatial and temporal processes affecting foundation design for the facilities interacting with the seafloor
- Analyzing, evaluating, and recommending optimum foundation locations and types to avoid geo-constraints and costly installation and foundation risks
- Documenting the geologic setting and processes that produced the geotechnical properties associated with each subsurface soil layer
- The subsurface assessment required mapping the relevant horizons and noting stratigraphic variations existing throughout the project area.

Bryant et al. (1995) previously described a realistic framework for defining the geotechnical stratigraphy of the sediments on the Continental Slope in the Gulf of Mexico. The framework for defining the geologic model is to interpret the stratigraphy in terms of the distribution and succession of the physical properties of the sediment layers that have been measured in each core, as shown in Figure 2.63. Each sediment layer must be defined in terms of its succession and sedimentation history in order to define the mode of origin and their geologic and geotechnical history. The resulting geologic/geotechnical stratigraphy will define the normal sequence of geotechnical properties and map the deviations caused by mass-wasting, erosion, creep, faulting, and slumping events observed in the seismic profiles. The end result was a series of geologic/geotechnical maps showing horizon isopachs, structure, soil provinces, and geologic features that might influence foundation installation and pose risk for drilling future wells. Areas exhibiting gas-charged sediments or areas prone to shallow water flow were also identified to allow adjustments for proposed well sites or placement of final drill center.

2.8.9.1 Chrono-stratigraphic Markers

The chronology of sea level/climatic history during the last 30,000 years (Slowey et al. 2003) of the Quaternary period played a very important role relative to the changes in sediment properties and lithology around the Mad Dog and Atlantis developments. The sediment accumulation during this period changed dramatically from the glacial low-stand to the present Holocene high-stand. Thus, the age dating information obtained from the C_{14} radiocarbon method was correlated with the regionally persistent seismic reflectors, as shown in Figures 2.55 and 2.64.

An example of a chrono-stratigraphic marker horizon was identified and named the Triplet. This marker horizon as described by Young et al. (2003) covers an extensive area along the Sigsbee Escarpment in the deepwater Gulf of Mexico. This marker horizon actually consists of three, sometimes four, silt-rich layers that were deposited between 18,900 and 20,200 years ago, near the end of the last sea level low-stand, as shown in Figure 2.53. The characteristics of this triple-kick signature are clearly evident in the MSCL velocity and bulk density logs. In the area around

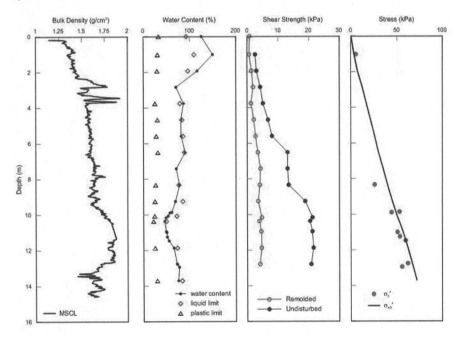

Figure 2.63 Typical properties from offshore plateau regions. Figure provided courtesy of Dr. Niall Slowey, Professor with Texas A&M University.

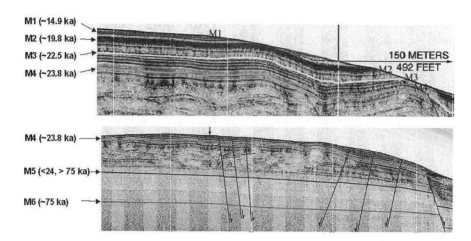

Figure 2.64 Chrono-stratigraphic marker horizons including Triplet © 2003 OTC (Slowey et al. 2003): Reproduced with permission of owner. Further reproduction prohibited without permission.

the Mad Dog and Atlantis Developments, the Triplet was encountered at about 6 m below the seafloor in normally consolidated depositional areas. It served as an excellent marker horizon and provided very accurate litho-stratigraphic correlation for understanding the timing of Pleistocene and Recent processes. The centimeter accuracy in stratigraphic consistency between Cores CSS1 and CSS2 is shown in Figure 2.65. These two cores are located about 2700 m apart.

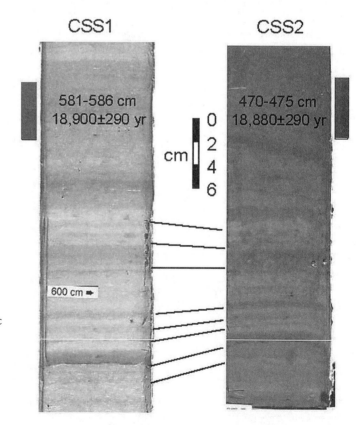

Figure 2.65 Lithostratigraphic correlation of triplet between Cores CSS1 and CSS2 © 2003 OTC (Slowey et al. 2003): Reproduced with permission of owner. Further reproduction prohibited without permission.

The sub-bottom profiles were correlated with age dating measurements such as C_{14} radiocarbon dates to assess the depositional history as described by Brand et al. (2003). For example, age dating was correlated with each stratigraphic horizon, as shown in Figure 2.64, to develop time markers of stratigraphic deposition. A number of time markers associated with stratigraphic deposition/erosion were used to identify areas prone to erosion and areas accumulating more sediment. This information was also used to compute the net sedimentation rates over different time periods, as illustrated in Figure 2.66. In addition, the age dating was used to measure the rate of seafloor erosion associated with deepwater currents.

Figure 2.67 shows a fence diagram prepared using sub-bottom profiler sections from the Upper Continental Rise at the Atlantis Development. The Triplet and other marker horizons indicate the uniform stratigraphic sequence over the area along with the areas where slump deposits were encountered. Mapping these shallow stratigraphic horizons also provided the basic framework for developing soil province maps. Each stratigraphic horizon was correlated with the geotechnical data to define the link between depositional history and soil properties.

With laboratory results and geologic data in hand, the geotechnical engineer and geologist could then fulfill their very important roles in terms of defining the potential variability of soil conditions and selecting the shear strength profile used in the foundation analyses. A key benefit was that three different types of data were used to compare subsurface soil conditions and

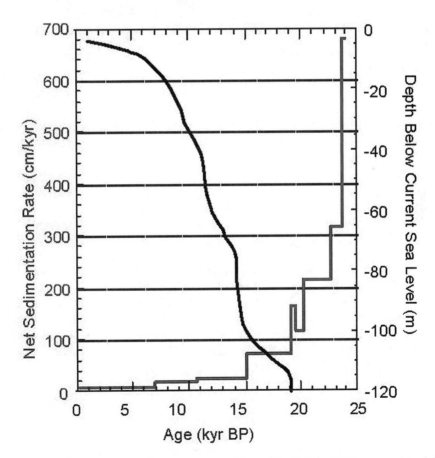

Figure 2.66 Net sediment accumulation rates computed from Atlantis Core CSS1 compared to global sea level change © 2003 OTC (Slowey et al. 2003): Reproduced with permission of owner. Further reproduction prohibited without permission.

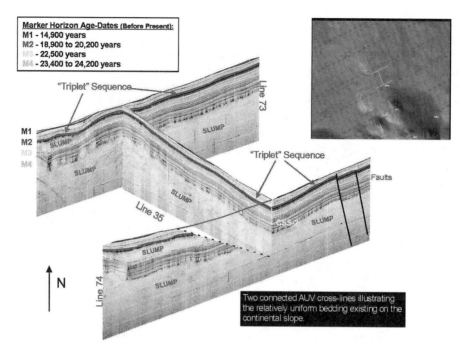

Figure 2.67   Fence diagram showing marker horizons on sub-bottom profiler sections along Upper Continental Rise at Atlantis development © 2003 OTC (Brand et al. 2003a): Reproduced with permission of owner. Further reproduction prohibited without permission.

to evaluate the spatial variability over the entire area of the development versus the variability throughout only the area of a specific foundation.

2.8.10 Site Favorability Assessment

The different production scenarios included placement of anchors, pipelines, and various subsea facilities that utilize different foundation concepts. The seafloor and subsurface conditions had to be fully understood to select the preferred locations for these structures and to analyze their foundation performance. Thus, the geosciences team prepared site favorability maps that defined the installation and operational criteria appropriate for each type of planned facility/foundation. An example of different site favorability criteria that might be considered is presented in Table 2.9.

The geologist then used the final geologic/geotechnical model to develop individual site favorability maps for each criteria listed in the assessment. Examples of site favorability maps prepared for the West Nile Project offshore Egypt is shown in Figure 2.68 as described by Moore et al. (2007). Maps such as these can be used in an iterative process to select the most optimum site for each individual facility. The following sections will describe the process used to assess the risk associated with each of the geo-constraints. In addition, the type of data that is required to accurately understand the key factors that must be considered in evaluating each of the geo-constraints will be described.

2.8.10.1 Slope Gradient/Reversal (Irregular Seafloor Topography)

All facilities placed on the seafloor require an evaluation of the slope gradient and seafloor topography throughout the foundation footprint. Different types of facilities have different tolerances in terms of the maximum slope gradients or nature of irregular seafloor topography on which they

Table 2.9 Site favorability assessment criteria

Site favorability assessment criteria			
Potential geo-constraint	Mudmat	Suction caisson	Pipeline
Steep slope gradient	3°	15°	20°
Slope reversal (irregular seafloor topography)	3°	10°	15°
Fault displacement/offset	< 0.1 m for 10-3 annual recurrence	< 1 m for 10-3 annual recurrence	< 3 m for 10-3 annual recurrence
Shallow-seated slope instability	Slope < 5° FS > 1.5	Slope < 15° FS > 1.25	Slope < 18° FS > 1.2
Deep-seated slope instability	Slope < 5° FS > 1.5	Slope < 15° FS > 1.25	Slope < 18° FS > 1.2
Debris/turbidity flows	Avoid	Limited	Avoid
Highly variable soil conditions	Avoid	Possible	Acceptable
Gas/fluid expulsion shallow water flow	Avoid	Avoid	Limited

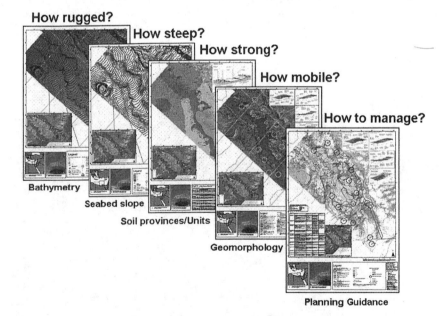

Figure 2.68 Example of site favorability maps for West Nile Project offshore Egypt (Moore et. al. 2007): Reprinted with permission from Dr. Roger Moore with Halcrow Group Ltd., Nat Usher, and Trevor Evans with BP.

can be placed. A mudmat needs to be placed on a seafloor with very little slope variation throughout the footprint to ensure that all portions develop full contact with the seafloor. For example, a mudmat with jumpers may require a seafloor slope less than 3°, whereas a suction caisson may be placed on steeper seafloor slopes as indicated in Table 2.9. The suction caisson has a higher tolerance for slope reversals and can probably tolerate up to about 15°. Seafloor lineaments such as pipelines are much more flexible and may be placed on slopes up to 20° or even greater if hold back piles are installed to provide more resistance to restrain pipeline movements.

To evaluate areas acceptable for these different facilities' criteria, a slope gradient map can be prepared using the swath bathymetry data. For example, areas fully satisfying the criteria can be color-coded to illustrate in a favorable color such as green. Yellow is used to indicate marginal areas that might be acceptable if special engineering design can be used to accommodate the marginal conditions. Of course, areas to be avoided will frequently be colored red, meaning they are off-limits. An example showing the planned layout of different facilities to accommodate the site favorability criteria in terms of slope gradient or reversal is shown in Figure 2.69.

2.8.10.2 Fault Displacement/Offset

In areas such as the Sigsbee Escarpment, numerous faults exist due to the movement of the salt underlying the rafted sediment, as shown in Figure 2.49. The large-scale effects of the salt movement include significant deformation, faulting, extension of the overlying supra-section of sediments, and over-steepening and failure of sediments on the face of the escarpment. The complex interaction between the salt and stability of the steep slopes requires a number of analytical studies to understand the risk that each of the geo-constraints represent to the planned development. We will first describe the risk of fault displacement to various planned facilities and then address the risk of slope stability.

The methods as described by Angell et al. (2003) were used to assess fault displacement hazard at the Mad Dog and Atlantis Developments. The methodology classified as a probabilistic fault displacement hazard analysis (PFDHA) involves determining the annual frequency of occurrence for the different size of individual fault movements per event. The input parameters are the selection of representative marker horizons, their age, the cumulative offset, and the average displacement per event.

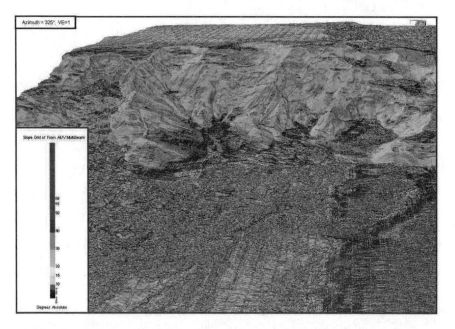

Figure 2.69   Site favorability map for slope gradient/reversal © 2003 OTC (Young et al. 2003): Reproduced with permission of owner. Further reproduction prohibited without permission.

There are two approaches considered in evaluating the fault displacement as a conditional probability of exceedance as described by Youngs et al. (2003). The first approach involves estimating the displacement per event, assuming that the fault displacement occurs as one large discrete event (>1 m). The second approach is based on the fault displacements occurring as creep movements or a large number of small events (0.1 to 1.0 m). To conduct a reliable assessment one must have good stratigraphic age control for key marker horizons as described earlier in this chapter. In addition, an accurate measurement of fault displacement of each marker horizon requires that high-resolution sub-bottom profiler data is available.

The results of the PFDHA are presented as a fault displacement hazard curve, as shown in Figure 2.70. A hazard curve represents the hazard for an individual fault crossing in terms of displacement per single event as compared to the frequency of exceeding the specified displacement. The field layout can be planned to accommodate the criteria set for each type of facility by evaluating the risk of fault displacement of each of the planned facilities. For example, the hazard curve indicates the probabilistic recurrence of a 1 m event is typically less than 10^{-3}. As shown in Table 2.9 (Angell et al., 2003), the criteria specified for the probabilistic recurrence for the three types of facilities varies from 0.1 to 3.0 m for a 10^{-3} annual frequency of exceedance.

2.8.10.3 Slope Instability

Examples of offshore slope failures, typically called a submarine landslide, occur in many forms depending upon the local geology and type of trigger mechanism. Two papers (Hampton and Lee 1996; Mulder and Cochonat 1996) previously described the wide range of marine landslide types and the geologic processes that may trigger each type. In the Gulf of Mexico, there have been a

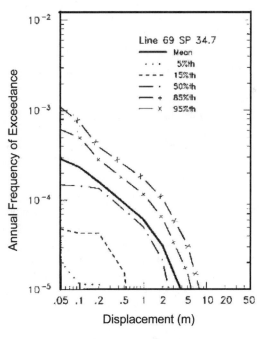

Figure 2.70 PFDHA hazard curve for Mad Dog fault © 2005 (Jeanjean et al. 2005): Reprinted with permission from Dr. Susan Gourvenec, Professor at University of Western Australia, Center for Offshore Foundation Systems, Perth.

wide variety of submarine slope instabilities, and many of the slumps/slides are associated with movement of the underlying salt masses covering very large areas of the continental slope as described by Silva et al. (2004). The methodology described in this chapter is covered in more detail relative to the field and laboratory methods used to investigate the possible trigger mechanism for relict and recent slope instabilities.

The supra-salt section undergoes a change in stress regime and loading due to salt movement and gravity to accommodate the increase in seafloor gradient. An increase in seafloor gradient moves the slope toward instability due to increases in the shear stresses in the slope. Thus, the soils on the slope will become progressively unstable as the slope gradient increases and vice-versa. In turn, the slope attempts to become more stable through progressively lengthening and reducing the seafloor gradient. Over-steepened slopes with unstable geometry try to reach equilibrium by changing their geometry. Shallow sediments shed early and decrease the slope as they try to maintain equilibrium. The stronger stiffer sediments attempt to stabilize by changing the stress regime via shear and translational sliding. These slope failures may be large and rapid or slow, long-term, progressive shear.

We will focus our discussion on the types of landslides that have been identified along the Sigsbee Escarpment in the area of the Mad Dog and Atlantis Developments that are also associated with salt movement. The escarpment lends itself to periodic slope failures and other mass movement processes due to the steep slope gradients and variable sediment properties. As previously described, the escarpment has experienced a number of landslide events, such as slumps, debris flows, mudflows, and turbidity currents as described by Young et al. (2003). Several processes have contributed to these instabilities and failure of the sediments on a given slope. Potential trigger mechanisms associated with the various processes include over-steepening of the slope angle, weak in situ sediment strengths, sediment loading, salt tectonic activity, excess pore pressure, and erosion.

In the following sections, we will describe the approach taken to evaluate the risk of slope instabilities for the two case study projects. We will first describe the methods used to evaluate the risk for shallow-seated slope failures, and subsequently, we will provide an overview of the methods to evaluate deep-seated slope failures.

2.8.10.4 Shallow-Seated Slope Failures

The placement of seafloor production infrastructure on or near the face of the escarpment requires a comprehensive assessment of the risk for shallow-seated slope instabilities. Many areas of the escarpment are covered by large areas of weak sediments on steep slopes that are prone to shallow-seated slope failures that can result in the formation of debris flows and turbidity currents. These mass soil movements can damage or destroy seafloor facilities in their path (Randolph et al. 2005). The extensive seafloor area occupied by these large developments makes it difficult to assess all areas with individual cores. An innovative method was developed for the Mad Dog and Atlantis projects as described by Brand et al. (2003b). Two correlations were developed between acoustic impedance vs. seismic amplitude and then shear strength vs. acoustic impedance and as illustrated in Figures 2.71 and 2.72, respectively. By combining these two correlations, it was possible to develop a correlation between seismic amplitude and soil shear strength, as illustrated in Figure 2.48. Since the seismic amplitude is a measure of the seismic reflection over the resulting tuning thickness of the seismic data (~10 m), then the average strength was estimated using the 3-D seismic data.

Shallow seated slope stability can then be assessed using the infinite slope theory proposed by Teunissen and Spierenburg (1995) and the estimated soil strengths from the 3-D seismic data. This method allows one to assess the average soil strength within the upper 6 to 8 m of shallow sediment over the entire seafloor area. Then the slope angle can be computed from the water depths picked from the seismic volume. The infinite slope analyses can be used to compute the factor of

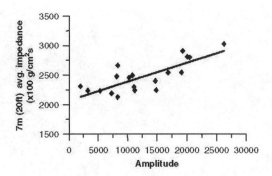

Figure 2.71   Seismic impedance vs. amplitude correlation © 2003 OTC (Brand et al. 2003): Reproduced with permission of owner. Further reproduction prohibited without permission.

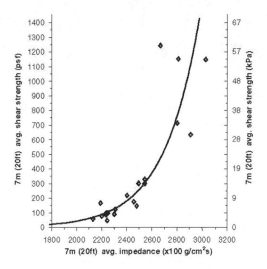

Figure 2.72   Seismic impedance vs. soil shear strength © 2003 OTC (Brand et al. 2003): Reproduced with permission of owner. Further reproduction prohibited without permission.

safety for areas of interest, and then the areas with the lowest factor safety can be flagged to verify that the geo-constraint, shallow-seated slope failure does not pose a risk. The critical high risk areas can be identified by cross-referencing the shear strength calibrated impedance values with seafloor gradients to locate steep slope areas posing potential instability (factor of safety less than 1.0), as shown in Figure 2.73.

2.8.10.5 Deep-seated Slope Failures

Extensive evidence exists throughout the area of the Mad Dog and Atlantis Developments confirming that massive slope failures have occurred in the past, as illustrated by Slump 8 (see Figures 2.49 and 2.50). These slope failures are a result of extensive deformation occurring in the steep slopes along the escarpment related to movements of the underlying salt. Most of the features showing the evidence for slope instability are readily observed at the seafloor. Other indicators existed deep within the slope that included changes in pore pressure and smaller

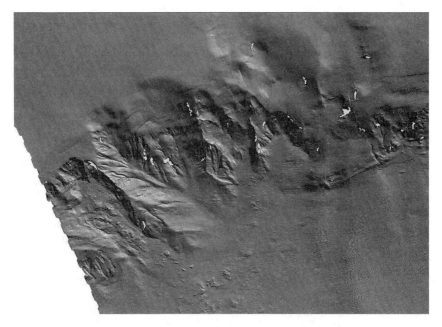

Figure 2.73 Critical slope areas identified from shallow-seated slope analyses posing potential risk to future production facilities © 2003 OTC (Brand et al. 2003): Reproduced with permission of owner. Further reproduction prohibited without permission.

internal movements. A complex pore pressure regime strongly influenced the stability of the slopes, as described by Orange et al. (2003). The goal of the slope stability analyses is to understand the factors that most significantly influence the slope condition as it approaches failure as described by Mikkelsen (1996). Unfortunately, not every parameter influencing a potential slope failure can be measured as easily and with equal success as others due to physical and economic constraints.

Slope stability analyses to evaluate the risk for deep-seated slope failures require an extensive amount of information on the soil properties throughout the soil profile (Duncan 1996). The field investigation included soil borings at optimum locations to obtain the data needed for the detailed slope stability analyses. In addition, a number of piezoprobe tests were performed within the soil profile to measure the resulting pore pressure regime. The laboratory testing previously described provided critical parameters based on the SHANSEP method utilizing DSS test data to establish undrained strength profiles.

Static limiting equilibrium method based on Morgenstern and Price (1965) was conducted to analyze circular and noncircular surfaces, as described by Nowacki et al. (2003). Since a trigger mechanism was not clearly evident, drained analyses was conducted to accompany the undrained analyses to cover all possible causes of slope failure. There were seven slope profiles analyzed that yielded minimum safety factors ranging from 1.1 to 1.3 for the undrained and drained analyses, respectively.

The numerous fault planes occurring in the soil profiles raised the question about the effects on the computed safety factors discussed above. A sensitivity study was conducted as described by Nowacki et al. (2003) to understand what a reduction in vertical and horizontal planes as represented by faults would have on the computed deterministic safety factors. The results, as shown in Figure 2.74, confirm that a reduction in strength on a horizontal weak plane has a much greater impact than a reduction on a vertical weak plane.

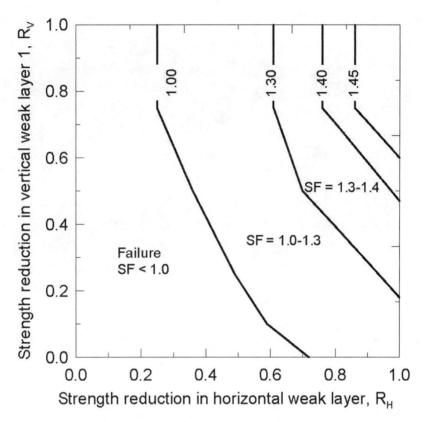

Figure 2.74 Fault plane influence on computed slope stability © 2003 OTC (Nowacki et al. 2003): Reproduced with permission of owner. Further reproduction prohibited without permission.

2.8.10.6 Debris Flows/Turbidity Currents

Many offshore areas exist where seafloor instabilities have triggered mass gravity flows, as described by Niedoroda et al. (2000) and Mulder and Cochonat (1996). Mass gravity flows are a significant geo-constraint that may pose significant risk to seafloor facilities positioned within their path of travel.

Many seafloor features exist on and near the base of the Sigsbee Escarpment that formed as a result of mass gravity flows. The steep highly faulted slopes reveal a rugged relief that are characterized by the excavation of failed soils that moved rapidly down the slope as the sediment and water are propelled by gravity. The mass gravity flows are categorized as debris flows, mudflows, and turbidity currents. Figure 2.75 show a schematic of the two broad classifications that are governed by different flow regimes and require different flow models, as described by Niedoroda et al. (2000).

Once the slope stability analyses were completed as described in the previous section, then areas on the escarpment exhibiting potential unstable slopes were analyzed to determine the size and soil characteristics of the failed soils. Potential debris flows and turbidity currents were analyzed, as described by Niedoroda et al. (2003). As part of the geo-constraint risk evaluation, the integrated team assessed the project areas for the following conditions:

1. Cataloged past mass gravity flows
2. Evaluated the causes for various events

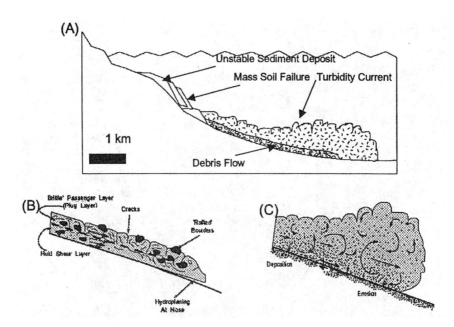

Figure 2.75   Schematic of debris flows and turbidity currents © 2000 OTC (Niedoroda et al. 2000): Reproduced with permission of owner. Further reproduction prohibited without permission.

3. Characterized the kinematics (speed, dimensions, run-out distance of each flow)
4. Compared the past flows to potential flow exposures

The models for debris flows included the size of the source sediment traveling down the slope until it came to rest as the potential energy was dissipated by friction. The turbidity current models must include the gravitational energy that drives the flow of suspended sediment and ambient water down the slope.

The analyses required full use of all integrated data sets to model seafloor conditions. Diagnostic modeling was first performed to demonstrate that the models replicated observations of past mass gravity flows that would overrun any planned facilities. The calibrated numerical models were then used to predict likely flow paths and the kinematics of potential flows. The results then helped establish the criteria to withstand their action in the design of seafloor production and transportation facilities.

2.8.10.7 Highly Variable Soil Conditions (Spatial Soil Provinces)

The data files for field and laboratory tests were then used to plot results from individual cores in a group to distinguish core data in regions exhibiting similar soil properties. By plotting the data for each core in different colors, the team could visually compare various properties on different plots as a function of depth, as shown in Figure 2.76. Soil variability became clearly evident by comparing certain soil properties, such as moisture content, submerged unit weight, in situ vertical effective stress, preconsolidation pressure, undisturbed and remolded miniature vane shear strength, and sensitivity. Trends in the different data sets improved our understanding of the various geologic processes and depositional history such as erosion or debris flows that could have influenced the spatial, epistemic, and aleatory variability throughout the project area.

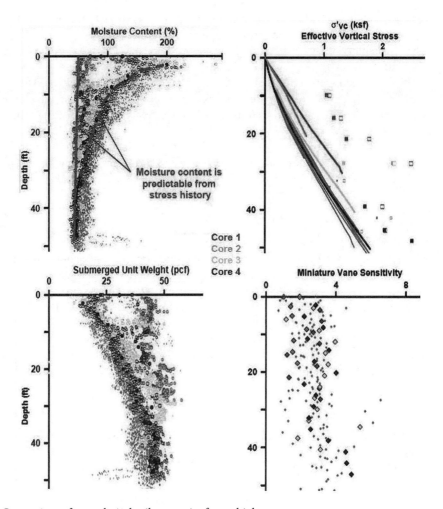

Figure 2.76　Comparison of geotechnical soil properties for multiple cores.

Trends in one data type such as moisture content were clearly evident in corresponding data sets that shared the same stress history such as pre-consolidation pressures, submerged unit weight, and miniature vane strength tests. For example, the removal of overburden pressure due to erosion appeared on the stratigraphic section that corresponded to lower values of moisture contents and higher values of miniature vane strength, submerged unit weight, vertical effective stress, and preconsolidation pressure.

The trends in laboratory data were then compared with the subsurface stratigraphic sections to identify areas with similar stress history. The stress history is probably the most important factor available for selecting geotechnical engineering parameters appropriate for preliminary foundation design. The generalized design parameters based on stress history generally bound the range of soil properties suitable for preliminary selection of preferred foundation types and geometry.

2.8.10.8　Shallow Water Flow

Problems setting the conductors in 1996 for the Ursa development, as described by Eaton (1999a, 1999b), clearly identified the risk of shallow water flow in certain deepwater areas of the Gulf of

Mexico. Shallow water flow occurs when water from a saturated sand aquifer flows up the casing string eroding the sediment providing support to the casing. As described by Pelletier et al. (1999), instability of the casing caused by shallow water flow can result in structural damage and loss circulation.

This list of geo-constraints that were evaluated as part of the integrated study for Mad Dog required an evaluation of the potential risk for shallow water flow for the planned wells. The assessment was made using conventional 3-D seismic data as used for exploration purposes, as described by Berger et al. (1998). High-resolution 3-D data also provided even more detail for evaluating the risk of shallow water flow as described by McConnell and Campbell (1999). Thus, evaluation of the subsurface conditions to depths of 500 to 600 m below the seafloor also played an important role in selecting final well locations relative to other planned infrastructure.

2.8.11 Foundation Site Evaluation

A number of site assessment factors were evaluated as the project design team laid out the seafloor field architecture and evaluated each proposed foundation site. The geophysical and geotechnical data was used to evaluate the risks associated with installation and performance of each proposed foundation type. A paper by Young et al. (2009) proposed a site assessment risk matrix for assessing the risk for anchoring problems within an offshore mooring spread. This site assessment matrix was not used at Mad Dog, but it can be used to assess the potential risks with other foundation types. The site assessment risk matrix included the following factors: (1) seafloor topography, (2) difficult soil conditions, (3) spatial soil variability, (4) type of seafloor strength profile, (5) foundation experience within the project area, (6) confidence in computed foundation capacity, and (7) maturity of the foundation concept.

The coauthors believe that this site assessment approach provides an improved framework for highlighting the risks of installing various type foundations in areas with complex geology and exhibiting highly variable soil conditions. The integrated geophysical and geotechnical data should be evaluated by experienced geotechnical engineers and geologists to assess potential foundation risks. When the site assessment risks are high as listed above, then the geosciences team should pinpoint new foundation sites or alternative foundation types where the likelihood of achieving the desired installation and foundation performance would be acceptable, as described in Chapter 3.

2.8.11.1 Foundation Selection

Once the results of the integrated geologic/geotechnical model were complete, the project team could perform their significant role in terms of selecting the most suitable foundations for the soil conditions existing throughout the project area. The principal risks in foundation design are associated with uncertainties in selection of design parameters, determination in foundation loads, and design method reliability. Each of these uncertainties should be evaluated by an experienced expert with a full understanding of the foundation concept and knowledge of subsurface conditions throughout the project area. Each foundation concept has a different level of reliability in terms of these foundation design uncertainties.

The types of foundations typically available for a typical deepwater development are described in Chapter 3. Deepwater foundations as described in this chapter are generally divided into two broad applications: (1) foundations (anchors) used for mooring and (2) foundations used to support seafloor facilities. The most suitable foundation type is generally dependent upon: (1) understanding all the sediment properties, (2) satisfying the available installation procedures, (3) providing required capacity to resist the most critical loading conditions, and (4) limiting foundation movements to satisfy the structure tolerances. Thus, the combined soil-foundation system must be designed to resist the maximum foundation loading applied during the facilities operating life without experiencing soil failure or excessive movement.

Final selection of the most appropriate foundation type was highly dependent upon conducting a high quality site investigation that defined the complex nature of the important physical and engineering properties of the sediments. The seismic profiles played an important role in terms of selecting the preferred locations of various foundation types.

The final locations for each foundation were selected by using the seismic profiles to confirm that soil stratigraphy was consistently uniform across its footprint. The Jeanjean et al. (2003) paper indicates the variability in subsurface conditions in complex geological settings may appear overwhelming. The challenge is to recognize the most crucial features and understand their nature and potential impact relative to the physical location of the proposed field architecture. Soil province maps played a very important role in defining the spatial variability that existed within and around the area for each planned facility. The marine geologist used the high-resolution sub-bottom data along with geotechnical data to ground truth the seismic data and to prepare maps showing different soil provinces.

2.9 Objective and Benefits of Integrated Geosciences Studies

The history of integrated geosciences studies over the last twenty years clearly demonstrates their benefits throughout the life of the project. As deepwater projects have moved into deeper water, the economic benefits have become more pronounced. The real economic benefit is associated with eliminating future risks to the development. This means that all risks need to be characterized and evaluated in a consistent manner. As described by Jeanjean et al. (2003), evaluation of risk requires two important considerations: (1) the annual probability of occurrence and (2) a measure of the consequences of all risks in terms of damage to health, environment, capital investment, and company reputation.

The final assignment during an integrated geosciences study is to develop criteria for evaluating geo-constraints in terms of risk acceptance, risk avoidance, and risk mitigation. To accomplish this assignment, as illustrated in Figure 2.77, the integrated team must quantify the frequency and magnitude of each geo-constraint event upon the impact on all planned seafloor infrastructure. The impact must be evaluated as described by Moore et al. (2007) in terms of the capacity of a facility (e.g., manifold) to resist the loading condition imposed by the event. There will be cases where the design capacity of the facility cannot be engineered to achieve risk free installation in a cost effective way. In this case, the facility must be moved to avoid the geo-constraint with an adequate buffer distance.

The risk evaluation requires that the geologic/geotechnical model is in final form and the regional geologic setting and environmental controls that have and may influence the area are clearly understood. The final interpretation of geophysical data has been integrated over the area by mapping the major seismic horizons, faults and other geologic features such as landslide detachment surfaces. The bathymetry data has been processed to provide detailed maps of the seafloor typography and geomorphic features that can be used by designers to model the potential constraints relative in their design of planned seafloor facilities. The spatial framework for geo-constraints is included as maps in the GIS system that can be used to select the final location of proposed facilities included in the field architecture.

The maps within the GIS system will show the nature and distribution of sediments, sedimentation rates, and the spatial distribution of key seafloor instabilities such as landslides and turbidity currents. Soil province maps are available based on the seismic data calibrated by the geotechnical sample data. This information provides a reasonable understanding of the soil conditions thus

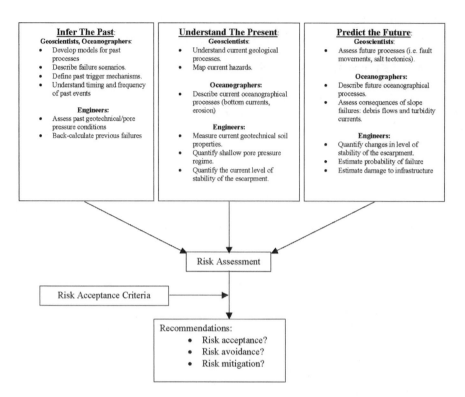

Figure 2.77　Purpose of integrated studies and contributions of individual disciplines © 2003 OTC (Jeanjean et al. 2003): Reproduced with permission of owner. Further reproduction prohibited without permission.

allowing estimates of geotechnical parameters such as spatial changes in undrained strength. Age dating on the soil samples supplements the mapped geologic features by providing an understanding of the frequency of past events. This chronological framework can then be used in the risk assessment to investigate the potential return periods of various geologic processes such as debris flows.

For example, each geo-constraint, such as slope instabilities, can be evaluated in terms of the annual probability of consequences. Then the consequence of any potential slope failure can be addressed relative to the risk of debris flows and turbidity currents causing damage to the infrastructure. The fault hazard can be appraised in terms of the annual displacement rates along fault planes and the potential damage of such displacement to the infrastructure. Thus, the frequency of each geo-constraint can be evaluated as part of the final risk assessment to determine the probability of future events over the lifetime of the development.

The final integrated geologic/geotechnical assessment will include a list of geo-constraints and the criteria used to evaluate their impact. The types of data frequently used to properly assess the cause, history, and potential risk imposed by various types of geo-constraints is shown in Table 2.5. Various methods available to analyze the potential risk of each geo-constraint are shown in Table 2.6 as recommended by Horsnell et al. (2009). Those imposing unacceptable risk based on these different types of analyses are identified on a suite of thematic maps, as shown in Figure 2.68, as described by Moore et al. (2007). These maps can then be used to move the facility to a favorable location or to plan mitigation measures. In the case of pipeline or shallow foundation design, these measures might include the addition of pin piles to provide greater lateral resistance against a turbidity flow. The other form of mitigation can include moving the facility to a safer area with a safe standoff or buffer zone to avoid the risk of a debris flow run-out.

All geo-constraints are evaluated, as indicated in Figure 2.77, as part of the integrated study to understand their important role in terms of avoiding damage to the environment, protecting investors and personnel potentially at risk on the anchored floating platforms, and assuring the regulators and the public that the facilities can be safely sited and operated.

Other economic benefits for the operator that have emerged are listed below:

1. Maximizes the use of the geophysical data to reduce the scope of the geotechnical investigation
2. Uses geophysical and geotechnical data to the maximum extent possible
3. Optimizes the type and amount of additional geophysical/geotechnical data acquired
4. Decreases field acquisition time thereby reducing field costs
5. Provides information for design at an early stage and supports changes throughout the lifetime of the project including installation
6. Allows data base management and supports project changes and contingencies throughout the project by using the GIS system
7. Combines expertise and cutting-edge technology to define critical design constraints and reduce overall project costs
8. Supports the selection of the most appropriate foundation type for the site conditions

All these benefits clearly highlight the significant commitment of personnel and equipment resources that are needed when planning and executing an integrated geosciences study. The combined experience of all team professionals and coordination of their work activities is a critical ingredient in executing the project and meeting project requirements. Finally, an integrated study can satisfy its objective and provide the myriad of benefits if one recognizes the key to success, as stated by Jeanjean et al. (2005).

"The key to a successful integrated study might reside in gathering outstanding high-resolution geophysical and geotechnical data early in the project, but not too early, hire world-class technical specialists, give them as much time and space as possible, and ensure that they communicate and interact appropriately."

2.10 Conclusion

The move into deepwater over the last twenty years has introduced a number of challenges with respect to the variety of geologic settings and complexity of geologic conditions. Taut-leg mooring systems and a variety of new anchor systems as described in Chapter 3 have emerged to moor the exploration rigs and production platforms in deepwater. Technological improvements have emerged that now allow high quality geophysical and geotechnical data to be acquired even in extreme water depths exceeding 3000 m.

Integrated geosciences studies now play a very significant role in allowing one to achieve the major objectives of planning and installing a deepwater development. Integrated geosciences studies are now carried out in discrete phases to more efficiently collect data needed to achieve reliable site selection, successful foundation design and installation, and safe operations of planned seafloor supported facilities.

The initial phase involves using available data to build a conceptual geologic model that defines the seafloor features, sequence of subsurface conditions, and geologic processes. This model is essential to plan the scope of the geophysical survey and to confirm that data coverage is adequate for the planned seafloor architecture.

AUVs can now operate in water depths exceeding 3000 m. AUVs include a suite of multi-sensory acoustic systems with different transducers, receivers, processors, and recorders. AUVs can simultaneously acquire high-quality data including: (1) multi-beam swath bathymetry, (2) side-scan sonar imagery, and (3) sub-bottom profiler seismic sections. A team of experienced geotechnical engineers and marine geologists can interpret these data and develop an integrated geologic model to plan the proper scope of the geotechnical investigation.

Specialized geotechnical drillships are now available that can drill soil borings to penetrations of 600 m below the seafloor in water depths up to 3000 m. New sampling and testing methods have evolved that dramatically improve the quality of the samples and reliability of in situ data. Seabed sampling systems are available to obtain continuous cores of sediments to depths up to 30 m. Down-hole sampling systems are available to obtain soil samples in a soil boring to the full depth that can be drilled by the geotechnical drillships. Innovative seafloor platform and down-hole in situ testing tools are now available to acquire reliable CPT and VST data to the same depths as the sampling systems. A new series of seabed systems have evolved over the last few years that allow deep borings and in situ testing to depths up to 100 m without the use of a deepwater drillship, as shown in Figure 2.4.

After integrating the results from the field and laboratory programs into the final geologic/geotechnical model, the multi-disciplinary team has all the high quality data needed to select sites where subsurface conditions are clearly defined and well understood. Geotechnical engineers now have data with the quality and coverage required to select appropriate parameters for the geologic complexity of the area and remain confident of their appropriateness for the foundation design methods described in Chapter 3. All team members must remember that more data may further improve the usefulness of the geologic/geotechnical model. However, the geologic model should never be considered final. Mr. John Lorenz, President of AAPG, recently made this quote to fellow geologists:

"Models are useful because they provide an outline, a standard by which to measure activities, process, and concepts. But one must always be aware that they are only a representation of reality."

The quote by Lorenz (2010) clearly describes the importance of all team geologists and geotechnical engineers working closely together to develop a geologic/geotechnical model that represents reality as closely as possible. They must recognize any uncertainty in the model and evaluate the impacts of any deviations during the risk evaluation.

Case studies for the Mad Dog and Atlantis Developments demonstrate a number of geo-constraints that were considered as part of the site favorability criteria to locate the final production field architecture. The major geo-constraints that were evaluated included: (1) steep slope gradients, (2) slope reversals and/or irregular seafloor topography, (3) amount of fault displacement and seafloor offset, (4) risk of shallow-seated slope instabilities, (5) risk of deep-seated slope instabilities, (6) potential for debris/turbidity flows, (7) extent of highly variable soil conditions, (8) presence of gas/fluid expulsion features, and (9) risk of shallow water flow. Each of the geo-constraints were evaluated as part of the owners' need to perform a comprehensive risk assessment.

An integrated geosciences study as stated by Jeanjean et al. (2003) must fulfill three main objectives. "Infer what happened in the recent geologic past, understand the current situation, and predict the evolution of the hazard in the near future."

A clear understanding of the global risks associated with these projects requires the work of a multidisciplinary team with the combined skills to conduct a comprehensive risk assessment. The capability of the geosciences team to fulfill their important role begins with the initial development plans and continues until after the foundations are installed and first loaded. The geoscientist experts must explain their role to the key decision makers and describe the importance and benefit of an integrated geosciences study to achieve overall success of a project. In turn, deepwater projects must be properly managed to allow the current state of the design practice to be employed and the preferred foundation concepts to be selected and properly installed. These are not quick decisions, and sufficient time is required to allow all interactions to occur that will produce a truly integrated study and viable geologic/geotechnical model.

The coauthors believe that integrated studies have evolved to a stage where the geosciences team has the capability to select the most appropriate foundation type for the existing soil conditions within the proposed development area. This capability has greatly reduced the risks associated with foundation installation and performance by relying upon high quality data to complete a geo-constraint risk assessment that includes risk acceptance, risk avoidance, and risk mitigation.

2.11 Acknowledgments

The coauthors greatly appreciate the opportunity to be a part of numerous integrated geosciences studies involving many project partners over the last decade. We are deeply appreciative to BP and their partners, Chevron, BHP, Devon, and Statoil for the opportunity to work on the case study projects, Mad Dog and Atlantis Developments. The coauthors could not have written this chapter without the support during the projects of many companies, universities, and individuals. We are especially indebted to Dr. Philippe Jeanjean with BP; Mr. Leon Holloway with ConocoPhillips; Mr. Earl Doyle; and Mr. Daniel Spikula and Mrs. Dianna Phu with GEMS for reviewing the text and providing helpful instructions for improvement.

We are also grateful to the professional staffs of TDI-Brooks International, C&C Technologies, Fugro-McClelland, Norwegian Geotechnical Institute (NGI), AOA Geophysics, and Texas A&M University. The technical contributions of the following individuals and their organizations, including Dr. Philippe Jeanjean, Dr. Ed Clukey, and Dr. Eric Liedtke with BP; Dr. Niall Slowey, Dr. Jean Louis Briaud, and Dr. Bill Bryant with Texas A & M University; Mr. Mike Angell and Mr. Dan McConnell with Fugro-William Lettis & Assoc.; Mr. Fritz Nowacki and Mr. Knut Schroeder with NGI; Mr. Bob Bruce, Mr. Shawn Williams, and Mr. Steve Taylor with BHP; Mr. Clarence Ehlers and Mr. Graeme Lobley with Chevron; Mr. Dan Orange with AOA Geophysics; Mr. Steve Garmon with C&C Technologies; Dr. Jim Brooks, Dr. Bernie Bernard, and Ms. Cassie Rutherford with TDI Brooks International, were critical to the success of these integrated studies.

We want to acknowledge the support of many current and past GEMS employees who worked diligently with us on these projects, including: Mr. Mike Kaluza, Mr. Dan Spikula, Mr. Dan Lanier, Mrs. Dianna Phu, Dr. John Brand, Mr. Bill Berger, Mrs. Jill Rivette, Mr. John Dobias, Mr. Chris Medere, and Mr. Cary Hongenan. We are also highly indebted to two long-term friends and past coworkers, Mr. Lowell Babb with Detail Design and Mr. Ron Boggess with Gregg Drilling, for their important contributions in developing many of the innovative geotechnical tools described in this chapter. Finally, we appreciate the long-term contributions of Dr. James D. Murff in guiding past research, developing numerous analytical methods, and preparing many industry guidelines that have become the backbone of the current offshore geotechnical engineering practice.

In summary, we would like to dedicate our chapter to a group of highly respected pioneers in geotechnical engineering who passed away within the last year. The coauthors are greatly indebted to our dedicated mentors, including: Dr. Lymon Reese, Mr. Bramlette McClelland, Dr. Bill Cox, Mr. Bill Emrich, Mr. John A. Focht, Dr. Wayne Dunlap, and Mr. Carl Fenske. The coauthors and our professional colleagues greatly appreciate the efforts of these pioneers to educate and help train numerous young geotechnical engineers over many decades.

2.12 References

Aas, G., Lacasse, S., Lunne, T., and Hoeg, K., 1986. Use of in situ tests for foundation design on clay, In *Proceedings of the ASCE Specialty Conference In Situ '86: Use of In Situ Tests in Geotechnical Engineering*, Blackburg, ASCE, 1-30.

Al-Khafaji, Z., Young, A. G., Degroff, B., Nowacki, F., Brooks, J., and Humphrey, G., 2003. Geotechnical properties of the Sigsbee Escarpment soil borings and jumbo piston core, In *Proceedings of the Annual Offshore Technology Conference*, Houston, TX, OTC 15158.

ASTM, 2005. Standard test method for mechanical cone penetration test of soil, ASTM D3441-05.

ASTM, 2008. Standard test method for field vane shear test in cohesive soil, ASTM D2573.

Andreson, A., Berre, T., Kleven, A., and Lunne, T., 1979. Procedures used to obtain soil parameters for foundation engineering in the North Sea, *Marine Geotechnology* 33:201-266.

Angell, M., Hanson, K., Swan, B., and Youngs, R., 2003. Probabilistic fault displacement hazard assessment for flowlines and export pipelines, Mad Dog and Atlantis Field developments, In *Proceedings of the Annual Offshore Technology Conference*, Houston, TX, OTC 15202.

ASTM, 1995. *Standard test method for electronic friction cone and piezocone penetration testing of soils*, ASTM D-5778.

ASTM International, 2007. ASTM Standards on Disc, 4(8) Soil and Rock I: D 420—D 5611, 4(9), Soil and Rock II: D 5714.

Baligh, M. M., Vivatrat, V., and Ladd, C. C., 1980. Cone penetration in soil profiling, *Journal of Geotechnical Engineering Division*, ASCE, 106GT4, 447-461.

Begemann, H. K. S., 1953. The friction jacket cone as an aid in determining the soil profile, In *Proceedings of the 6th International Conference on Soil Mechanics and Foundation Engineering* 1:213-217 Montreal.

Begemann, H. K. S., 1965. The friction jacket cone as an aid in determining the soil profile, In *Proceedings of the 6th International Conference on Soil Mechanics and Foundation Engineering* 1:17-20, Montreal, Quebec, Canada.

Berger, W. J., Kaluza, M. J., and Usher, N. F., 1998. The use of very high resolution 3-D seismic data in conjunction with 3-D data in evaluating the potential for water flow and other shallow hazards, In *Proceedings of the Annual Offshore Technology Conference*, Houston, TX, OTC 8593.

Berger, W. J., Lanier, D.L., and Jeanjean, P. (2006). Geologic Setting of the Mad Dog Mooring System, In *Proceedings of the Annual Offshore Technology Conference*, Houston, TX, OTC 17914.

Bingham, D., Drake, T., Hill, A., and Lott, R., 2002. The application of AUV technology in the oil industry—vision and experiences, TS4.4 Hydrographic Surveying, FIG XXII International Congress, Washington, D.C.

Bjerrum, L., 1973. Geotechnical problems involved in foundations of structures in the North Sea, *Geotechnique* 23(3): 319-358.

Blondel, P., 2009. *The handbook of sidescan sonar*, Chichester, UK: Praxis Publishing.

Boggess, R. and Roberton, P. K., 2010. CPT for soft sediments and deepwater investigations, *CPT'10: 2nd International Symposium on Cone Penetration Testing*, Huntington Beach, CA.

Bolstad, P., 2002. *GIS fundamentals*, White Bear Lake, MN: Eider Press, 411.

Borel, D., Puech, A., Dendani, H., and de Ruijter, M., 2002. High quality sampling for deepwater geotechnical engineering: the STACOR experience, Conference on Ultra Deep Engineering and Technology, Brest.

Brand, J., Angell, M., Lanier, D., Hanson, K., Lee, E., and George, T., 2003a. Indirect methods of dating seafloor activities: regional stratigraphic markers, and seafloor current processes, In *Proceedings of the Annual Offshore Technology Conference*, Houston, TX, OTC 15200.

Brand, J. R., Lanier, D. L., Berger, W. J., Kasch, V. R., and Young, A. G., 2003b. Relationship between near seafloor seismic amplitude, impedance, and soil shear strength properties, and use in prediction of shallow seated slope failures, *In Proceedings of the Annual Offshore Technology Conference*, Houston, TX, OTC 15161.

Briaud, J. L. and Meyer, B., 1983. *In situ tests and their application in offshore design, geotechnical practice in offshore engineering*, ed. S. G. Wright, ASCE, Austin, TX, 244-266.

Bryant, W. R., Liu, J. Y., and Ponthier, J., 1995. The engineering and geological constraints of intraslope basins and submarine canyons of the northwestern Gulf of Mexico, *Gulf Coast Association of Geological Societies Transactions* 45:95-101.

Campbell, K. J. and Hough, J. C., 1986. *Planning and design of fixed offshore platforms*, New York: Van Nostrand Reinhold, 176-223.

Campbell, K. J., Humphrey, G. D., and Little, R. L., 2008. Modern deepwater site investigation: getting it right the first time, In *Proceedings of the Annual Offshore Technology Conference*, Houston, TX, OTC 19535.

Campbell, K. J., Quiros, Q. W., and Young, A. G., 1988. The importance of integrated studies to deepwater site investigation, In *Proceedings of the Annual Offshore Technology Conference*, Houston, TX, OTC 5757.

Carter, J. P., Davies, P. J., and Krasnostein, P., 1999. The future of offshore site investigation—Robotic drilling on the seabed, *Australian Geomechanics* 343:7-84.

Chandler, R. J., 1988. The in situ measurement of the undrained shear strength of clays using the field vane, *Vane Shear Strength Testing of Soils: Field and Laboratory Studies*, ASTM STP 1014:13-45.

Christian, J. T., 2003. Geotechnical engineering reliability: how well do we know what we are doing? The 39th Karl Terzaghi Lecture, ASCE, *Journal of Geotechnical Engineering*, October.

ConocoPhillips, 2009. *Upstream engineering standards and practices CPMS—STR—EP—007*, Offshore Shallow Geophysical Surveys.

Douglas, B. J. and Olsen, R. S., 1981. Soil classification using electric cone penetrometer, *Symposium on Cone Penetration Testing and Experience, Geotechnical Engineering Division*, 209-227, St. Louis, MO.

Doyle, E. H., McClelland, B., and Ferguson, G. H., 1971. Wire-line vane probe for deep penetration measurements of ocean sediment strength, In *Proceedings of the 3rd Annual Offshore Tech. Conference* 1:21-32, Houston, TX.

Doyle, E. H., Smith, J. S., Tauvers, P. R., Booth, J. R., Jacobi, M. C., Nunez, A. C., Diegel, F. A., and Kaluza, M. J., 1996. The usefulness of enhanced surface renderings from 3-D seismic data for high resolution geohazards studies, MMS Gulf of Mexico 16th Information Transfer Meeting, New Orleans.

Doyle, E. H., 1998. The Integration of Deepwater Geohazard Evaluations and Geotechnical Studies, In *Proceedings of the Annual Offshore Technology Conference*, Houston, TX, OTC 8590.

Duncan, J. M., 1996. Soil slope stability analysis, In Landslides: Investigation and mitigation, special report 247:337-371, Transportation Research Board, National Research Council, eds. A. K. Turner and R. L. Schuster.

Eaton, L. F., 1999a. Drilling through deepwater shallow water flow zones at Ursa, *SPE/IADC Drilling Conference*, SPE/IADC 52780, Amsterdam.

Eaton, L. F., 1999b. The Ursa template failure, In *Proceedings of the 1999 International Forum on Shallow Water Flow*, League City, TX.

Ehlers, C. J. and Lobley, G. M., 2007. Exploring soil conditions in deepwater, *Geo-Strata, Publication of Geo-Institute-ASCE*, Reston 8(3): 30-33.

Ehlers, C. J., Young, A. G., and Focht, J. A., 1980. Advantages of using in situ vane tests for marine soil investigation, International Symposium on Marine Soil Mechanics, Mexico City, Mexico.

Emrich, W. J., 1970. Performance study of soil samplers for deep penetration marine borings, *Sampling of Soil and Rock*, ASTM STP 483:30-50, American Society for Testing and Materials.

Groshong, R. H. Jr., 2006. *3-D structural geology: a practical guide to quantitative surface and subsurface map interpretation*, 2nd ed., Springer-Verlag Berlin Heidelberg.

Hampton, M. A., Lee, H. J., and Locat, J., 1996. Submarine landslides. *Reviews of Geophysics* 34(1): 33-49.

Healy-Williams, N., 1984. *Principles of Pleistocene stratigraphy applied to the Gulf of Mexico*, International Human Resources Development Corporation, Boston, MA.

Horsnell, M. R., Little, R. L., and Campbell, K. J., 2009. The Geotechnical challenges of active geohazards in the design of deepwater facilities, In *Proceedings of the SUT Annual Conference*, Perth, Western Australia.

ISO 17776:2000 E, 2000. *Petroleum and natural gas industries—Offshore production installation—guidelines on tools and techniques for hazards risk assessment*.

ISO/FDIS 22476-1, 2006. *Geotechnical investigation and testing—Field testing—Part 1: Electrical cone and piezocone penetration tests*, International Organization for Standardization.

ISSMFE, 1989. International reference test procedure for cone penetration test CPT, *Report of the ISSMFE technical committee on penetration testing of soils*—TC 16, with Reference to Test Procedures, Swedish Geotechnical Institute, Linköping, Information 7:6-16.

ISSMGE, 1999. International reference test procedure for the cone penetrometer test CPT and the cone penetration test with pore pressure CPTU, Report of ISSMGE TC16 on Ground Property Characterisation from In situ Testing, In *Proceedings of the 12th European conference on Soil Mechanics and Geotechnical Engineering* 3:2195-2222.

ISSMGE Technical Committee 1, 2005. *Geotechnical & geophysical investigations for offshore and nearshore developments*, International Society for Soil Mechanics and Geotechnical Engineering, London.

ISO, 2006. *Geotechnical investigation and testing—Field testing—Part 1: Electrical cone and piezocone penetration tests*, ISO/FDIS 22476-1.

Jeanjean, P., Hill, A., and Taylor, S., 2003. The challenges of siting facilities along the Sigsbee Escarpment in the southern Green Canyon area of the Gulf of Mexico, Framework for Integrated Studies, In *Proceedings of the Annual Offshore Technology Conference*, Houston, TX, OTC 15156.

Jeanjean, P., Liedtke, E., Clukey, E. C., Hampson, K., and Evans, T., 2005. An operator's perspective on off-shore risk assessment and geotechnical design in geohazard-prone areas, In *Proceedings of the International Symposium on Frontiers in Offshore Geotechnics*, Perth.

Kjekstad, O., Lunne, T., and Clausen, C. J. F., 1978. Comparison between in situ cone resistance and laboratory strength for overconsolidated North Sea clays, *Marine Geotechnology* 31:23-36.

Kolk, H. J. and Campbell, K. J., 1997. Significant developments in offshore geosciences, *Boss '97: Behavior of Offshore Structures*, vol. 1, ed. J. H. In Vugts, 3-40.

Kolk, H. J. and Wegerif, J., 2005. Offshore site investigations: new frontiers, In *Proceedings of the International Symposium on Frontiers in Offshore Geotechnics*, Perth.

Kulhawy, F. H. and Mayne, P. H., 1990. *Manual on estimating soil properties for foundation design*: Report EL-6800 Electric Power Research Institute, August.

Ladd, C. C., 1991. Stability evaluation during staged construction, 22nd Terzaghi Lecture, *Journal of Geotechnical Engineering* 117(4): 540-615.

Ladd, C. C. and Foott, R., 1974. New design procedures for stability of soft clays, *ASCE, Journal of the Geotechnical Engineering Division* 100(GT7): 763-786.

Ladd, C. C., Foott, R., Ishihara, K., Schlosser, F., and Poulos, H. G., 1977. Stress-deformation and strength characteristics, In *Proceedings of the 9th International Conference of Soil Mechanics and Foundation Engineering* 2:421-494, Tokyo.

Lorenz, J. C., 2010. Geology, models, and AAPG, President's Column, *AAPG Explorer* 31(2).

Lunne, T., Berre, T., Strandvik, S., Andersen, K. H., and Tjelta, T. I., 2001. Deepwater sample disturbance due to stress relief, In *Proceedings of the OTRC International Conference on Geotechnical, Geological, and Geophysical Properties of Deepwater Sediments*, OTRC, 64-85, Austin, TX.

Lunne, T. and Kleven, A., 1978. Role of CPT in North Sea foundation engineering, *Session at the ASCE National Convention: Cone Penetration Testing and Materials*, ASCE, 76-107, St. Louis, MO.

Lunne, T., Robertson, P. K., and Powell, J. J. M., 1997. *Cone penetration testing in geotechnical practice*, 312, Blackie Academic, EF Spon/Routledge, New York.

Lunne, T., Tjelta, T. I., Walta, A., and Barwise, A., 2008. Design and testing out of deepwater seabed sampler, In *Proceedings of the Annual Offshore Technology Conference*, Houston, TX, OTC 19290.

McClelland, B., 1972. Techniques used in soil sampling at sea, *Offshore* 323:51-57.

McClelland, B. and Ehlers, C. J., 1986. Offshore geotechnical site investigations, In *Planning and Design of Fixed Offshore Platforms*, 224-265, Van Nostrand Reinhold, New York.

McConnell, D. and Campbell, K. J., 1999. Interpretation and identification of the potential for shallow-water-flow from seismic data, In *Proceedings of the 1999 International Forum on Shallow Water Flow*, League City, TX.

Mikkelsen, P. K., 1996. Field instrumentation in landslides: Investigation and mitigation, In the special report 247, *Transportation Research Board*, National Research Council, ed. A. K. Turner and R. L. Schuster, 278-316.

Moore, R., Usher, N., and Evans, T., 2007. Integrated multidisciplinary assessment and Mitigation of West Nile Delta geohazards, In *Proceedings of the 6th International Offshore Site Investigation and Geotechnics Conference: Confronting New Challenges and Sharing Knowledge*, London, UK.

Moore, T. C. and Heath, G. R., 1978. Sea floor sampling techniques, *Chemical Oceanography*, 2nd ed., 7:75-126, Eds. J. P. Wiley, and R. Chester, Academic Press, London.

Morgenstern, N. R. and Price, V. E., 1965. The analysis of the stability of general slip surfaces, *Geotechnique* 15(1): 79-93.

Mulder, T. and Cochonat, P., 1996. Classification of offshore mass movements, *Journal of Sedimentary Research* 66(1): 43-57.

Niedoroda, A. W., Reed, C. W., Parsons, B. S., Breza, J., Forristall, G. Z., and Mullee, J. E., 2000. Developing engineering design criteria for mass gravity flows in deep-sea slope environments, In *Proceedings of the Offshore Technology Conference*, Houston, TX, OTC 12069.

Niedoroda, A. W., Reed, C. W., Hatchett, L., Young, A. G, Lanier, D., Kasch, V., Jeanjean, P., Orange, D., and Bryant, W., 2003. Analysis of past and future debris flows and turbidity currents generated by slope failures along the Sigsbee Escarpment in the Deep Gulf of Mexico, In *Proceedings of the Annual Offshore Technology Conference*, Houston, TX, OTC 15162.

NORSOK, 2004. Marine soil investigations, *NORSOK Standard G-001*, Rev. 2.

Nowacki, F., Solheim, E., Nadim, F., Liedtke, E., and Andersen, K., 2003. Deterministic slope stability analyses of the Sigsbee Escarpment, In *Proceedings of the Annual Offshore Technology Conference*, Houston, TX, OTC 15160.

Oil & Gas Producers (OGP), 2011. Guidelines for the conduct of offshore drilling hazard site surveys. International Association of Oil and Gas Producers, Report No. 373-18-1, 1-31.

Orange, D., Angel, M., Brand, J., Thompson, J., Buddin, T., Williams, M., Hart, B., and Berger, B., 2003a. Shallow geological and salt tectonic setting of the Mad Dog and Atlantis fields: relationship between salt, faults, and seafloor geomorphology, In *Proceedings of the Annual Offshore Technology Conference*, Houston, TX, OTC 15157.

Orange, D., Saffer, D., Jeanjean, P., Al-Khafaji, Z., Riley, G., and Humphrey, G., 2003b. Measurements and modeling of the shallow pore pressure regime at the Sigsbee Escarpment: successful prediction, In *Proceedings of the Offshore Technology Conference*: Houston, TX, OTC 15201.

Peck, R. B., 1962. Art and science in sub-surface engineering, *Geotechnique* 12(1): 60-62.

Pelletier, J. P., Ostermeier, R. M., Winker, C. D., Nicholson, J. W., and Rambow, F. H., 1999. Shallow water flow sands in the deepwater Gulf of Mexico: Some recent shell experience, In *Proceedings of the 1999 International Forum on Shallow Water Flow*, League City, TX.

Penetration testing, 1988. In *Proceedings of the 1st International Symposium on Penetration Testing ISOPT-1*, De Ruiter ed. Balkema, Rotterdam.

Prior, D. B. and Doyle, E. H., 1984. Geological hazard surveying for exploratory drilling in water depths of 2000 meters, In *Proceedings of the 16th Annual Offshore Technology Conference*, Houston, TX, OTC 4747.

Peterson, L. M., Johnson, G. W., and Babb, L. V., 1986. High-quality sampling and in situ testing for deepwater geotechnical investigation, *ASCE Specialty Conference*, In Situ Test Methods in Geotechnical Engineering Practice, 913-925, Blacksburg, VA.

Preslan, W. L. and Babb, L. V., 1979. Piezometer measurement for deep penetration marine applications, In *Proceedings of the Annual Offshore Technology Conference*, Houston, TX, OTC 3461.

Prior, D. B., Doyle, E. H., Kaluza, M. J., Woods, D. D., and Roth, J. W., 1988. Technical advances in high-resolution surveying, deepwater Gulf of Mexico, In *Proceedings of the 20th Annual Offshore Technology Conference*, Houston, TX, OTC 5758, 109-117.

Prothero, D. R. and Schwab, F., 1996. *Sedimentary geology*, W. H. Freeman and Company, New York.

Quiros, G. W., Little, R. L., and Garmon, S., 2000. A normalized soil parameter procedure for evaluating insitu undrained shear strength, In *Proceedings of the Annual Offshore Technology Conference*, Houston, TX, OTC 12090.

Quiros, G. W. and Young, A. G, "Comparison of Field Vane, CPT, and Laboratory Strength Data at Santa Barbara Channel Site." *Vane Shear Strength Testing in Soils: Field and Laboratory Studies. ASTM STP 1014.* A. F. Richards, Ed., American Society for Testing and Materials, Philadelphia, 1988, pp. 306-317.

Quiros, G. W., Young, A. G, Pelletier, J. H., and Chan, H. C., 1983. *Shear strength interpretation for Gulf of Mexico clays, geotechnical practice in offshore engineering*, ASCE, Ed. S. G. Wright, Austin, TX, 144-165.

Randolph, M. F., Cassidy, M. J., Gourvenec, S. M., and Erbrich, C., 2005. Challenges of offshore geotechnical engineering, In *Proceedings of the 16th International Conference of Soil Mechanics and Geotechnical Engineering* 1:123-176, Osaka.

Randolph, M. F., Martin, C. M., and Hu, Y., 2000. Limiting resistance of a spherical penetrometer in cohesive material, *Geotechnique* 50(5): 573-582.

Randolph, M. F., Hefer, P. A., Geise, J. M., and Watson, P. G., 1998. Improved seabed strength profiling using the T-bar penetrometer, Proc. Int. Conf. Offshore Site Investigation and Foundation Behavior—"New Frontiers", Society for Underwater Technology, London, 221-235.

Randolph, M. F., 2004. Characterisation of Soft Soils, Geotechnical and Geophysical Site Investigation: In *Proceedings of the Second International Conference on Site Characterization* ISC-2, Porto, Portugal, Eds. V. Viana da Fonsecca, and P. W. Mayne, Millpress, Rotterdam, 209-231.

Roberts, H. H., Doyle, E. H., Booth, J. R., Clark, B. J., and Kaluza, M., 1996. 3-D-seismic amplitude analysis of the seafloor: An important interpretative method or improved geohazard evaluation, In *Proceedings of the 28th Annual Offshore Technology Conference*, Houston, TX, OTC 7988.

Robertson, P. K., 1990. Soil classification using the cone penetration test, *Canadian Geotechnical Journal* 271:151-158.

Robertson, P. K., 2009. Interpretation of cone penetration tests—a unified approach, *Canadian Geotechnical Journal* 4611:1337-1355.

Schjetne, K. and Brylawski, E., 1979. Offshore soil sampling in the North Sea, In *Proceedings of the International Symposium on Soil Sampling*, Singapore, 139-156.

Schmertmann, J. H., 1974. Penetration pore pressure effects on quasi-static cone bearing, In *Proceedings of the European Symposium on Penetration Testing* 2(2): 345-51, Stockholm.

Schmertmann, J. H., 1976. An updated correlation between relative density, d_r and fugro-type electric cone bearing, Contract report DACW 39-76 M 6646, Waterways Experiment Station, Vicksburg, Miss.

Schnaid, F., 2009. *In situ testing in geomechanics. The main tests*, Taylor Francis, London and New York.

Schultheiss, P. J. and Weaver P. P. E., 1992. Multi-sensor core logging for science and industry, *Ocean92, Mastering the Oceans through Technology* 2:608-613.

Semple, R. M. and Johnston, J. W., 1979. Performance of stingray in soil sampling and in situ testing, In *Proceedings of the International Conference of Offshore Site Investigation*, 169-182, London.

Silva, A. J., Baxter, C. D. P., LaRosa, P. T., and Bryant, W. R., 2004. Investigation of mass wasting on the continental slope and rise, Ed. B. V. Elsevier, *Marine Geology* 203:355-366.

Skempton, A. W., 1985. Residual strength of clays in landslides, folded strata, and the laboratory, *Geotechnique* 35(1): 3-18.

Slowey, N., Bryant, B., Bean, D. A., Young, A. G, and Gartner, S., 2003. Sedimentation in the vicinity of the Sigsbee Escarpment during the last 25,000 years, In *Proceedings of the Offshore Technology Conference*, Houston, TX, OTC 15159.

Spencer, A., 2007. Rovdrill—The development and application of a new ROV operated seabed drilling and coring system, In *Proceedings of the 6th International Offshore Site Investigation and Geotechics Confererence, Confronting New Challenges and Sharing Knowledge*, 533-540, London.

Spencer, A., 2008. Rovdrill and the Rovdrill "M" Series, pushing the limits of the offshore geotechnical investigation, In *Proceedings of the Rio Oil & Gas Expo and Conference*, Rio de Janeiro.

Spencer, A. G., Remmes, B. and Rowson, I., 2011. A fully integrated solution for the geotechnical drilling and sampling of seafloor massive sulfide deposits, In *Proceedings of the Annual Offshore Technology Conference*, Houston, TX, OTC 21439.

Stark, T. D. and Eid, H. T., 1997. Slope stability analysis in stiff fissured clays, ASCE, *Journal of Geotechnical and Geotechnical Engineering* 120(5): 856-871.

Sterling, G. H. and Strohbeck, E. E., 1973. The failure of the south pass 70 B platform in Hurricane Camille, In *Proceedings of the 5th Annual Offshore Technology Conference*, Houston, TX, OTC 1898.

Stewart, D. P. and Randolph, M. F., 1994. T-bar penetration testing in soft clay deposit, ASCE 12012, *Journal Geotechnical Engineering Division*, 2230-2235.

SUT Offshore Site Investigation and Geotechnics Group, 2005. Guidance notes on site investigations for offshore renewable energy projects, revision, *Society for Underwater Technology*, London.

Sweirz, A. M., 1992. Seismic stratigraphy and salt tectonics along the Sigsbee Escarpment, southeastern Green Canyon region, *CRC Handbook of Geophysical Exploration at Sea*, 2nd ed., Ed. R. A. Geyer, CRC Press, Boca Raton, Florida, 227-294.

Templeton, J. S., Murff, J. D., Goodwin, R. H., and Klejbuk, L. W., 1985. Evaluating soils and hazards in the Mississippi Canyon. In *Proceedings of the Annual Offshore Technology Conference:* Houston, TX, OTC 4964.

Terzaghi, K. and Peck, R. B., 1948. *Soil mechanics in engineering practice*, Wiley, New York.

Teunissen, J. A. M. and Spierenburg, S. E. J., 1995. Stability of infinite slopes, technical note, *Geotechnique* 45(2): 321-323.

TDI-Brooks International, 2010. *CPT stinger—Deepwater static cone penetrometer*, Technical Service Sheet, College Station, TX.

The unified soil classification system, 1953. Waterways Experiment Station, Corps of Engineers, U. S. Army, Technical Memorandum, vol. 1-3, 3-357, Vicksburg, MS.

Watson, P. G., Newson, T. A., and Randolph, M. F., 1998. Strength profiling in soft offshore soils, In *Proceedings of the 1st International Conference on Site Investigations* 2:1389-1394, Atlanta.

Weller, J. M., 1947. Relations of the invertebrate paleontologist to geology, *Journal of Paleontology* 21:570-575.

Williamson and Associates, Inc., 2010. *AMS-120 Sonar Mapping System, Technical Specification.*

Wissa, A. E. Z., Martin, R. T., and Garlanger, J. E., 1975. The piezometer probe, In *Proceedings of the ASCE Specialty Conference on In Situ Measurement of Soil Properties* 1:536-545, Raleigh, NC.

Wroth, C. P., 1984. The interpretation of in situ soil tests, Rankine Lecture, *Geotechnique* 4.

Young, A. G, Babb, L. V., and Boggess, R. L., 1988. Mini-probes: a new dimension in offshore in situ testing, In *Proceedings of the Oceans '88 Conference*, 423-427, Baltimore, MA.

Young, A. G, Bernard, B. B., Remmes, B. D., Babb, L. V., and Brooks, J. M., 2011. "CPT Stinger"—An innovative method to obtaining CPT data for integrated geosciences studies, In *Proceedings of the Annual Offshore Technology Conference*, Houston, TX, OTC 21569.

Young, A. G, Honganen, C. D., Silva, A. J., and Bryant, W. R., 2000. Comparison of geotechnical properties from large diameter long cores and borings in deepwater Gulf of Mexico, In *Proceedings of the Annual Offshore Technology Conference*, Houston, TX, OTC 12089.

Young, A. G, McClelland, B., and Quiros, G. W., "In Situ Vane Shear Testing at Sea," *Vane Shear Strength Testing in Soils: Field and Laboratory Studies. ASTM STP 1014.* A. F. Richards, Ed., American Society for Testing and Materials, Philadelphia, 1988, pp. 46-67.

———, 1988. In situ vane shear testing at sea, *Vane Shear Strength Testing in Soils*, Ed. Adrian F. Richards, ASTM STP 1014, Baltimore, MD, 46-67.

Young A. G, Phu, D. R., Spikula, D. R., Rivette, J. A., Lanier, D. L., and Murff, J. D., 2009. An approach for using integrated geoscience data to avoid deepwater anchoring problems, In *Proceedings of the Offshore Technology Conference*, Houston, TX, OTC 20073.

Young, A. G, Quiros, G. W., and Ehlers, C. J., 1983. Effects of offshore sampling and testing on undrained soil shear strength, In *Proceedings of the 15th Annual Offshore Technology Conference*, 193-204, Houston, TX.

Young, A. G, Slowey, N., Bryant, B., and Gardner, S., 2003 Age dating of past slope failure events from C_{14} and nanno-fossil analyses, In *Proceedings of the Annual Offshore Technology Conference*, Houston, TX, OTC 15204.

Youngs, R. R., Arabasz, W. J., Anderson, R. E., Ramelli, A. R., Ake, J. P., Slemmons, D. B., et al., 2003. A methodology for probabilistic fault displacement hazard analysis PFDHA, *Earthquake Spectra* 1(1): 191-219.

Zang, Z.J., and McConnell, D. (2010). Backseater Characterization of Seep-Associated Seafloor Features in the Vicinity of Bush Hill, Northwest Green-Canyon, Gulf of Mexico, In *Proceedings of the Annual Offshore Technology Conference*, Houston, TX, OTC 20662.

Zuidberg, H. M., 1975. Seacalf: A submersible cone penetrometer rig, *Marine Geotechnology*, 111:15-32.

Zuidberg, H. M. and Windle, D., 1979. High capacity sampling using a drillstring anchor, In *Proceedings of the International Conference of Offshore Site Investigation*, 193-204, London.

3

Deepwater Foundation Design

Alan G Young, PE
Vice President, Geoscience Earth & Marine Services

James D. Murff, PE, PhD
Geotechnical Consultant

Jill A. Rivette, PE
Geotechnical Engineer, Geoscience Earth & Marine Services

3.1 Background

In this chapter, we will describe the current state of design practice for foundation concepts employed in deepwater. For the purposes here, we will define deepwater as approximately 450 m since this is roughly the practical limit of the use of fixed bottom production platforms.

It may be useful to briefly recount the evolution of offshore concepts to gain an appreciation of current deepwater practice. Of course, hand-in-hand with the evolution of foundation concepts has been the evolution of site investigation and characterization tools and methods. This latter subject is discussed in Chapter 2 and will not be detailed here.

The earliest platforms were template type structures resembling scaffolding to some degree. Small 0.3-meter-diameter pipe piles supported these platforms. The piles were driven through the vertical legs of the template, as shown in Figure 3.1. As the desire to drill and produce in deeper water mounted, piles became larger diameters and more deeply embedded. In addition, batter was introduced to hold down the size of the platform base footprint, as overturning became an important consideration. During this period, pile driving, axial pile capacity, and lateral pile performance were the central concerns of the foundation engineer. Shallow foundations played a secondary role in that jacket structures set on the bottom required temporary support prior to pile installation. This role was filled using shallow footings or *mudmats* affixed to the structure's base (see Helfrich et al. 1980). This innovation was required to deal with the soft soils that typically occur at the seafloor.

There were also foundation issues associated with temporary drilling structures such as submersibles and jack-ups as well as with pipelines. Anchors for mooring floating drilling rigs were also a concern in water depths too large to use bottom founded drilling systems. The latter, however, were usually installed on an ad hoc, trial and error basis. As water depths increased, the foundation concepts remained largely the same. The major changes were primarily a matter of scale—that is, larger loads and bigger and deeper piles. Of course, the costs escalated dramatically, and the need for increased understanding and improvements to achieve efficient design became

Figure 3.1   Early offshore platforms. Figure provided courtesy of Dr. James D. Murff, retired Professor at Texas A&M University.

more important. This need led to major innovations, such as new pile capacity models supported by large-scale load tests and new equipment such as the underwater pile-driving hammer.

Once water depths exceeded the range of conventional fixed bottom structures, industry turned to floating systems for permanent production facilities. This required a rather dramatic shift in foundation concepts employed. While the basic principles underlying the design of these new foundation types remained largely the same, there were many new design aspects that confronted the foundation engineer. A paper by Aubeny et al. (2001) described many of the issues that previously challenged deepwater foundation design, construction, and installation. The following sections describe how the geotechnical engineering community has faced these challenges and developed novel solutions for foundations in these deepwater frontiers.

3.2 Deepwater Foundation Applications

The move of oil and gas exploration and production activities into deepwater over the last 20 years has resulted in two broad categories of foundation applications—mooring anchors and foundations for on-bottom facilities. The field architecture for these large deepwater developments cover an extensive area of the seafloor, as illustrated in Figure 3.2. The Thunder Horse Development in the Gulf of Mexico is its largest development. It is located in about 1845 m of water and covers a seafloor area of over 24 sq km. The production drilling quarters (PDQ) platform at Thunder Horse is the largest production semi-submersible ever built with a topside area equal to three football fields. The PDQ has the capability to process and export a quarter of a million barrels of oil per day.

Deepwater developments of this type typically include:

1. A major floating structure such as a large diameter deep-draft floating caisson (SPAR) or tension leg platform (TLP)
2. An extensive number of seafloor structures such as pipeline end terminals (PLET), pipeline end manifolds (PLEM), riser bases, manifolds, and holdback piles
3. A wide variety of anchor and foundation types
4. A large suite of mooring lines, pipelines, flow lines, umbilicals, risers, steel catenary risers, etc.

Figure 3.2 Deepwater field architecture. (Courtesy of FMC Technologies, Inc.)

To select appropriate foundation types and locations, the seafloor and subsurface conditions must be interpreted and mapped over the entire area that will encompass the field architecture. In addition, the depth of foundation interest below the seafloor must be investigated with borings, in situ testing, and/or cores in order to analyze/design all foundation types that may be considered.

A number of international specifications and guidelines have been developed by the American Petroleum Institute (API) and the International Standards Organization (ISO) that provide extensive details covering the analytical procedures appropriate for the different design applications of these offshore foundation types (API 2000; ISO 2007). Thus, this chapter will not repeat all the details of the methods associated with the current design practice. Rather, our intent is to present an overview of deepwater foundation practice and current design methods, including their limitations. We will also highlight key technical improvements in the analytical methods and important innovations influencing installation.

Deepwater foundations are generally divided into two broad applications: foundations (anchors) used for mooring and foundations used to support seafloor facilities. Table 3.1 shows many of the types of foundations (anchors) used for various mooring applications required for deepwater oil and gas activities, and Table 3.2 shows foundation types used for support of seafloor facilities.

3.2.1 Foundations for Moorings

The type of mooring system used for floating drilling and production vessels has evolved with the move into deepwater. To reduce the size of the mooring footprint, the conventional catenary mooring system has been largely replaced with taut-leg mooring systems as described by D'Souza et al. (1993) and illustrated in Figure 3.3. The TLP that employs vertically loaded piles is another option to resist large uplift loads with a minimal footprint.

Table 3.1 Foundations (anchors) used for mooring

| Mooring application | Foundation loading conditions | | | |
Foundation type	Vertical compression	Vertical tension	Horizontal	Oblique lateral tension
Suction caisson	Yes	Yes	Yes	Yes
Driven pile	Yes	Yes	Yes	Yes
Vertical loaded plate anchors	No	Yes	Yes	Yes
Drag embedment	No	Some	Yes	Yes
Suction embedded plate anchor	No	Yes	Yes	Yes
Dynamically penetrating anchor	No	Some	Yes	Yes

Table 3.2 Foundations used for seafloor facilities

| Seafloor facilities | Foundation loading conditions | | | |
Foundation type	Vertical compression	Moment	Horizontal	Torsion
Suction caisson	Yes	Yes	Yes	Yes
Driven pile	Yes	Yes	Yes	Yes (limited)
Drilled and grouted pile	Yes	Yes	Yes	Yes (limited)
Jetted pile	Yes (limited)	Yes (limited)	No	No
Shallow foundation	Yes	Yes	Yes	Yes

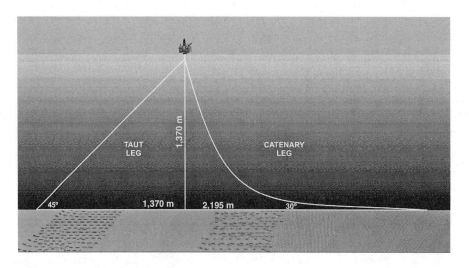

Figure 3.3   Types of deepwater mooring systems.

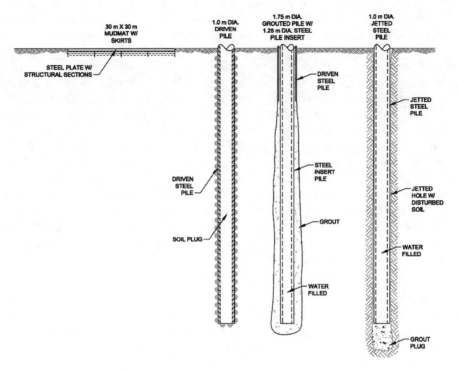

Figure 3.4 Deepwater foundation types.

For inclined moorings, the angle of the mooring line with respect to the horizontal mudline has increased from 25° to 30° to angles ranging from 40° to 45°. The steeper inclination angles result in a larger component of vertical load being applied to the anchor by the mooring line as compared to the mooring load from a catenary mooring system. In most cases, the vertical holding capacity or uplift resistance controls the design of the anchors for taut-leg mooring systems, whereas lateral resistance plays the dominant role for most catenary moorings.

3.2.2 Foundations for Seafloor Facilities

Many foundation types have evolved to support seafloor facilities for deepwater applications. The classical types of foundations that have accommodated the move into deepwater are shown in Figure 3.4. These include: (1) mats (frequently called mudmats), (2) driven or drilled and grouted pipe piles, and (3) jetted piles. In addition, the suction caisson, which was developed primarily for mooring applications, has been found to be an effective means for supporting on-bottom facilities.

3.3 Foundation Design Requirements

The philosophy behind deepwater foundation design is similar to that for other offshore foundations in that the foundation must be appropriately sized to resist the applied loading without experiencing excessive deformation. For most mooring applications, the vessel is not sensitive to the small deformations of an anchor, and hence, *excessive deformation* typically translates to ultimate capacity. In most codes and standards, a factor of safety is applied to both an operating and a maximum loading condition. The operating condition is one that has a relatively high probability of being exceeded during the system's life and a larger safety factor is specified.

The maximum loading condition is usually associated with a relatively rare event, such as a major storm or hurricane. A factor of safety of 2.0 is typically applied to the operating loads and 1.5 to the maximum storm loading. Seafloor facilities, on the other hand, are sometimes sensitive to displacements and rotations, and excessive deformation may have a more literal meaning. The actual deformation limits for the facilities set the target. Predicting displacements is a much more difficult problem with larger uncertainties and hence, more conservatism may be warranted. This may be implicitly accomplished by using a higher safety factor on ultimate capacity.

The API guidelines for production mooring systems require that the individual lines and anchors be designed for a factor of safety ranging from 1.0 to 1.5. For loading conditions based on a 100-year storm event, there are two different load cases to be analyzed. For a completely intact mooring system (intact case), a factor of safety of 1.5 is required, whereas a factor of safety of 1.0 is required for the most-heavily-loaded line missing condition (damaged case). According to Gilbert et al. (2005), the damaged case usually governs the design, although in most instances the design capacities are about the same for the two cases.

The major conclusion of the Gilbert study is as follows:

"Foundation design for floating production systems may be excessively conservative. The values showing the probability of failure being achieved are several orders of magnitude smaller than industry-recommended targets for components (single line or foundation) in a mooring systems. The practical concern with such excessive conservatism is that installation can become unnecessarily costly and problematic."

Their study also shows that where a lower bound or minimum available capacity can be established, it can have a significant, positive effect on the foundation reliability.

For foundations supporting seafloor facilities, storm loading is not a factor because of the large water depths. For these foundations, installation and operating loads such as those arising from hardware and mat weights, pipeline pull-in loads, and thermal expansion loads, all of which may generate moments, usually control the design. This further complicates the problem, as the installation and operating loads are themselves very uncertain. The various guidelines listed provide specific recommendations for the different foundation types and loading conditions. Of course, the experience and engineering judgment of the design team is important to confirm that all uncertainties are covered in the final selected factor of safety.

A further consideration for design of seafloor facilities is the fact that larger foundation displacements are required for a mat to mobilize its ultimate foundation capacity than for a pile. Thus, designers must take this into consideration when connecting various structural members, such as piping or jumpers to these foundations. The methods used to predict foundation deformation for the various foundation types also must be addressed even though it is not covered in this chapter.

3.3.1 Analytical Procedures

Various analytical procedures are available to assess the performance of each foundation/anchor type for the different applications and loading conditions. The level of sophistication depends upon the complexity of loading conditions, vertical variation in soil properties, and the time rate of change in drainage conditions. Analytical methods vary from limit equilibrium procedures such as classical bearing capacity theory as outlined in various recommended practices (API 2000; ISO 2007) to 3-D finite element analyses (FEA) using complex soil models.

3.3.2 Interpreted Soil Properties

During the course of a geotechnical site investigation, as described in Chapter 2, numerous soil properties will be measured either in situ or in the laboratory on recovered samples. The most important soil properties that will be required for the analyses of the different foundation types are listed in Table 3.3.

In all cases, there are a number of soil properties that must be properly interpreted throughout the appropriate zone of influence for each foundation type. For a clay profile, the soil parameters that must be interpreted include: (1) undrained shear strength, (2) remolded shear strength, (3) moisture content, (4) unit wet weight, and (5) the Atterberg limits. These soil properties are routinely measured and presented in graphical form on an offshore boring log, as shown in Figure 3.5. An experienced geotechnical engineer should review and determine if all appropriate soil parameters have been measured and can be interpreted for the required foundation analyses.

In this case, we have selected a best estimate profile for each soil parameter. In the authors' opinion, the designer should select a best estimate interpretation instead of lower and upper bound profiles that show variations often caused during the field sampling, in situ, or laboratory testing activities (Young et al. 1983). With this approach, the computed factor of safety covers the

Table 3.3 Soil properties used in analyses

Soil property	In situ testing	Laboratory testing
Soil stratigraphy	Sub-bottom profiler	Multi-Sensor Core Logger (MSCL)
	Piezo CPT (PCPT)	CT scans
	Gamma logger	Core photographs
Soil classification	PCPT	Grain size
	Gamma logger	Moisture content
		Atterberg limits
Soil strength (undrained)	PCPT	Tovane
	In situ vane	Miniature vane
	T-bar	Unconsolidated Undrained (UU) triaxial
	Ball cone	Consolidated Undrained (CU) triaxial
		Direct simple shear
Shear strength (drained)	PCPT	CU triaxial
		Direct simple shear
		Tilt table
		Ring shear
Sensitivity	PCPT	Miniature vane
		UU-triaxial
Stress history (consolidation characteristics)	PCPT	Direct simple shear
		CU-triaxial
		Constant Rate of Strain (CRS) consolidation
Cyclic strength	T-bar	Cyclic simple shear
		Cyclic triaxial
Permeability	Piezoprobe	CRS consolidation
	PCPT	Permeability

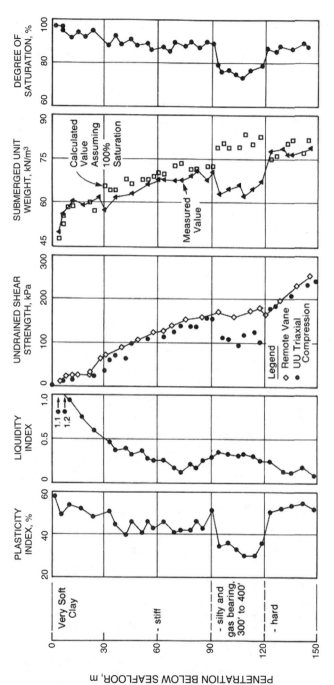

Figure 3.5   Typical offshore boring log. Figure provided courtesy of Mr. Clarence J. Ehlers.

best estimates of uncertainty in the entire foundation design process and does not include individual bias that may disguise inherent errors caused by an improper sample handling or testing procedures. Of course, the use of the continuous in situ data, such as the CPT profile shown on Figure 3.6, helps avoid sampling disturbance and human handling problems caused by inexperienced drillers, technicians, and geotechnical engineers and gives a better relative profile than individual data points.

3.3.3 Soil Strength Profile

The soil conditions at most deepwater sites generally consist of clays or highly plastic silts that are suitable for most foundation types described here. Of course, there is potential for extreme variation in spatial soil conditions existing within different geologic settings due to various processes, as described in Chapter 2. The potential uncertainty in soil conditions needs to be recognized and considered during the foundation design process.

In many instances, these deepwater sites consist of normally or slightly under-consolidated clays that have a linearly increasing undrained strength profile, as shown in Figure 3.7. The soil

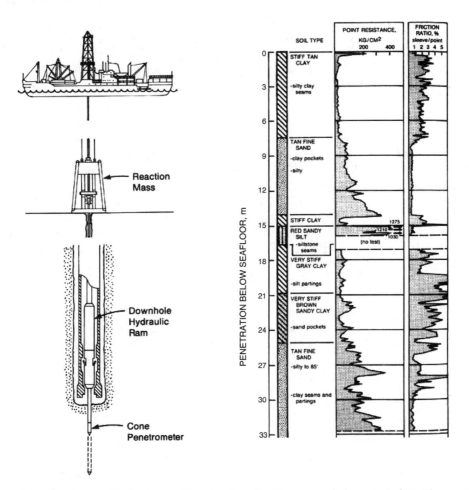

Figure 3.6 Cone penetration testing equipment and results. Figure provided courtesy of Mr. Clarence J. Ehlers.

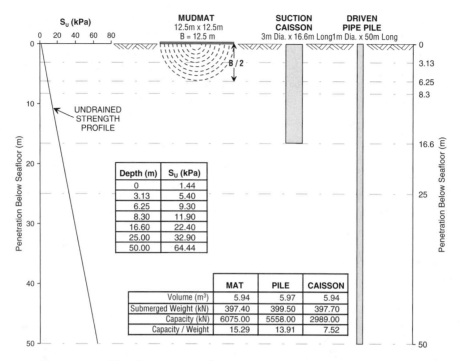

Figure 3.7   Comparison of foundation types and capacities.

strength profile for this example has an undrained strength of 1.44 kPa at the seafloor and in-creases at a rate of 1.26 kPa per m of depth below the seafloor.

Foundations develop their capacity by mobilizing the soil resistance throughout the soil region being affected by the loading. A mudmat develops its soil resistance throughout the applied pressure bulb, as shown in Figure 3.7. This zone is equivalent to about one-half the width of the mud-mat base. A driven pile or caisson develops its capacity along pile wall (typically within a few cm) but over a much deeper foundation depth. The advantage of the pile foundation over the mudmat is that the pile develops its capacity deeper in the stratigraphy where the soil strength is greater.

To illustrate the benefit of increasing strength with depth, we will use the three foundation types with equivalent areas and volumes of steel to demonstrate the relative efficiencies in terms of material cost. For this example, the bearing capacity of a square mat foundation is compared to the axial compressive capacity of a driven pipe pile and a suction caisson. The dimensions of these foundations types are included in Table 3.4. Each foundation type is assumed to have a uniform steel wall thickness of 3.8 cm. The total steel volume is 5.95 cu m and a total submerged steel weight is 398 kN. Table 3.4 shows the ultimate foundation capacities for the three different foundation types using current recommended API/ISO procedures (API 2000; ISO 2007) and a normally consolidated clay profile of 1.44 kPa at the seafloor and linearly increasing to 64.4 kPa at 50 m depth.

The ratio of foundation capacity to submerged steel weight, an indicator of material cost efficiency, is included in Table 3.4. Based on these results, the mudmat is the most efficient in terms of foundation capacity. The driven pile appears as a close second in terms of the ratio of foundation capacity to submerged steel weight. What is not included in this simplified example of the mat are the steel bracing members necessary for the structural integrity. The additional submerged weight of this steel could be as much as the submerged weight of the 12.5 × 12.5 m plate. Therefore, the ratio for the mudmat is overstated because it neglects this additional weight. In reality, the mud-

Table 3.4 Ultimate foundation capacities

Foundation Type	Computed ultimate capacity for different type loadings	
	Axial compressive capacity (MN)	Ratio of foundation capacity to submerged steel weight
Mudmat 12.5m × 12.5m with 0.6m skirt	6.08	15.3
Driven pipe pile 1.0m dia. × 50m long	5.56	13.9
Suction caisson 3.0m dia. × 16.6m long	2.99	7.5

mat will most likely provide about the same capacity to weight ratio as the suction caisson, and the pipe pile would be the most efficient.

The suction caisson is the most inefficient since it provides the lowest capacity, and the steel weight neglects the weight of the structural cap. The benefit of the driven pile is that it mobilizes the skin friction over a much deeper depth, so the average strength used in the pile computations is 32.9 kPa as compared to 11.9 and 5.4 kPa for the suction pile and mudmat, respectively.

This comparison in steel costs does not include any difference for fabrication or installation costs that will also vary for the different foundation sizes and shapes. For example, the fabrication costs for the mudmat and the suction caisson will generally be greater than for a driven pile. However, the installation cost of the driven pile can be much greater than the suction piles depending on equipment availability and location. This comparison demonstrates that there are many variables that need to be investigated to verify the most economic foundation alternative, including the cost of installation.

3.4 Deepwater Anchors

A number of different mooring concepts have evolved with the move into deepwater over the last two decades to reduce costs and uncertainties. One example is shown in Figure 3.3. The taut-leg mooring system is a major innovation that offers many advantages, including a stiffer response to platform/rig motion, a significant reduction in the area of the mooring footprint, and potential reduction in the tensile loads on the platform/rig.

The taut-leg mooring system imposes a much larger component of uplift capacity as compared with the more traditional catenary mooring systems. Previously, conventional drag anchors have been the most widely used anchor type and generally provide adequate capacity for catenary type loading, although uncertainties in anchor placement location and load carrying capacity are still significant. This uncertainty, along with the requirement for adequate uplift capacity, has increased the use of other seafloor anchor types such as piles, vertical-loaded anchors (VLA), and the now common suction caisson.

In addition to these more conventional anchors, a number of new anchor types have been developed:

- SEPLA—Suction-embedded plate anchor (Dove et al. 1998)
- DPA—Deep penetrating anchor (Lieng et al. 1999)
- Torpedo anchor (Medeiros 2001)
- SPEAR—Soil penetrating embedment arrow radial anchor (Zimmerman et al. 2005)

These anchor types are illustrated in Figure 3.8. Ehlers et al. (2004) presented a table (see Table 3.5) showing the advantages and disadvantages of these four anchor types. The table also highlights

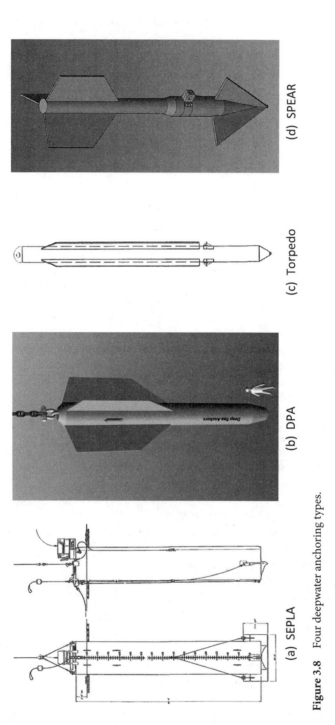

Figure 3.8 Four deepwater anchoring types.

Table 3.5 Comparison of four anchor types (modified after Ehlers et al., 2003)

Anchor	Advantage	Disadvantage
Suction caisson anchor	Simple to install accurately with respect to location, orientation, and penetration Leverage design experience with driven piles Well-developed design and installation procedures Anchor with the most experience in deepwater for mooring Mobile Offshore Drilling Units (MODUs) and permanent facilities	Heavy—derrick barge may be required Large—more trips to shore to deploy full anchor spread Requires Remote Operated Vehicle (ROV) for installation Requires soil data from advanced laboratory testing for design Concern with holding capacity in layered soils
Drag Embedment Plate (Vertical Loaded Anchors, VLA)	Lower weight Smaller—fewer trips to transport the full anchor spread to a site Well developed design and installation procedures	Requires drag installation, keying, and proof testing—limited to bollard pull of installation vessel; requires 2 or 3 vessels and ROV No experience with permanent floating facilities outside of Brazil Difficult to assure installation to and orientation at design penetration
SEPLA	Uses proven suction caisson installation methods Cost of anchor element is the lowest of all the deepwater anchors Provides accurate measure of penetration and positioning of anchor plate Design based on well developed design procedure for plate anchors	Proprietary/patented installation Installation time about 30% greater than suction caisson and may require Dynamic Postitioned (DP) vessel Requires keying and proof testing—limited to bollard pull of installation vessel; also requires ROV Limited field load tests and applications limited in number to MODU only
Torpedo anchor/ deep penetrating anchor	Simple to design—conventional API RP 2A pile design procedures used for prediction of capacity; thus, capacity calculations are likely to be readily acceptable to verification agencies Simple and economical to fabricate Robust and compact design makes handling and installation simple and economical using one vessel and no ROV Accurate positioning with no requirements for specific orientation and proof testing during installation	Proprietary/patented design and no formal design guideline Verification of penetration and verticality Limited experience outside Brazil Mooring line attachment point has been at the top of the anchor, which is not at the optimum location for holding capacity Max. design holding capacity to date is 7.56 MN; required design holding capacity for deepwater anchor in GoM may be as great as 13.34 MN

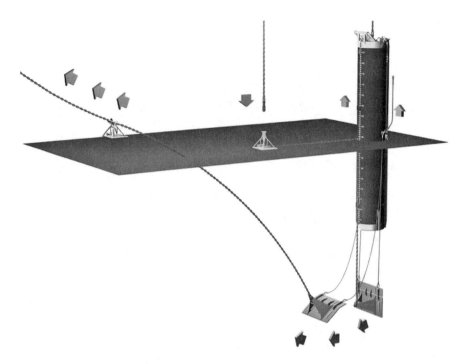

Figure 3.9   Installation of SEPLA anchor. Figure provided courtesy of Mr. Bob Wilde with InterMoor.

the state of practice associated with their unique installation requirements. After the appropriate anchor is selected for the facility and site conditions, the installation vessel and its support equipment need to be carefully evaluated to confirm that it can properly deploy the anchor. Figure 3.9 illustrates the process of installing the SEPLA.

The Ehler paper, along with an earlier paper (Eltaher et al. 2003), provides a complete overview of the design practice for these anchor types including their limitations in terms of foundation design. These two papers and more recent international guidelines describe the general basis for design of these foundation types. This chapter will provide an overview of current analysis/design procedures for the various anchor types, including improvements in the design practice that have emerged since publication of these previous references.

3.5 Anchor Line

The flexible anchor line is a common feature of all deepwater mooring systems except the TLP (Paulling and Horton 1970). Vertical pipes called tendons moor the TLP as described in Chapter 4. The tendons are pre-tensioned by partially submerging the floating vessel while connecting them to the seafloor foundations and subsequently removing the ballast. Tendons are usually connected to piles (Doyle 1999) but in some cases have been connected to suction caissons. Flexible anchor lines are used for all other applications at present. The applications for flexible anchor lines will be discussed under their respective sections, but general anchor line characteristics are described in the following.

Flexible anchor lines are composed of various materials including chain, wire rope, and polyester or combinations of these and are usually connected to the anchor at a padeye below the

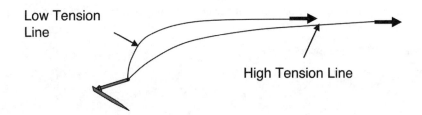

Schematic of Anchor Lines with Increasing Load

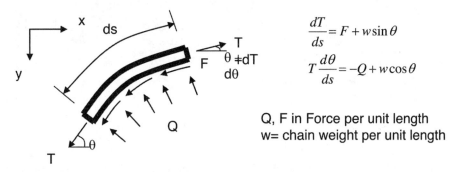

$$\frac{dT}{ds} = F + w\sin\theta$$

$$T\frac{d\theta}{ds} = -Q + w\cos\theta$$

Q, F in Force per unit length
w= chain weight per unit length

Figure 3.10 Schematic of anchor line installation forces. Figure provided courtesy of Dr. James D. Murff, retired Professor at Texas A&M University.

mudline. There are two primary methods of anchor line deployment: tensioning the line after the anchor is installed, such as for a pile or suction caisson, and tensioning the line in concert with the anchor installation, such as for a drag embedment anchor. These methods are discussed in more detail in subsequent sections, but first a general overview of anchor line mechanics is presented.

Consider an anchor line fixed to a given point below the mudline that is tensioned at the mudline at some small angle to the horizontal, as illustrated in Figure 3.10. As the line is tensioned, it cuts through the soil, developing both shear and normal resistance all along the line. The line is flexible—that is, it does not develop any moment resistance such that the line forces are strictly tensile. The soil resistance depends on the line geometry/composition and the soil strength.

Reese (1973) and Gault and Cox (1974) developed early models for the anchor line behavior as they are tensioned. Vivitrat et al. (1982) derived a more general form of the governing equations of a short segment of the anchor line by considering all the forces acting on the line. The forces on a line segment are shown in Figure 3.10 along with the differential equations governing equilibrium of the line. Since the equations are nonlinear, the authors proposed an iterative finite difference solution that is relatively simple to implement. Neubecker and Randolph (1995) developed a closed form solution to the equations by neglecting the weight of the anchor line itself (the nonlinear terms) and showed that the error introduced by this assumption is usually small. Details of their solution for general application are given in their paper, but a special case of interest is given here for convenience. For the case of a horizontal pull at the mudline, as for a catenary with small line angles below the mudline, the relationship between the tension τ_a at the anchor and the angle with the horizontal at the attachment point θ_a is illustrated in Figure 3.11 where it is shown that:

Recipe for CatenaryAnchor Line Analysis

(After Neubecker and Randolph, 1996)

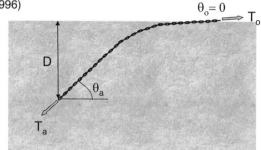

$$T_a = \frac{2D\overline{Q}}{\theta_a^2}$$

$$T_o = T_a \exp(\mu\theta_a)$$

where:

T_a = anchor holding capacity

T_o = mudline line load

Q = average bearing resistance of line per unit length = $bN_c s_{uavg}$

b = effective width of line

N_c = bearing capacity factor

s_u = undrained shear strength

μ = "friction" coefficient (approx 0.4)

Figure 3.11   Horizontal pull at the mudline as for a catenary and for small line angles below the mudline. Figure provided courtesy of Dr. James D. Murff, retired Professor at Texas A&M University.

$$\tau_a = \frac{2D\overline{Q}}{\theta_a^2} \qquad\qquad (3.1)$$

In this case, $\overline{Q}$ is the average bearing resistance (per unit length of anchor line) from the mudline to the anchor attachment point, and D is the depth below the mudline of the attachment point. It has been found that the solution is surprisingly robust even for angles of 35° to 40°.

3.5.1 Post Tensioning of the Anchor

For pile and suction caisson anchors, the anchor line is generally attached to the side of the anchor from the attachment point (padeye) to the mudline while the anchor is being installed. The line is released from the side once the pile or caisson is at final penetration depth and then tensioned. As discussed previously, the line angle at the mudline may range from zero degrees for a catenary line up to 40° or more for a taut line. As tensioning proceeds, the line cuts through the soil until the design value of line tension is achieved. Figure 3.12 shows an example of an anchor line configuration during various levels of tensioning. During this process, the line takes on a reverse catenary-like shape and maintains some curvature even at very high-tension values. The consequence is that the line is always at a higher angle at the attachment point than it is at the mudline, and hence there is always some vertical component of load at the attachment point. For catenary systems, this angle may be modest depending on the depth of the padeye, but for taut line moorings, it is significant. Thus, there is always some anchor line inclination and some vertical-horizontal load interaction to consider in the anchor design. It is very important to have a reasonable estimate of this angle to properly design the anchor.

3.5.2 Interaction Tensioning of the Anchor

Installation of anchors for exploration activities has normally been a trial and error process using crude charts showing anchor capacity versus anchor weight for several soil classification

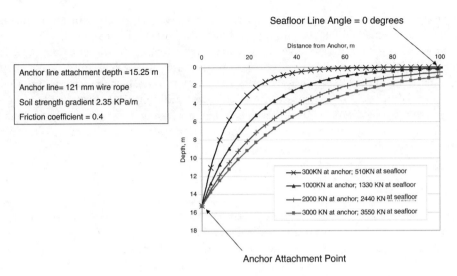

Figure 3.12   Anchor line configuration as a function of tension. Figure provided courtesy of Dr. James D. Murff, retired Professor at Texas A&M University.

descriptions. These charts introduce significant uncertainty, and most operators are unwilling to rely on such a process for permanent production facilities.

For drag embedment anchors and vertical loaded anchors (VLA) installed by drag embedment, there is an important interaction between the anchor line and the anchor as it is dragged into place. The anchor will initially rest on the mudline or at a relatively shallow depth below the mudline. As it is dragged horizontally, the fluke will bite into the soil and will gradually rotate downward, as shown in Figure 3.13. Once the attachment point is below the mudline, the soil will resist the anchor line and tend to increase its horizontal angle. The angle will continue to increase as the anchor penetrates; thus, there is a continuous interaction between the line and the anchor

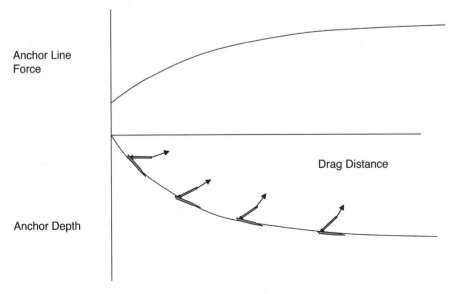

Figure 3.13   Schematic of anchor orientation during drag embedment. Figure provided courtesy of Dr. James D. Murff, retired Professor at Texas A&M University.

during its installation. As the angle increases, the line will apply a larger and larger moment on the fluke that will eventually rotate the fluke upward toward a horizontal orientation.

There are several methods available for incrementally estimating the trajectory of drag embedment anchors (Neubecker and Randolph 1995; Thorne 1998; Aubeny et al. 2005). These methods are an improvement over the empirical charts but still involve considerable uncertainty. This has been a major drawback to widespread use of anchors, including VLAs, installed by drag embedment. This will be discussed further in the section describing drag embedment anchors and VLAs. There remains significant economic incentive to improve the reliability of drag embedment anchor placement and hence design capacity.

3.6 Pipe Piles

The driven pile has been most widely used for support of jacket structures due to its numerous advantages. One of the major advantages is the capability to mobilize its capacity at a deeper penetration below the seafloor where the soil strength is much greater. Another advantage is the ability to install the driven pile with a minimum amount of soil disturbance. Even this disturbance is eventually recovered over time to develop the pile's full capacity as a result of soil setup.

In deepwater, one major drawback of driven pipe piles has been the limits of operating water depth for underwater hammers. Today, there are hydraulic hammers that have overcome some of this drawback, such as the Menck MHU-900T that can operate in water depths approaching 2200 m. On the other hand, the use of these hammers in deepwater requires an expensive construction barge. Other anchor types may be installed with a smaller, less expensive installation vessel. A recent project offshore Brazil (van Noort et al. 2009) demonstrated that a large underwater hydraulic hammer could be operated from an anchor handling vessel (AHV) to drive conductors in water depths ranging from 1600 to 1950 m. The use of an AHV can significantly reduce the overall cost of using deepwater hammers to drive piles.

Driven, jetted, and drilled-and-grouted piles, as illustrated in Figure 3.4, are all still being used in deepwater since they offer many benefits in terms of foundation performance that are independent of water depth. The design practice for each of these pile foundations types will be described in subsequent sections along with the advantages and disadvantages associated with their use.

3.6.1 Driven Pipe Piles

The design practice for offshore pile foundations is mature for most soil types. An earlier paper by Pelletier et al. (1993) described in detail the historical development of the practice of driven pile design for axial loads. The design procedures for laterally loaded piles are detailed in API (2000) based on work by Matlock (1970), Reese et al. (1974 and 1975), O'Neill and Murchison (1983), and others.

Recent experience as reported by Chen et al. (2009) during unprecedented hurricanes occurring in 2004 and 2005 within the Gulf of Mexico confirm that very few foundation failures occurred due to axial or lateral loading of driven pipe piles. The pile foundations performed better than expected as compared to the jacket structure. Subsequent studies indicate that driven pipe piles in clay developed at least their predicted ultimate axial capacity while the ultimate axial capacity in granular soils may be greater than predicted by the currently available design methods.

The design of piles for axial and lateral loading is generally treated as an uncoupled problem. This is reasonable since the majority of axial resistance is mobilized at depth whereas lateral resistance is mobilized nearer the mudline. As shown on the left in Figure 3.14, for a typical pile, less than 10% of the axial capacity is developed in the top 30% of the pile. As shown on the right,

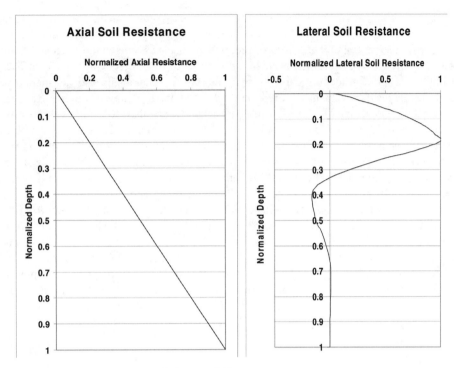

Figure 3.14   Axial and lateral loading behavior of piles.

the resistance to lateral loads and moments is mobilized in the near surface soils. In this case, approximately 85% of the lateral resistance is developed in the top 30% of the pile. For the purposes here, these two aspects of behavior will be discussed in separate sections.

3.6.2 Axial Capacity

The current methods for designing driven pipe piles to resist axial loading has been continually evolving, and updates in the procedures have been made as a result of API and other agency funding over the last forty years. The history of development of the different design methods to predict the axial pile capacity was previously documented in a number of references, such as Pelletier et al. (1993), Toolan and Hornell (1992), and Focht (1983).

The static method of analyses has been traditionally used to compute the ultimate axial capacity in compression by adding the shaft resistance Q_f to the end bearing capacity Q_p. The shaft resistance is computed by integrating unit skin friction f for each soil layer penetrated over the entire surface area of the pile. The end bearing capacity (Q_p) is computed by integrating the unit end bearing Q over the total end area of the pile. The total axial capacity of the pile Q_t is then

$$Q_t = \pi D \int_{z=0}^{z=L} f(z)\,dz + \pi D^2 q/4 \qquad (3.2)$$

where D = pile diameter
              $f(z)$ = unit skin friction as a function of depth,
                q = unit end bearing

The parameters f and q are discussed in more detail for different soil types in subsequent sections.

The static method assumes the maximum values of the two components of capacity are mobilized at the same time. In reality, unit skin friction develops at very small displacements (as little as 10 to 20 mm) whereas the unit end bearing requires tip displacements in the order of 10% of the pile diameter. This means a pile tip movement in the order of 100 mm for a 1 m diameter. The flexibility of the pile and degradation of the soil/pile load transfer relationship also play major roles in terms of the pile head movement required to mobilize the maximum axial compressive capacity.

The procedure developed by (Coyle and Reese 1966) can be used to simulate the entire pile/soil interaction, resulting in a pile top load versus displacement curve. The method combines the different load versus displacement responses along the sides of the pile with the pile tip reaction, along with the pile deformation, into a realistic solution as compared to the simple static pile capacity computation. The evolution of the different methods used to compute the different component of axial pile capacity is described in the following paragraphs.

A large number of different methods have been proposed for computing the unit skin friction, but the method most widely accepted and recommended is the current API (2000) design guidelines. This method is based on a correlation between α, the fraction of shear strength mobilized, and the strength ratio (S_u/σ_v') as recommended by McClelland (1974) and Wroth (1972). The correlations later proposed by Semple and Rigden (1984) and then Randolph and Murphy (1985) were used as the basis for a new design method that relies upon measurable physical properties such as soil over-consolidation and pile stiffness. RP2A incorporated a somewhat simplified version of that proposed by Randolph and Murphy, specifically:

$$f = \alpha S_u \tag{3.3}$$

where α = a dimensionless factor that must be
       S_u = undrained shear strength of the soil at the point in question

The factor α can be computed by the following equations:

$$\alpha = 0.5\psi^{-0.5} \text{ for } \psi \leq 1.0 \tag{3.4}$$

$$\alpha = 0.5\psi^{-0.25} \text{ for } \psi > 1.0 \tag{3.5}$$

where $\psi = S_u/\sigma_v'$ for the point in question
       σ_v' = effective overburden pressure at the point in question (kPa)

We need to emphasize that the capacities computed with the above equations are applicable only after full setup occurs after pile installation. Cavity expansion theory as proposed by Whittle et al. (2001) and Randolph (2003) has been used to estimate the time rate of consolidation for driven piles and suction caissons. This approach indicates that 90% consolidation may take up to 500 to 700 days to occur after pile installation. We believe that these estimates are on the conservative side, as indicated in a case study by Bogard et al. (2000) that shows 90% of soil setup occurs within a period of a year or less.

The two studies used to develop the unit skin friction correlations recommended a pile to soil stiffness ratio be used to assess the effects of progressive strain softening at the pile-soil interface. The current API/ISO guideline (API 2000) does not include an explicit equation to correct for the pile *flexibility* effect. The procedures recommend that load-transfer curves (t-z) include a degraded or residual value beyond the peak resistance to account for this effect. We should emphasize that the ultimate pile capacity curve typically provided by the geotechnical consultants for selecting the final pile penetration does not include this effect. Thus, the designer should include this step in their soil-structure analyses to verify the pile can develop the required design capacity. This can be accomplished by conducting load displacement simulations (Coyle and Reese 1966) using softened t-z curves to represent the soil skin friction versus displacement behavior.

The unit end bearing Q_p in clay soils has traditionally been computed based on the following equation:

$$q = N_c S_u \qquad (3.6)$$

where N_c = bearing capacity factor usually assumed equal to 9.0

S_u = undrained shear strength averaged immediately below the pile tip

The unit end bearing is integrated over the entire base area of the pile based on the assumption that, at deeper penetrations, the pile will plug under static loading.

The traditional pile capacity curve for our earlier example using a 1 m diameter pipe pile driven to 50 m depth in clay is shown in Figure 3.15. The total end bearing resistance is usually small; in this case, it provides about 8.0% of the ultimate axial compressive capacity.

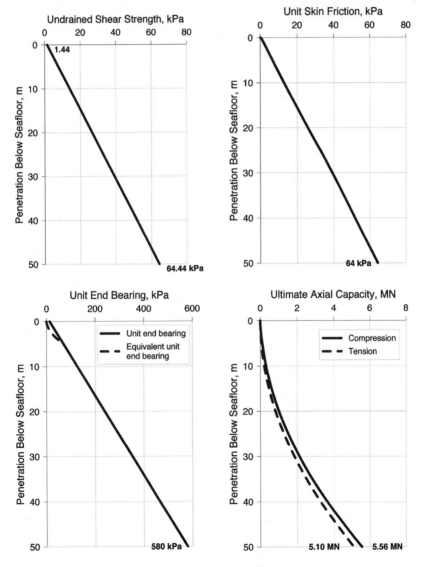

Figure 3.15   Axial capacity results.

The most common method used to compute the two components of axial capacity in sand, the shaft resistance Q_f and the end bearing capacity Q_p, is described in the current API/ISO guideline (API 2000) and is similar in form to the static method presented in the first RP 2A edition. The unit skin friction f and unit end bearing q are based on the following equations:

$$f = k\sigma'_v \tan\delta \le f_{max} \tag{3.7}$$

where k = dimensionless earth pressure coefficient
            σ'_v = effective overburden pressure
            δ = friction angle between the pile wall and the soil
            f_{max} = limiting value for unit skin friction

Table 3.6 shows the current limiting values f_{max} for the various types of granular soils. The appropriate parameters from this table were assigned primarily based on the grain size (sieve) distribution and blow count information obtained during sample collection for each soil layer encountered in the boring.

The unit end bearing in granular soils is based on the following equation:

$$q = \sigma'_v N_q \le q_{max} \tag{3.8}$$

where σ'_v = effective overburden pressure
            N_q = bearing capacity factor
            q_{max} = limiting value for unit end bearing

The values of N_q and q_{max} are again based on the gradation and relative density of the granular soil as determined by the sampler blow count data.

In recent years, the inaccuracy of assigning design parameters based on sampler blow count data has been recognized, as reported by Gilbert et al. (2010). Their recent study suggests that the limiting values and parameters assigned based on sampler blow count data may be very conservative based on hind casts of observations of damage to platforms/piles during hurricanes in the Gulf of Mexico. In addition, earlier studies, such as by Hossain and Briaud (1993), conclude that the limiting end bearing values recommended in the API/ISO guidelines may under predict the actual field value by a factor of 2 for dense sands. A number of studies reported by Fugro (2004) support the conclusion that the API/ISO axial pile design method for granular soils is conservative in dense to very dense sands when applying the limiting values of end bearing and skin friction.

To address these shortcomings, a number of different methods have been proposed that rely on CPT data in order to improve the API practice for computing unit end bearing and skin friction (Lehane et al. 2005; Jardine et al. 2005; Clausen et al. 2005; Kolk et al. 2005a). These methods have been included in the commentary of the API RP. In most cases, the new CPT methods provide a significant increase in computed ultimate axial capacity as compared with the API method. Unfortunately, there are only a few large-scale pile tests such as the Euripides load tests (Kolk et al. 2005b) available to calibrate the new methods, and there is still much debate regarding these procedures. It should be pointed out that the API method, in spite of possible shortcomings, has been successfully used for design of offshore foundations for some fifty years.

3.6.3 Lateral Capacity

The practice of designing driven piles to resist lateral loading is also a relatively mature practice. Piles can be used as anchors to resist lateral loads or inclined loads although they are generally not as effective in this role as they are in resisting axial loads. For conventional design applications, the pile-soil system is represented as shown in Figure 3.16. The pile is modeled as a linearly elastic

Table 3.6 Limiting values for granular soils (API 2000)

| Design parameters for cohesionless siliceous soil* | | | | | |
Density	Soil description	Soil-pile friction angle δ	Limiting unit skin friction values f_{max}	N_q	Limiting unit end bearing values q_{max}
Very loose	Sand				
Loose	Sand-silt**	15°	47.8 kPa	8	1.9 kPa
Medium	Silt		(1.0 ksf)		(40 ksf)
Loose	Sand				
Medium	Sand-silt**	20°	67.0 kPa	12	2.9 kPa
Dense	Silt		(1.4 ksf)		(60 ksf)
Medium	Sand				
Dense	Sand-silt**	25°	81.3 kPa	20	4.8 kPa
			(1.7 ksf)		(100 ksf)
Dense	Sand				
Very dense	Sand-silt	30°	95.7 kPa	40	9.6 kPa
			(2.0 ksf)		(200 ksf)
Dense	Gravel				
Very dense	Sand	35°	114.8 kPa	50	12.0 kPa
			(2.4 ksf)		(250 ksf)

*The parameters listed in the table are from API (2000). Where detailed information such as in situ cone tests, strength tests on high quality samples, model tests, or pile driving performance is available, other values may be justified.

**Sand-silt includes those soils with significant fractions of both sand and silt. Strength values generally increase with increasing sand fractions and decrease with increasing silt fractions.

beam column. The soil is represented as continuous, uncoupled, nonhomogeneous, nonlinear soil springs. These springs are idealized such that they resist the movement of the pile in the lateral direction. The characteristics of these springs are based on empirical procedures that are calibrated with test data for site-specific soil property types (Matlock 1970; Reese et al. 1974; O'Neill and Murchinson 1983). General recommendations for characterizing soil springs for clay and sand deposits are detailed in API RP2A. This model is used to design the pile steel for both working stress design and load and resistance factor design.

The basic governing equation for the pile soil system is a generalized form of the equation for a beam on elastic foundation:

$$EI\frac{d^4y}{dx^4} + k(x,y)y = 0 \qquad (3.9)$$

where EI = bending stiffness of the pile
        x = distance along the pile length
        y = lateral displacement
        k = soil spring stiffness, which is a nonlinear function of x and x

This equation along with the boundary conditions can be used to calculate the pile displacements and stresses for specified loading conditions. The solution to the governing equations has been implemented in a number of widely available *beam column* programs, most of which employ

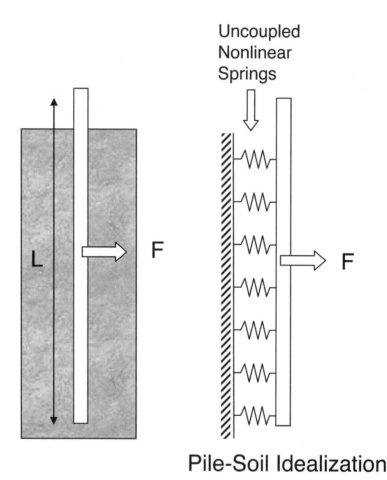

Pile-Soil Idealization

Figure 3.16 Laterally-loaded pile-soil system. Figure provided courtesy of Dr. James D. Murff, retired Professor at Texas A&M University.

finite difference approximations to the governing equations. There are both proprietary and non-proprietary versions available, many of which have been written with industry sponsorship.

The conventional design approach for laterally loaded piles is somewhat different from the design for axial load. Rather than designing for ultimate load as for axially loaded piles, the piles are designed for allowable stresses based on beam column analysis. As an example, Figure 3.17 shows the calculated normalized moment diagrams for a 1.22 m diameter pile anchor with a lateral load applied at the top and at 15.2 m below the mudline in soft, normally consolidated clay. The moment for the top loaded case is over three times larger than the below mudline case. Clearly, attaching the anchor line well below the mudline is much more effective than attaching it to the pile top. Attachment below the mudline does not come free since it involves complications for both fabrication and installation.

The design moment generally decreases as the depth of the attachment point increases, but the anchor line inclination angle increases such that the load becomes more vertical, increasing the possibility that vertical loading may govern the design. For mooring applications, the pile displacements generally have a negligible effect because the mooring lines are themselves so compliant. There is some merit in designing such anchors based on ultimate load rather than allowable stresses or at least using ultimate load as an additional check. In this regard, plastic limit

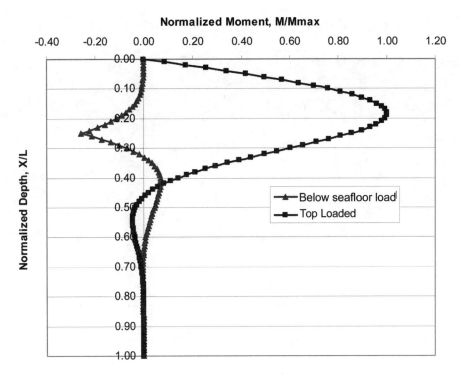

Figure 3.17 Comparison of moment diagram for different loading positions. Figure provided courtesy of Dr. James D. Murff, retired Professor at Texas A&M University.

analysis methods can be used to estimate the ultimate capacity of the pile soil system under lateral loading.

An exact solution can be obtained for some simple cases, as shown in Figure 3.18. For a top loaded pile, the failure mechanism involves the formation of a plastic hinge at some depth below the mudline. The solution for a uniform soil resistance and uniform pile cross section is shown here on the left where R_0 is the soil resistance per unit length and M_p is the plastic moment capacity of the pile. When the load is applied at a large depth, the mechanism involves the formation of three plastic hinges. In this case, the capacity is approximately three times that for the top loaded case. For highly variable soil resistance profiles, the critical mechanisms can be more complex, and numerical methods may be required to determine a solution.

3.6.4 Jetted Piles

The incentive to use jetted piles is that the piles can be installed using conventional deepwater drilling technology—that is, the same processes that are used in exploration and production drilling in deepwater. These processes include jetting in the so-called *structural casing*, which is the first section of casing set from the mudline down to perhaps 60 m or more. For very light loads, the use of jetted piles may have economic advantages, but there are serious problems associated with installing piles by jetting. The major disadvantage is that the soil surrounding the pile is usually highly disturbed and hence does not develop a large degree of the available soil strength as do soils adjacent to driven piles. In the case of normally consolidated clays, the driven pile develops close to its full potential skin friction resistance ($\alpha = 1$) due to soil setup over time. Full soil setup develops over different time periods, varying from about 90 days to longer periods approaching one year, depending on the soil type and pile diameter.

Pile Failure Modes Under Lateral Loading

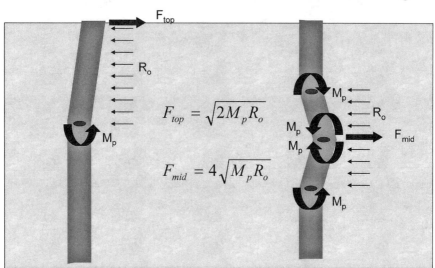

$$F_{top} = \sqrt{2M_p R_o}$$

$$F_{mid} = 4\sqrt{M_p R_o}$$

Figure 3.18　Simplified cases of pile system under lateral loading. Figure provided courtesy of Dr. James D. Murff, retired Professor at Texas A&M University.

On the contrary, a number of studies (Beck et al. 1991; Jeanjean 2002) confirm that a jetted pile develops a much lower percentage of the skin friction resistance due to soil disturbance, including a substantial increase in moisture content. The unit skin friction used for jetted pile design is totally dictated by the quality of the jetting procedures and the weight on bit during the jetting operation. The unit skin friction used for jetted pile design may be only 7 to 15% of the values used for driven pile design after complete setup. Even this modest soil resistance is highly dependent on the installation procedures, mud control, and the care applied during the jetting operations. Hydraulic fracture is a major consideration that is very dependent upon proper control of the mud weights, as described by Andersen et al. (1994). The studies by Beck et al. (1991) and Jeanjean (2002) describe some of the most important procedures that must be implemented to develop consistent methods for computing the ultimate axial capacity.

Using the same pile dimensions as in the previous example for a driven pile, the ultimate pile capacity for a jetted pile would range from about 0.39 to 0.84 MN, as compared to 5.56 MN for a driven pile. The state of practice relative to our understanding of the potential setup of the jetted piles that may occur over various time periods is a very immature technology with data available only for time periods varying from 1 to 10 days (Jeanjean 2002). There is not any empirical data available for setup periods greater than 10 days, so the ultimate axial capacity may not be predicted with confidence considering the many variables associated with installation procedures. Given the uncertainties involved in the installation process and hence the pile capacity, the authors believe that jetted piles should remain a niche technology. These pile types should be used only for very lightly loaded piles and only then with very carefully controlled installation procedures.

3.7　Drilled and Grouted Piles

Drilled and grouted piles have many of the same drawbacks of jetted piles. Drilling and grouting, however, can potentially result in a higher quality foundation that does not cause significant disturbance to the soil if carefully controlled. One of the major drawbacks though is quality control—for example, assessing the grout-soil interface. Field measurements are critical to confirm that the grout weight is sufficient to properly displace the drilling mud without causing hydraulic fracture of the soil formation during installation. There have been a number of major projects, such as the Ekofisk and Thistle Platforms in the North Sea as described by Ehlers and Bowles (1973) and Young et al. (1975b), respectively, where drilled-and-grouted piles were successfully installed and withstood maximum storm loads during 100-year storm conditions.

There are some soil conditions where drilled and grouted piles may serve as a preferred substitute for driven piles. For example, there have been a significant number of field studies reported by Kraft and Lyons (1974), O'Neill and Hassan (1994), and Murff (1987) that confirm that drilled and grouted piles provide a more suitable foundation when driving problems with piles are anticipated. Other studies (Angemeer et al. 1973) demonstrate a significant increase in skin friction resistance over driven piles in lightly cemented to uncemented calcareous sand. There are other geologic provinces where driven piles may be difficult to install due to rock layers or dense cementation. The drilled and grouted pile may prove to be the only good option for these difficult foundation conditions.

3.8 Suction Caisson

The suction caisson has become one of the most widely used anchor types for deepwater mooring applications. The first offshore industry application (Senpere and Auvergne 1982) involved the installation of twelve suction caissons in 1981 for mooring a Catenary Anchor Leg Mooring (CALM) buoy to offload oil onto tankers. The installation owned by Shell Oil was in the Gorm Field located in the North Sea in about 40 m water depth with a sandy seabed.

The suction caisson did not come into widespread use until 10 to 12 years later when its advantages for deepwater moorings became apparent. Since that time, the installation experience using this technique has been excellent in a wide variety of soil types, although most installations have occurred in normally consolidated clays. This section will describe the evolution of the suction caisson as a preferred anchor alternative and the critical design and operational procedures that must be employed to achieve satisfactory performance.

3.8.1 Suction Caisson Design Geometry

A typical suction caisson, as shown in Figure 3.19, consists of a steel cylindrical shell with a top plate and various fittings that allow water to be pumped into or out of the shell. The caisson has an open bottom that allows soil to enter the internal volume of the caisson. Many of the caissons used to date have internal stiffeners to prevent buckling of the caisson wall during installation and to distribute the anchor line load from the attachment point.

Murff and Young (2007) state that suction caissons have large diameters ranging from 2.4 m to over 6 m (8 to 20 ft) and length to diameter (L/D) ratios typically in the range of 4 to 8. House et al. (1999) states that theoretical computations indicate that suction caissons with L/D ratios as high as 10 may be installed with suction, but their tests indicate that plug failure may be initiated at much lower aspect ratios, in the range of 5 to 7.

The current trend is to use thicker walls (greater than 3.5 cm) to avoid the use of internal rings or plate stiffeners. The thicker wall has the added advantage of causing less soil disturbance during installation than internal stiffeners, thereby providing a greater and more reliable vertical capacity.

For mooring applications, the padeye (anchor line attachment point) is frequently located about two-thirds down the length of the embedded pile, as shown in Figure 3.19, to mobilize the

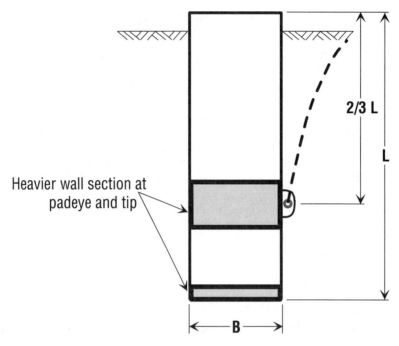

Figure 3.19   Typical suction caisson geometry. Figure provided courtesy of Mr. Alan G Young.

maximum lateral capacity. However, where vertical loading is critical, the padeye for mooring line attachment can be made at any point along its length.

3.8.2 Key Design Factors

The key design objective, according to Murff and Young (2007), is the appropriate selection of a pile size that develops the required holding capacity and withstands the under pressure during suction embedment. Thus, the pile dimensions should be carefully selected to achieve full penetration without buckling the top or walls of the caisson or causing soil plug failure (upward movement of the soil plug without further penetration of the caisson). For each design, a number of computations are carried out that include:

1. Penetration resistance with depth
2. Required under pressure with depth
3. Allowable under pressure with depth
4. Self-weight penetration depth
5. Maximum recommended penetration depth
6. Anchor line geometry especially the inclination at the attachment point
7. Required holding capacity
8. Recommended optimum attachment point of padeye if used

In addition, the design needs to address a number of other factors that include: (1) the setup time required to develop full shaft resistance, (2) the effects of the cyclic mooring loads, and (3) whether a gap may form at the backside of the caisson during loading. There are also a number of installation considerations that should be addressed to verify that the suction caisson maintains the specified orientation and verticality.

3.8.3 Design Procedures

The design procedures for suction caissons are generally divided into two broad groups that cover the installation phase and the loading phase, as described by Randolph et al. (1998) and Andersen and Jostad (1999). The methods used to investigate these requirements are, in many ways, qualitatively similar to those employed in the design practice, developed over many decades, for installation and loading of driven piles. The types of advanced laboratory testing required for defining the soil parameters used in the suction caisson analyses are described by Jeanjean et al. (1998).

The soil resistance to penetration consists primarily of shear stresses distributed along the internal and external walls of the steel shell and along any internal longitudinal stiffeners. Minor soil resistance develops along the bottom edge of the steel shell as well as on stiffener edges. The shaft resistance is computed as the sum of the remolded shear strength integrated over the embedded surfaces of the caisson. The tip resistance is equal to the end areas times the undrained shear strength times a bearing capacity factor equal to about 9.0.

The vertical holding capacity Q_t of a suction caisson is computed as the skin friction resistance on the external caisson surface and the end bearing capacity computed over the total bottom area of the caisson. The equation for predicting the vertical holding capacity in clay soils is similar in form to the equation for estimating axial pile capacity previously discussed, specifically:

$$Q_t = \pi DLS_{u(\text{ave})}\alpha + N_c S_{u(\text{tip})}\frac{\pi D^2}{4} \tag{3.10}$$

where D = diameter of the suction caisson
               L = length of the suction caisson
               S_u = average DSS shear strength over the caisson embedment depth
               α = ratio of skin friction to undrained shear strength
               N_c = bearing capacity factor

The α factor and the bearing capacity factor N_c are two important factors that must be carefully selected since the maximum skin friction and end bearing capacity do not reach their maximum values at the same displacement. The API method indicates that the α value would be close to 1.0 for normally consolidated clay. Past design practice (Andersen and Jostad 2002) recommended the use of α values ranging from 0.65 to 0.70, depending upon how the suction caisson was embedded.

The findings from centrifuge testing as reported by Chen and Randolph (2005) found that α values were the same whether the suction caisson was jacked or installed with suction. Their testing indicates an N_c factor of 12 and an α value of 0.76. Subsequent centrifuge testing as described by Jeanjean et al. (2006) confirms the findings of Chen and Randolph relative to the influence of the method of embedment. The Jeanjean study recommends an N_c of 9.0 and an average α value equal to 0.85 to compute the two components of holding capacity at compatible displacements. The maximum skin friction resistance develops at much smaller displacement as compared to the 10% displacement required to mobilize full reverse end bearing.

Analytical methods for predicting holding capacity should take into account the coupling of lateral and vertical resistance, as described by Aubeny and Murff (2005), that is usually ignored for driven piles. This interaction diagram shown in Figure 3.20 illustrates variations in the lateral and vertical resistance. The interaction diagram shows that the suction caisson under inclined load may have vertical and lateral components that are smaller than the pure vertical or horizontal failure load.

For mooring applications, the mooring line can be attached at any point along the caisson's length. The objective is to place the padeye at the point where a failure would be translational—that

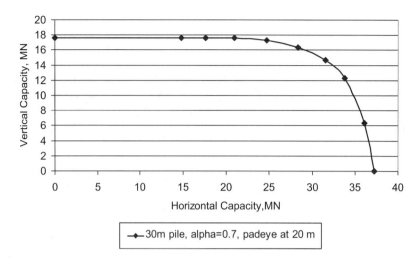

Figure 3.20   Interaction diagram of suction caisson loading. Figure provided courtesy of Dr. James D. Murff, retired Professor at Texas A&M University.

is, occurs without rotation. This point provides the largest holding capacity and is referred to as the optimal load attachment point. For mooring applications, the external padeye is then located such that the load application point on the caisson centerline is at or near the centroid of the lateral soil resistance profile over the pile depth. For inclined loads, this means that the padeye is placed at some distance above this point.

To illustrate the importance of the load application point, Figure 3.21 shows the variation in lateral capacity under purely horizontal load for two different soil strength profiles with the same average resistance over the caisson depth. For a 6.1 m diameter suction caisson with a length of 30.5 m, the holding capacity is almost 44.4 MN for both cases. The optimum load point is about two-thirds of the embedment depth for the linearly increasing resistance profile and one-half the embedment depth for the uniform resistance profile. If the lateral load is applied above the optimum attachment point, the suction caisson rotates forward and develops lower soil resistance.

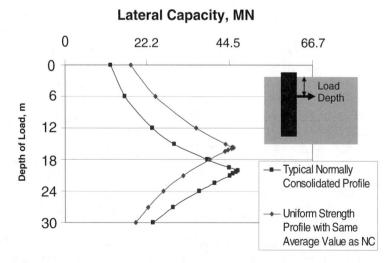

Figure 3.21   Variation of lateral capacity versus load application depth. Figure provided courtesy of Dr. James D. Murff, retired Professor at Texas A&M University.

If loads are applied below this location, the caisson tends to rotate in a reverse direction, and the capacity is also reduced.

The methods used to predict holding capacity are generally more complex than for conventional piles since the failure mechanism needs to address the type of loading, loading components and point of application, type of soil profile, and caisson dimensions (L/D). Many of the limit equilibrium methods are plane strain models that do not accurately model asymmetrical failure of a cylindrical geometry. These methods are adjusted based on model tests and 3-D finite element analyses to correct for the inherent limitations.

Murff and Hamilton (1993) published a paper that incorporated a true 3-D failure mechanism in a plastic limit analysis (upper bound) solution. This collapse mechanism, as illustrated in Figure 3.22, is capable of estimating the horizontal resistance for a general soil profile but requires optimization of the solution using four parameters that describe the mechanism geometry. To simplify the computationally intensive procedure, Aubeny et al. (2001) proposed a simpler equivalent mechanism, as shown in Figure 3.23. This simplification is based on an empirical fit to solutions using the full model with varying depth to diameter ratios and soil profiles of uniform and linearly increasing strength. The simplification involves only one optimization parameter, the depth of the center of rotation of the caisson, and hence is easily carried out in a spreadsheet format.

The model developed for the suction anchors for inclined loading is an extension of the lateral loading model described previously and is again based on plastic limit analysis procedures. In this case, finite element solutions were used as the base model, and simplified solutions were developed by curve fitting the finite element results. These solutions incorporate the interaction between lateral load and vertical load. Aubeny et al. (2003a) presents a comprehensive parametric

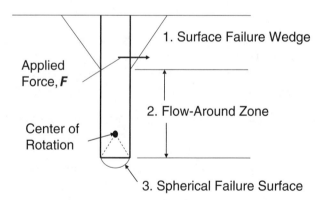

Figure 3.22 Murff and Hamilton collapse mechanism. Figure provided courtesy of Dr. James D. Murff, retired Professor at Texas A&M University.

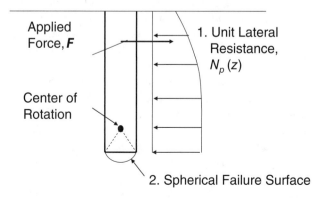

Figure 3.23 Simpler equivalent mechanism. Figure provided courtesy of Dr. James D. Murff, retired Professor at Texas A&M University.

study of inclined load capacity of suction caissons based on this procedure. Interaction diagrams have been developed for other conditions, including nonuniform strength profiles (Aubeny et al. 2003a) and variable adhesion factors (Aubeny et al. 2003b).

A comprehensive study comparing calculated holding capacity using different methods (Andersen et al. 2005) was conducted by: (1) the Offshore Technology Research Center, (2) the Centre for Offshore Foundation Systems, and (3) the Norwegian Geotechnical Institute. The study compared the various methods of predicting holding capacity by comparing the vertical load versus horizontal load interaction diagrams calculated by each method. The agreement among the methods was good considering the wide range of conditions. Among the important conclusions reached in the study was the following:

> "The interpretation of in situ and laboratory test data to establish a design shear strength profile may often be one of the major uncertainties in anchor design."

The authors also want to emphasize the importance of correctly measuring soil sensitivity (S_t). Soil sensitivity has many useful applications when used appropriately in computing the penetration resistance of a suction caisson and the driving resistance of driven piles. On some projects, bad decisions have been made when evaluating soil sensitivity that resulted in foundation installation problems.

A pioneering paper by Skempton and Northey (1952) clearly describes how to measure soil sensitivity and thixotropy and use them in our geotechnical engineering practice. Soil sensitivity is a simple dimensionless parameter defined by the following equation:

$$S_t = \frac{S_{und}}{S_{rem}}$$
(3.11)

where S_{und} = undisturbed undrained shear strength
             S_{rem} = remolded undrained shear strength

Soil sensitivity is also a useful parameter for evaluating sample quality. For example, typical values of soil sensitivity for deepwater Gulf of Mexico clay range from 3.0 to 4.0. If measured values are less than typical values, the sensitivity indicates that samples disturbance may have occurred during field acquisition.

The finite element method is another technique that has been used to determine more solutions to this complex problem. This is a more rigorous method requiring fewer assumptions and has been used to calibrate other methods (e.g., Randolph and House 2002), and/or to benchmark centrifuge testing (e.g., Clukey et al. 1995; Clukey and Phillips 2002; Clukey et al. 2003; Clukey et al. 2004).

As pointed out by Templeton (2002), 3-D FEA have become practically tractable in recent years providing a number of benefits over more conventional methods such as: (1) detailed predictions of load versus displacement behavior, (2) including critical 3-D effects and highly variable soil properties, and (3) more accurate modeling of the load path interaction associated with inclined loading. Thus, FEA has enjoyed more widespread use for suction caisson design, which has resulted in greater accuracy.

Another advantage is that the soil reactions obtained from the geotechnical analysis based on FEA can be used directly as input for the structural model. A disadvantage of FEA is that, in spite of vast improvements, it remains rather labor and computer resource intensive, especially compared to the more conventional tools such as plastic limit analysis. In particular, it requires more start-up effort than some of the simpler methods. The strategy preferred by the authors involves taking advantage of the strengths of each approach. For example, limit equilibrium or plastic limit analysis can be used to carry out extensive parameter studies that are useful for preliminary screening and sensitivity studies, whereas FEA can be employed for detailed analysis and design once a final configuration has been determined.

3.8.4 Installation Procedure

The suction caisson is installed using under pressure within the caisson to develop the driving force used to embed it into the seafloor. The concept of using a hydrostatic head to install suction caissons was first introduced in the early 1960s, as described in a paper by Goodman et al. (1961). Subsequent papers by Brown et al. (1971), Wang et al. (1975), and Helfrich et al. (1976) presented methods for predicting the vertical holding capacity and described potential practical applications. Shell initiated an extensive testing program in 1976 to verify the offshore suitability by performing very small suction pile experiments (Hogervorst 1980) in a lake located in The Netherlands.

The suction caisson is launched from the installation vessel using a variety of methods. Those with large diameter and long lengths are frequently placed in the water with a large derrick barge, as shown in Figure 3.24. Smaller suction caissons such as those used to moor exploration-drilling rigs are often over-boarded over the stern of an AHV using a fairly simple procedure, as illustrated in Figure 3.25. This latter technique has been routinely used as described by Dupal et al. (2000) to moor MODUs using a semi-taut mooring system. The approach has a number of benefits, but the key benefits that are achieved include the following:

1. The anchors may be pre-installed prior to arrival of the construction vessel to reduce the time and cost for anchor installation.
2. The mooring spread footprint may be smaller.
3. The anchors can be installed in a more confined area without dragging them to the final site.

Figure 3.24   Large-diameter suction caisson launching.

Figure 3.25   Small-diameter suction caisson launching. Figure provided courtesy of Mr. Evan Zimmerman with Delmar Systems, Inc.

After the AHV over-boards the suction caisson over the stern rollers, it then lowers the suction caisson down near the seafloor with the large cable operated from the vessel winch. Before embedment, the vessel and remotely operated vehicle (ROV) properly orient the suction caisson at the proposed site. The cable is further lowered until the suction caisson embeds under its self-weight at a controlled rate of penetration. Trapped water is expelled through a valve on the top cap of the caisson. Once self-weight penetration is fully achieved, the ROV closes the valve and uses a pump connected to valves on its top to withdraw water, creating an under-pressure (referred to, albeit incorrectly, as suction) within the suction caisson. Once the suction caisson achieves its desired embedment, the ROV pump is removed and all valves closed. Closure of the suction caisson ensures that during uplift loading, *suction* under pressure will allow development of the full end bearing resistance of the soil plug.

   Although the procedures for suction caisson installation are fairly standardized and have been routinely used, a written installation plan is highly desirable to verify that the suction caisson achieves the intended performance. Typical measurements that are made and carefully monitored during the installation process include:

1. Applied under-pressure
2. Penetration
3. Penetration rate
4. Pump rate
5. Tilt
6. Plug heave
7. Orientation

These measurements should be plotted as they are recorded in a graphical form, as illustrated in Figure 3.26. All these measurements should be observed in real time to avoid problems during the installation. A quick review will confirm that the final installation conditions satisfy the acceptance criteria. In addition, these measurements help to confirm that the soil conditions are consistent with the interpreted design conditions. They also help to calibrate the design procedures by providing empirical data that can be compared with their predicted installation performance.

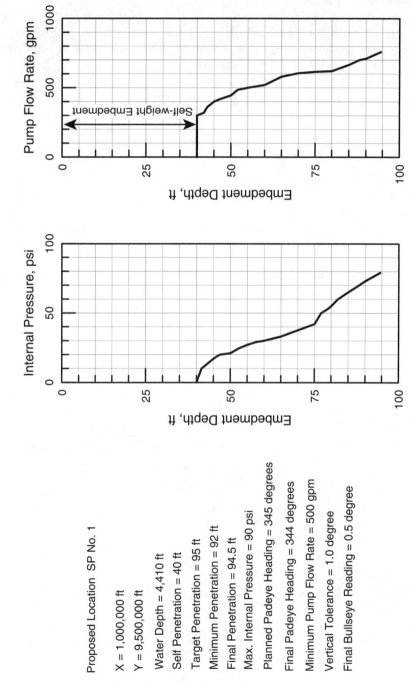

Figure 3.26 Field measurements during suction caisson installation. Figure provided courtesy of Mr. Alan G Young.

The penetration rate, pump rate, and the measured under-pressure are critical during suction embedment. The measurement of under-pressure should be made inside the caisson instead of at the pump to avoid errors due to fluid friction losses. The pump must be sized to have sufficient capacity compared to the total volume of the suction caisson, which can be very large, to achieve an adequate penetration rate. Suction caissons with large length to diameter ratios (L/D) penetrating strong soils near their final design depth need to be carefully monitored. The suction caisson must be designed to achieve its planned penetration without exceeding its structural collapse capacity. When suction caissons are planned for difficult soil conditions, then a backup ROV may be warranted to avoid the risk of soil setup. Time delays caused by mechanical breakdowns during the installation process will result in increased penetration resistance that can be detrimental to achieving the design requirements.

3.9 Vertical Loaded Anchors (VLA)

Drag embedment anchors, as shown in Figure 3.27, have been used for many years to moor offshore drilling vessels and offer numerous advantages for this application. These applications are typically limited to a few months duration and are relatively inexpensive, quick to deploy, and recoverable even though there is considerable uncertainty in their load carrying capacity. These anchors continue to be widely used in exploration activities, because of their economic and practical advantages. For example, in many exploration operations, the large footprint required by catenary mooring systems does not constitute a significant problem. The uncertainty in placement location and hence capacity has made these anchors more problematic for use in permanent production systems where reliability requirements are significantly higher. Another major disadvantage is the fact that these anchors have marginal uplift resistance that is required for the taut line systems that are so advantageous where extensive infrastructure is in place.

Figure 3.27 Drag embedment anchor (Stevpris Mk5). Figure provided courtesy of Dr. James D. Murff, retired Professor at Texas A&M University.

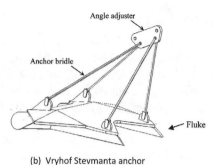

(a) Bruce Dennla anchor   (b) Vryhof Stevmanta anchor

Figure 3.28   Bruce Anchor's Dennla and Vryhof Anchor's Stevmanta. Figure provided courtesy of Dr. James D. Murff, retired Professor at Texas A&M University.

To address these disadvantages, new drag-in plate anchor concepts were introduced in the 1990s that were later called vertical loaded anchor (VLA). There were two anchor companies that conceived and developed the new VLAs: Bruce Anchor's Dennla and Vryhof Anchor's Stevmanta (2005), as shown in Figure 3.28. While there are significant differences in the details of these anchors, the general operating principles are very similar.

VLAs are installed by drag embedment similar to the conventional drag embedment anchor. However, after installation to the design depth, the fluke of the VLA is rotated from the installation position to a position normal or almost normal to the anchor line loading direction. Note that the anchors are actually loaded normal to the fluke and not necessarily vertically as the name might imply (Figure 3.29). This normal loading configuration significantly increases the holding capacity. This can be estimated using convention bearing capacity equations. Furthermore, these anchors generally achieve greater depth below the seafloor than conventional drag embedment anchors and hence are in stronger soils. Plates loaded in this manner develop the maximum capacity of the fluke. The capacity estimate Q is given simply as:

$$Q = S_u N_c A \tag{3.12}$$

where N_c is a bearing capacity factor typically taken as between 10 and 11.

The VLA over the last decade has been used quite extensively for mooring exploration rigs and on a limited basis for deepwater taut-leg mooring systems for floating oil and gas production facilities. For example, Petrobras has used 11 m² Stevmanta VLAs (Ruinen 2004) with an estimated holding capacity of 7.56 MN to moor the FPSO Fluminense. The taut-leg mooring system used a chain-polyester mooring line that was installed in 700 m of water in the Bijupira and Salema fields, offshore Brazil.

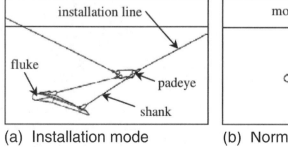
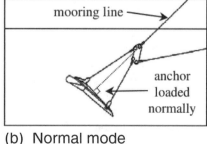

(a) Installation mode   (b) Normal mode

Figure 3.29   Vertically loaded plate anchor (VLA) installation. Figure provided courtesy of Dr. James D. Murff, retired Professor at Texas A&M University.

3.9.1 VLA Geometry

Figure 3.28 shows examples of the two different VLA concepts. The first concept, the Bruce Dennla, looks quite similar to a conventional drag embedment anchor with a flat plate as a fluke and a rigid shaft as a shank. The difference is that the shank, while fixed during installation, can be rotated after embedment such that its orientation is nearly normal to the fluke. The second concept, the Vryhof Stevmanta (Annevall 1997), has a rather unique shape, using a wire rope bridal instead of a shank. By shearing a pin in the anchor line attachment mechanism and tensioning the line, the fluke rotates into a position normal to the anchor line. There are other brands and configurations of anchors now available, but they generally employ similar concepts.

3.9.2 Key Design Factors

Successful installation and performance of a VLA depends upon the anchor being designed and installed properly for the soil conditions within the area of the proposed mooring spread. Earlier procedures developed for drag embedment anchors were modified for VLA applications as part of an industry JIP described by Dahlberg (1998). Based on the design procedures developed during this JIP, DNV (2000) published an international recommended practice for geotechnical design and installation. The design requires that the fluke area be properly sized to the installation vessel in order to achieve the required anchor depth and hence capacity. A listing of steps that typically must be taken to design the anchor, as well as the installation and keying procedure, is described in the Vryhof Stevmanta VLA User Guide (2005) as follows:

1. Determine the required ultimate holding capacity
2. Determine the geotechnical design profile
3. Determine the required anchor penetration depth as a function of fluke area
4. Determine the optimal VLA fluke area
5. Determine the required installation load for the selected VLA size

The anchor design involves two major considerations that include the measured installation load and the post-installation effects due to consolidation and cyclic loading. Anchor manufacturers indicate that the holding capacity may be 2.5 to 3.5 times the maximum installation load, although we believe that this may be somewhat optimistic. The actual holding capacity is dependent upon the installation vessel providing sufficient bollard pull to achieve the required penetration depth of the anchor and keying the anchor in the near normal alignment with the anchor line. The installation procedures critical to a successful installation in terms of the penetration path and anchor drag distance are described in the next section.

3.9.3 Installation Procedures

As mentioned, anchor installation of a VLA is very similar to that of a drag embedment anchor where the plate (fluke) is placed on the seafloor and dragged horizontally by an anchor-handling vessel. As the anchor is dragged, the fluke rotates downward, bites into the soil, and penetrates until the line load reaches a prescribed tension, and is then, ideally, at the desired target penetration. The horizontal force from the anchor line is applied through the shank or bridle/harness system that is set at a prescribed angle with the fluke, as shown in Figure 3.29. The angle of the shank or bridle to the fluke typically exceeds the 50° typical for a drag embedment anchor to achieve maximum penetration in soft soil formations.

Upon reaching a predetermined tension during installation, a pin, with a set capacity, shears, allowing the shank to rotate the plate and become normal or perpendicular to the anchor line loading. The process of orienting the fluke to a normal or near normal position to the anchor line is referred to as keying or setting the anchor. The anchor-handling vessel then increases its bollard pull, thereby increasing the tension in the line up to a proof load that is approximately equal to 80 to 100% of the maximum intact load for the mooring system design. In this case, the holding capacity is limited to the maximum bollard pull of the installation vessel.

3.9.4 Anchor Installation Plan and Performance

Developing an anchor installation plan requires a method for estimating the anchor performance. Several such methods are available, ranging from simple empirical charts to detailed numerical simulations of the anchor embedment process. Some of these approaches are briefly described in the following.

- Empirical charts: Many anchor suppliers have developed anchor specific charts that relate anchor-holding capacity to anchor weight and soil type. These methods are generally very rough and subject to much uncertainty. (Example references Vryhof 2006; Bruce 2010.)
- Limit equilibrium methods: These methods simulate the anchor embedment process in incremental steps. In each step, the soil forces acting on the anchor are assumed based on limiting equilibrium approximations along with the anchor line force and configuration. The anchor is then advanced a small increment, including rotation if any. The forces on the anchor and anchor line are then revised based on the new configuration, and the anchor is advanced another increment. This is continued until the anchor achieves its target or maximum practical depth. (Example references Neubecker and Randolph 1995; Thorne 1998; Kim et al. 2005; Dahlberg 1998.)
- Plastic limit analysis: These methods also involve an incremental approach to estimating the anchor trajectory. The major difference with limit equilibrium approaches is that the anchor is modeled as an interaction surface or plastic yield function in moment, normal (to the fluke) load, and parallel (to the fluke) load space. The increment directions are based on the concept that plastic displacements and rotations are normal to the yield surface. The solution for each increment is obtained by solving the anchor line equations interactively with the anchor yield surface equations. (Example references Bransby and O'Neill 1999; Yang 2009.)

The latter two methods have the advantage of more realistically incorporating details of the anchor, anchor line, and soil strength profile in the model of the simulation process. While they are an improvement over the empirical charts, the methods still result in significant uncertainty in the estimates of trajectory and holding capacity, as reported by Murff et al. (2005) based on API study.

Successful installation and hence operation of the mooring spread requires a plan that includes all design factors and verifies that design assumptions are met by monitoring the spread performance during its operating life. The installation plan should consider as a minimum the following:

1. Recommended VLA size (fluke area)
2. Required installation load to satisfy class requirements; AHV size
3. Installation load at seabed
4. Estimate of ultimate holding capacity
5. Estimates of required penetration depth and drag length for the applied installation load using the best prediction methods available
6. Inverse catenary calculations for the anchor line in concert with the anchor installation model to determine the embedded length of the anchor forerunner
7. Recommended installation procedures that incorporate the above features

There remains a strong incentive to use VLAs in the moorings for deepwater production systems. While considerable practical experience has been obtained in the last 15 years, and performance prediction methods have improved, there remains significant uncertainty in VLA design that is an impediment to widespread use in permanent mooring systems. The drag embedment installation still involves uncertainties regarding positioning that detract from their desirability as permanent anchors.

3.10 SEPLA

The SEPLA was introduced in 1998 and offers a novel approach for placing a plate anchor at a specified depth and location. The anchor is a hybrid design using the proven installation methods of the suction caisson and the simple VLA anchor geometry to accurately embed the plate anchor to a specified depth below the seafloor. The plate anchor is installed vertically in a slot at the bottom of a suction caisson, as shown in Figure 3.30. The caisson is then installed using standard procedures as described by Wilde et al. (2001), the plate is released, and the suction caisson is

Figure 3.30   SEPLA anchor (vertical plate in slot of suction caisson). Figure provided courtesy of Mr. Bob Wilde with InterMoor.

removed, leaving the plate anchor in place. The plate anchor is then keyed or rotated into place normal to the anchor line pull direction by tensioning the anchor line attached to the plate anchor shank, as illustrated in Figure 3.9. This innovation provides a potential reduction in the size of the anchor plate as well as a reduction in placement uncertainty.

The installation procedures and design methods used to predict penetration resistance will not be described in detail since the basic method is the same as described in the previous sections for suction caissons and VLAs. We should caution that the SEPLA is generally not the preferred anchor type for layered soil profiles where a strong layer exists near the final location of the anchor plate. The SEPLA must overcome the combined penetration resistance of the suction caisson and the plate anchor. This maximum penetration resistance will develop close to full embedment of the suction caisson when the maximum suction pressure is required. Thus, the suction caisson must be designed for an under-pressure that can overcome the combined soil resistance, including potential problems with soil setup.

The anchor plate must penetrate any deep stronger soil layer to a sufficient depth that it will not break out under the maximum mooring load, resulting in a significant loss of capacity. The study by Liu et al. (2005) shows that the pullout capacity of the anchor plate embedded into the stronger underlying clay depends upon the distance between the plate and the soil layer boundary and the soil layer strength ratios. For example, the anchor plate usually needs to penetrate the stronger layer a distance much greater than the anchor plate width.

3.11 Gravity-installed Anchors

A number of gravity-installed anchor types have evolved since Petrobras began development in 1996 of the Torpedo anchor, as illustrated in Figure 3.31. Petrobras has successfully installed more than 1000 Torpedo anchors for mooring deepwater flow lines and facilities offshore Brazil.

These anchor types typically feature a long cylindrical shaft and various fin geometries. The anchor is released above the seafloor and becomes completely buried by dynamic self-weight penetration after freefall to depths of 20 to 30 m below the seafloor in normally consolidated soils. These anchor types have the potential of providing significant holding capacity due to the

Figure 3.31   Torpedo anchor installation. Figure provided courtesy of Mr. Bob Wilde with InterMoor.

fact that the anchor embeds to penetrations where the soil strength is much greater than near the seafloor.

The Torpedo anchor features a cone tip and tail fins. Their past designs are 0.76 m to 1.07 m in diameter and are about 12- to 15-m long. The size of the fins depends on the mooring application. Large fins are used for MODU and FPSO moorings to increase holding capacity. The cylindrical shaft has a cast iron padeye attached to the rear of the anchor to provide an omni-directional attachment point for the mooring line.

To increase the submerged weight, the shaft is generally filled with lead ballast near the tip, a section of cast iron above the lead, and concrete above the iron. The size and submerged weight depends upon the application: T-24 torpedo piles typically are used for flowline restraint, T-43 for MODUs, and T-98 for FPSO permanent anchors. The number following the T indicates the dry weight of the anchor in metric tons.

Torpedo anchors are typically installed over the stern of an AHV as shown in Figure 3.31. The anchor is lowered to a set height above the seafloor with a lift line. The smaller torpedo anchor (T-24) is often released about 120 m above the seafloor. The heavier T-98 anchor requires a height of only 30 m above the seafloor to achieve tip penetrations of 20 to 30 m.

The DPA was developed as part of a JIP in Norway and operates on the same principles as the Torpedo anchor. The anchor is a similarly shaped, thick-walled steel cylinder with fins (flukes) attached to the upper end along with an anchor line attachment point, as shown in Figure 3.32.

The SPEAR anchor, as illustrated in Figure 3.33, was conceived by coauthor, Young, in 2004. A key design feature of this anchor, as described by Zimmerman et al. (2005), is the capability to embed deeper as greater anchor line loading is applied. This anchor is designed to freefall through the water column and to embed up to 15 to 20 m penetration below the seafloor. Key design features involve the shape and area of the nose and tail fin sections and the location of the mooring line attachment. This configuration develops less soil resistance on the nose section than tail

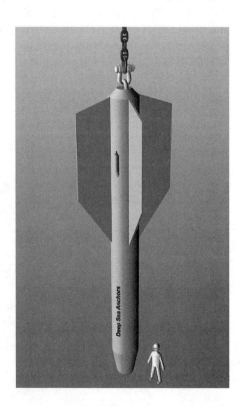

Figure 3.32   Deep Penetrating Anchor (DPA). Figure provided courtesy of Jon Tore Lieng with Deep Sea Anchors AS in Trondheim, Norway.

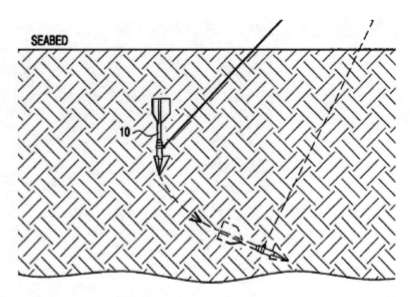

Figure 3.33   SPEAR anchor embedment with increasing line loading (Zimmerman et al. 2005): Figure provided courtesy of Mr. Alan G Young.

section, causing the anchor to rotate about the mooring line attachment point while penetrating deeper below the seafloor as line loading increases, as illustrated in Figure 3.33.

The advantages and disadvantages of the gravity-installed anchor are listed in Table 3.5, as described by Ehlers et al. (2004). The major benefit of all the gravity-installed anchors is that they can be deployed from an AHV in almost any water depth. These anchor types provide significant economic benefit since multiple anchors can be installed with a single vessel deployment, and simplicity of installation means less time is required to install each anchor. Other benefits include the facts that orientation is required during installation and unlike drag anchors, a specified drag distance or proof-load is not required to verify its holding capacity.

3.11.1 Installation Procedures

The installation and retrieval procedures of all gravity-installed anchors are fairly similar. The key requirement is that the anchor must achieve sufficient embedment to develop the vertical uplift capacity required for the taut-leg mooring system. Thus, the height of drop and submerged weight need to be selected to overcome the penetration resistance for expected soil strengths at the proposed anchor site. This means that the soil conditions and strength profile must be known within reason to model the dynamic processes occurring at the soil-anchor interface.

As mentioned, the anchors are typically installed from an AHV with a large deck space and high capacity winches with large drums for the cable used as the installation line, as shown in Figure 3.34. This approach is typically used to moor exploration drilling rigs but has been used in a few cases for permanent production rigs.

The anchors are typically deployed in a similar fashion as suction caissons over the stern roller and then lowered to the proper height above the seafloor, as shown in Figure 3.35. The permanent mooring line is also attached prior to installation with a special arrangement that allows the temporary installation line to be released in a controlled manner. At this point, an acoustic connector releases the anchor, allowing it to freefall and embed into the seafloor. The elastic rebound of the installation line is reduced by increasing the mass of the connector.

Figure 3.34 Anchor handling vessel with a large deck space and high capacity winches with large cable drums. Figure provided courtesy of Mr. Bob Wilde with InterMoor.

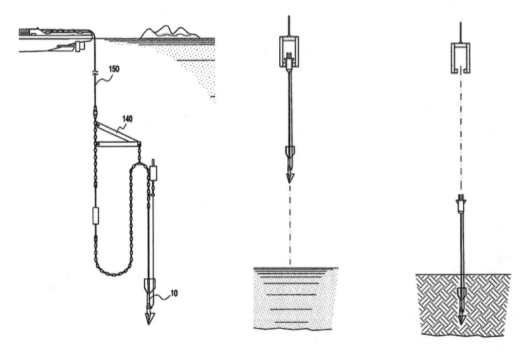

Figure 3.35 Deployment of a gravity SPEAR anchor (Zimmerman et al. 2005): Figure provided courtesy of Mr. Alan G Young.

3.11.2 Design Methods

Gravity-installed anchors rely upon analytical procedures to predict the final embedment depth and their final holding capacity. Like any of the other anchor types, the design methods require a thorough understanding of the soil stratigraphy and engineering properties throughout the zone to be penetrated by the anchor. Gravity anchors like suction caissons and VLAs are not suitable for all soil conditions. The three anchor types are most ideally suited for clay soils and should be avoided for dense sands and cemented soils.

A number of input parameters are required to predict gravity anchor embedment. The model for free-fall embedment must include the anchor velocity upon impact of the seafloor, the anchor mass and geometry, and soil sensitivity. There have been a number of studies aimed at developing methods to predict dynamic projectile penetration into the seafloor, so we will only reference a few such as (Young 1967) and (True 1976). A more recent study by O'Loughlin et al. (2004) describes the current procedures for predicting gravity installed anchor embedment and holding capacity in clay soils based on centrifuge model testing.

Vertical holding capacity is a combination of pile weight and soil resistance along the external surface of the pile and fins. The vertical holding capacity of the different gravity anchors will increase over time due to soil setup, as occurs with driven pipe piles. Most of the strength loss during installation will be regained in approximately 90 days, depending on the soil type (Bogard et al. 2000).

3.12 Foundations for Seafloor Facilities

Foundations used to support seafloor facilities are typically divided into two categories: shallow mats (or mudmats) and piles. In deepwater, the pile category includes both conventional pipe piles and suction caissons. Foundation design in deepwater generally must satisfy similar requirements and employ similar methods as those for shallow water. Nonetheless, there are major differences in the typical applications and hence major differences in the magnitude and type of loadings. Foundation design in shallow water is typically associated with exploration structures, such as submersibles and jack-ups, and production structures, primarily steel pile jackets. Foundation design for these structures is usually controlled by environmental loadings such as those developed from maximum storm conditions. Many of the foundation design considerations as described by Young et al. (1975) and Helfrich et al. (1980) for the shallow water mats (mudmats) are also applicable to the shallow foundations used in deepwater, but as mentioned, the loading conditions are extremely different.

Shallow foundations in deepwater are associated with on-bottom facilities such as PLETs, PLEMs, pipeline lift structures, and riser bases. These must be designed for loading conditions associated with operating, installation, and removal that are more complex and quite different than the maximum environmental loading in shallow water. Previous sections have described the design of piles and suction caissons for moorings. The design methods for these foundations are not remarkably different for seafloor applications. Therefore, in this section, we will focus on the use of shallow foundations, which are significantly different, to support seafloor facilities. We will describe the characteristics of this foundation type, describe the design procedures, and discuss the key factors that must be considered to achieve satisfactory performance.

3.12.1 Mudmat Design Geometry

A typical example of a mudmat used as a shallow foundation in deepwater to support a PLET is shown in Figure 3.36. The mat dimensions are 9 m by 12 m, providing 108 sq m of bearing area. The mat includes 0.5-m-long skirts consisting of steel plates that extend below the base of the mats in a grid arrangement, as shown in Figure 3.36. The skirt is added to improve foundation performance for a typical deepwater soil profile in two ways. The skirt penetrates the very weak soil at the mudline interface, resulting in an increase in the horizontal sliding resistance, and also improves the mat's ultimate bearing capacity.

Another key design feature of the mudmat shown in Figure 3.36 is the addition of numerous holes cut in the bottom plate of the mudmat. In this example, each skirt compartment has

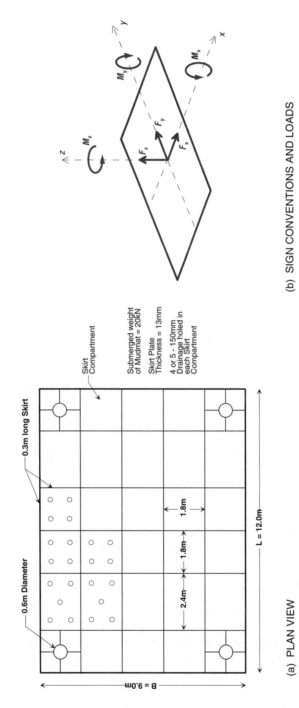

(a) PLAN VIEW

(b) SIGN CONVENTIONS AND LOADS

Figure 3.36 Mudmat design loading and geometry. Figure provided courtesy of Mr. Alan G Young.

four or five 150-mm diameter drainage holes. The drainage holes allow water to escape when the mat is lowered through the water column and also allow the extremely weak seafloor soils to be extruded from beneath the mudmat. The soil in direct contact with the mudmat base will hence be stronger than the soil supporting the mudmat without holes. The drainage holes also serve another useful purpose during removal of the mudmat by reducing the required pullout force and reducing the time required to eliminate soil suction. Finally, the drainage holes will help increase the ultimate bearing capacity over time due to consolidation since the drainage path for the pore water is shortened. In a typical case, the increase in soil strength beneath the base of a mudmat for a period of about 18 months has been measured to be over 10%, as verified by continuous CPT measurements. The percentage increase will vary depending on the structure weight and site-specific soil strength profile.

3.12.2 Key Design Factors

The primary objective of the mudmat design is the proper sizing of the mat area to provide an adequate factor of safety against foundation failure under the anticipated loading conditions. The stability of the mudmat depends on preventing a generalized bearing capacity failure and limiting the combination of vertical, horizontal, overturning, and rotational displacements. The loading during installation and operation will control the design and is typically quite variable. These loads include any set-down loads, structure weight, pipeline or jumper pull-in, and thermal expansion/contraction loads. This load combination will generally impose loads in all six degrees of freedom. These loadings are complex to predict and are difficult to design for since conventional design procedures are limited to idealized conditions.

The different types of foundation loads associated with mudmat design are shown in Figure 3.36. The loads consist of three forces and three moments applied in three-dimensional space with sign convention and axes, as shown in Figure 3.36. For design consistency, the force diagram with all load components should be referenced to the centroid of the mudmat area. For mudmats with skirts, the load application point may be transferred to the base of the skirt, and the skirt system should be designed to effectively transfer the loads to the skirt tips. This is discussed in more detail in subsequent paragraphs.

The design process must include the following steps:

1. Interpret the available shear strength data and select a design shear strength profile
2. Compute the factors of safety for generalized bearing capacity, including horizontal loads and overturning moments for a range of mudmat dimensions and loadings and select a final configuration
3. Analyze the mudmat and skirt geometry for the final installation, operating, and removal conditions
4. Compute soil springs for final loading, soil strengths, and mudmat dimensions to be used in structural analysis

The process often involves numerous iterations of the above steps to adequately complete the design. Because of the complex loading that the combined vertical and horizontal loads and three moments impose and the variable soil strength profile, using classical limiting equilibrium methods may be inadequate. Alternative methods include plastic limit analysis where failure mechanisms are consistent with the postulated soil and load conditions. In some cases, FEA methods are necessary, which require fewer assumptions and idealizations. The following section will describe the classical methods and point out their limitations along with adjustment factors that are frequently applied to accommodate unusual site conditions such as a sloping seafloor. Plastic limit analysis and FEA methods are also discussed in more detail.

3.12.3 Design Procedures

The assessment of foundation stability of the mudmat should take into account all of the loads acting on the mat. The design should also consider any limitations on rotations and lateral or vertical movement, including long-term settlement. A reasonable first step in this process is to size the mudmat to provide an adequate factor of safety with respect to bearing capacity, overturning, and sliding resistance of the supporting soil using conventional analysis methods. We recommend a factor of safety of 2.0 for these operating load conditions to accommodate uncertainties in the design parameters, including the possible effects of cyclic loading and creep movement.

Conventional analysis methods rely on classical bearing capacity theory to size the mudmat. Classical bearing capacity equations as proposed by Prandtl (1921) and Terzaghi (1943) have been used for vertically loaded strip foundations placed on a uniform soil profile. Meyerhof (1951) and Skempton (1951) later recommended shape factors should be applied for circular and rectangular shape foundations. These classical methods had design limitations since they neglected the effects of load inclination and eccentricity, embedment, seafloor slope, and nonuniform strength profiles. The basic bearing capacity equation was later modified by others such as Meyerhof (1953), Hansen (1970), and Vesic (1975) to include adjustment factors for these additional critical design factors.

The formula for the ultimate bearing capacity, including the recommended adjustment factors, is available in several excellent documents describing the recommended practice for shallow foundations. These design guidelines include API (2000), ISO (2007), DNV (1977) and a chapter of the *Foundation Engineering Handbook* written by Chen and McCarron (1991). In this section, we will summarize some of the factors associated with mudmat design for deepwater facilities.

We will first consider the ultimate bearing capacity equation proposed by Skempton (1951) that is applicable for undrained bearing capacity analyses for clay soils with a uniform or linearly increasing strength profile. The ultimate bearing capacity of the near seafloor soils supporting a mudmat can be estimated using the following equation:

$$q_{ult} = 5S_u \left(1 + 0.2\frac{D}{B}\right)\left(1 + 0.2\frac{B}{L}\right) + \gamma'D \tag{3.13}$$

where q_{ult} = ultimate bearing capacity of the foundation soils (kPa)
         S_u = average undrained shear strength of the soil over a depth of $B/2$ below the mudmat (kPa)
         D = mudmat embedment depth below the seafloor (m)
         B = width of the mudmat (m)
         L = length of the mudmat (m)
         γ' = submerged unit weight of the soil (kN/cu m)

This simple equation includes a few key variables like mudmat geometry, depth, and strength profile. Using this equation, the ultimate bearing capacity under centric loading with the linearly increasing strength profile (from Figure 3.15) can be computed in a straightforward manner. Likewise, a single safety factor can be applied to the vertical load to size the mat. However, the typical mudmat in these applications is loaded with six different components of loading. This combined loading condition, especially the overturning moments and load inclination, will substantially reduce the generalized bearing capacity of the mudmat and hence the computed factor of safety based only on vertical loading can be misleading. The classical methods may be used to account for eccentric loading as proposed by Meyerhof (1953) and Hansen (1970). These methods include the eccentricity effect by providing rules for reducing the effective area of the foundation available to resist the loading. Methods for accounting for load inclination have also been developed—for example, see Meyerhof (1953), Hansen (1970), and Vesic (1975). These methods are formulated in terms of correction factors that are multiplied by the bearing capacity equation.

We should mention that a very important consideration involving the bearing capacity equation is the selection of design shear strength. Skempton recommended averaging the shear strength over a depth below the foundation of two-thirds of the foundation width (with *B* being the foundation width) if the average shear strength does not vary by more than 50% over this depth. Helfrich et al. (1981) pointed out that the stipulation stated by Skempton is not achieved for most soft clay soil shear strength profiles or foundation sizes. Thus, they recommended averaging the shear strength over a depth equal to one-half of the foundation width.

Booker and Davis (1973) used rigorous plasticity solutions to develop a method for computing the ultimate bearing capacity that takes into consideration a linearly increasing shear strength profile. Jeanjean et al. (2010) compares bearing capacity of an example mudmat using shear strength averaged over different foundation widths ranging from one-fourth to two-thirds. The study suggests that the past practice of averaging shear strength over different widths can result in over-predicting the bearing capacity, compared to the Booker and Davis method. The extent of the over-prediction will vary with the mat width, mudline soil strength, and strength gradient. While the Booker and Davis method is considered to be the most accurate, Skempton's method with the strength averaged over a depth of one-half the foundation width remains a very useful approximation for quick estimates.

The effect of torsion is a somewhat special case. It has not previously been included in conventional analyses as it is not a common load component for shallow foundations. In general, torsion will reduce the sliding resistance of the foundation but does not have a significant effect on the vertical bearing capacity. Publications by Yun et al. (2009) and Murff (2010) provide a better understanding of how torsion can influence the overall load carrying capacity of a mat.

The ultimate bearing capacity computed with the limiting equilibrium equations using the effective area method and inclination factors is usually conservative, especially for large components of lateral load and an overturning moment acting in the same direction. However, with more than one load component, the factor of safety no longer has a unique interpretation. For example, consider the case of a surface footing under inclined, eccentric load. We can represent the bearing capacity solution as an interaction or failure diagram where lateral load is plotted against vertical load for varying moments, as shown in Figure 3.37. Note that in this case, the moment is due to the lateral load being applied at some height above the footing base. Each interaction curve represents a constant load application height. This is just one example of how safety factors can be applied for cases of multi-axial loading.

Considering the design load as shown in this figure, we can then define the safety factor in at least three ways. For example, the factor of safety represents the relative distance from the design load to the interaction diagram, assuming:

1. The horizontal load is increased to failure (appropriate where the horizontal load is the primary uncertainty).
2. The vertical load is increased to failure (appropriate where the vertical load is the primary uncertainty).
3. The loads are increased to failure proportionately (appropriate where both loads are of equal uncertainty).

The problem is compounded many fold when the footing is subject to loads in all six degrees of freedom. However, the use of interaction diagrams can be a great advantage in trying to understand the problem and the sensitivity of the solution to potential failure under the various loads.

If the moments (*M*) and lateral loads (*H* horizontal and *V* vertical) are resolved into single resultant components, the solution begins to take on a more tractable form. One can then plot various two-dimensional cross-sections of the interaction surface—for example, *M* vs. *V*, *M* vs. *H*, and *H* vs. *V*. Knowing which components are the most uncertain and which are likely to remain

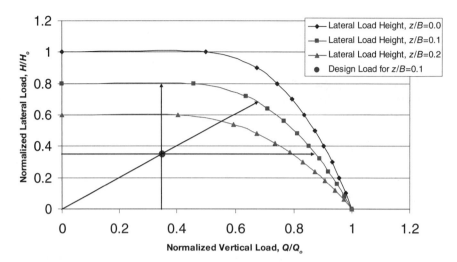

Figure 3.37   Bearing capacity solution as an interaction or failure diagram. Figure provided courtesy of Dr. James D. Murff, retired Professor at Texas A&M University.

constant (system weight, for example), can then provide the designer with tools to develop practical design criteria. As an example, Randolph et al. (2004) indicates that the ultimate limit state in the *V-H-M* space provides a more realistic prediction for undrained failure than the conventional methods. Since the more exact plasticity solution is only available for a few simple cases, geotechnical engineers will still need to rely upon the classical bearing capacity equations with appropriate adjustment factors.

In some cases, the large loads on the foundation or the weak soil conditions or both can require very large mudmats. For example, if wells have been predrilled, the very weak drill cuttings may accumulate on the ocean floor in thicknesses of 1 to 2 m or more. In these cases, it has been found that the size of the required mudmat is prohibitive, or at least very costly, for offshore handling and installation. For instance, the poor hydrodynamic characteristics of a large mat can make it very difficult to lower through the wave zone or may require an expensive construction spread.

One solution is to incorporate *wings*, or additional mats, that are hinged on both sides of the central mat, which can be folded up during installation and folded down to increase the overall mat size after the mat has been placed on-bottom. Another option is to incorporate pile slots in the mudmat corners where short *pin* piles can be inserted after the mudmat is set down. Such pin piles may be 8 m to 12 m long and still installed using a dead weight. These piles can significantly increase the foundation resistance in all failure modes and decrease the mat area by taking advantage of the additional capacity provided by the pin piles.

As stated, the primary function of the skirts is to mobilize stronger soils at depth for sliding resistance and, to a lesser degree, bearing capacity. The objective of skirt design is, therefore, to ensure that the skirt system is able to transmit the sliding and bearing loads to the skirt tips. This requires that the spacing between skirts is designed such that the full potential capacity in sliding and bearing is mobilized. Various analysis methods have been developed for carrying out this design, for example, Young et al. (1975) and Murff and Miller (1977).

The final skirt geometry must be designed with an arrangement of skirt elements with the proper width, length, spacing, and thickness to achieve full embedment for the submerged weight of the mudmat and the soil strength profile at the site. We recommend using a factor of safety of 2.0 on the skirt penetration resistance, as compared to the submerged unit weight of the mudmat. The penetration resistance of the skirt system during embedment into the soil is computed as skin friction along both sides of the skirt and bearing capacity resistance on the skirt plate thicknesses for the total skirt arrangement. The skirts have a narrow width compared with their length, so bearing capacity theory for a deeply embedded strip foundation can be used to estimate the end bearing resistance along the skirt tips.

Another important general analysis approach, plastic limit analysis, is available in situations where the classical bearing capacity method is clearly inadequate or where more rigorous analysis is desired. In this approach, special failure mechanisms are postulated that recognize the unique aspects of the problem, such as layered soil profiles, highly irregular mat geometries (multiple footings, for example), or difficult loading patterns. Rules for developing such mechanisms were developed by early workers in plasticity theory, for example by Drucker and Prager (1952). Designing appropriate limit analysis solutions for special problems is somewhat of an art form, so there have been many specialized solutions published for bearing capacity and related applications. While these are far too numerous to detail here, the books by Chen (1975) and Chen and Liu (1990) provide an excellent introduction to the topic.

Finally, we should mention that FEA of complex problems is always an option. This approach requires the fewest assumptions, but it is generally more labor and computer time intensive, especially for three-dimensional problems. In addition, solutions for specific complex problems require multiple computer runs to confirm a proper solution is achieved. Extensive parameter studies may be required, which are time consuming and expensive. In light of these difficulties, we strongly recommend that any problem be initiated with the simplest models possible, taking maximum advantage of classical and plastic limit approaches. Once the designer has developed a good *feel* for the problem, then confirmatory analysis of a specific design using FEA, if necessary, can be exploited to maximum advantage as described by Zdravkovic and Potts (2005).

3.13 Conclusion

Deepwater foundation design has evolved over the last couple of decades to become a mature design practice employing a wide variety of analytical procedures. The advances in equipment deployment and operation and the quality of data acquisition systems enable very high quality geophysical and geotechnical site investigations to be conducted in water depths up to 3000 m. The resolution of the data sets allows the geoscientist to develop a very comprehensive three-dimensional geostatistical model of geologic constraints and geotechnical soil properties. The ability to define spatial variability in soil conditions has dramatically improved. The geotechnical engineer now has the capability to select the most appropriate foundation type with a clearer picture of the existing soil conditions within the proposed development area. These improvements can reduce the risks associated with foundation installation and performance.

Previous studies (Ehlers et al. 2004) state that the principal uncertainties in foundation design include the proper selection of design parameters, accurate estimation of foundation loads, and reliable application of the analytical methods. The design team must appreciate the relative importance and potential variability of each of these uncertainties. The geoscience experts must work closely with the facility designers to confirm that all seafloor and subsurface conditions are thoroughly understood and applicable for the foundation types being considered.

The state of practice of foundation design for deepwater foundations has benefitted from the numerous contributions made by individual practitioners, companies, and universities. A comprehensive reference list, included in this chapter, shows the major contributions made over the last twenty years for the different foundation types. Analytical methods have improved to accommodate the more complex loading conditions, so we now have a broad suite of procedures, varying from the classic limit equilibrium methods to 3-D FEA employing very sophisticated soil models.

Centrifuge model testing has also played an important role in helping calibrate the analytical methods, as described in Chapter 6 of this book. We now have more confidence that the analytical methods truly model the failure mechanisms for the variety of loading conditions. Centrifuge testing and analytical methods now provide a clearer understanding of the deformation response of the soil foundation system for the different stages of loading, thus helping confirm that an acceptable factor of safety is being used.

One of the major contributions has been the evolution of plasticity methods that can be used to define the ultimate limit state under combined vertical, moment, and horizontal loading, which can be characterized using three-dimensional interaction diagrams in *V-H-M* space. The authors applaud Dr. Mark Randolph and Dr. Chuck Aubeny and their coworkers for their numerous contributions to solutions for these complex problems. Many other researchers have contributed greatly to the evolution of new foundation concepts and analysis methods used in our current design practice, as evidenced by the many papers in the reference list.

The authors are confident that the design practice, as described in this chapter, is at a level of maturity that can be effectively applied to deepwater foundation design. Further improvements will evolve and additional experience will refine the empirical basis for the procedures described within. The authors want to caution that design standards not become too prescriptive to supplant sound engineering judgment.

As pointed out by Dr. Jim Mitchell (2009), we may find ourselves working in other areas of the world where the soil conditions are very different from where we have most of our experience. He cautions that potential pitfalls await if we rely only on our prior experience and use theories and experience not justified for these new soil types. The role of the integrated geoscience team and the geotechnical engineer should not end when design of the facilities is completed. The geoscience experts should stay involved during the foundation installation to confirm that the installation data does not reveal extreme variations from the plan. Comparison of predicted versus actual installation performance will help identify the potential cause of foundation problems in the field and later provide a greater confidence in the computed foundation capacity. The installation records also provide *ground truth* to our geophysical and geotechnical data interpretation and help calibrate our empirical methods for the variety of soil conditions existing in various offshore regions.

Finally, we believe that the practice of foundation design will continue to serve its important role in developing our offshore resources in a safe and reliable manner.

3.14 References

Andersen, K. H. and Jostad, H. P., 1999. Foundation design of skirted foundations and anchors in clay, In *Proceedings of the Annual Offshore Technology Conference*, Houston, TX, OTC 10824.

Andersen, K. H. and Jostad, H. P., 2002. Shear strength along outside wall of suction anchors in clay after installation, Paper presented at the *12th International Society of Offshore and Polar Engineers Conference*, Ky'ushu, Japan.

Andersen, K. H., Murff, J. D., Randolph, M. F., Clukey, E. C., Erbrich, C. T., Jostad, H. P., Hansen, B., Aubeny, C., Sharma, P., and Supachawarote, C., 2005. Suction anchors for deepwater applications, In *Proceedings of the 2nd International Symposium on Frontiers in Offshore Geotechnics*, Perth.

Andersen, K. H., Rawlings, C. G., Lunne, T., and Trond, H., 1994. Estimation of hydraulic fracture pressure in clay, *Canadian Geotechnical Journal* 31:817-828.

Angemeer, J., Carlson, E., and Klick, J. H., 1973. Techniques and results of offshore pile load testing in calcareous soils, In *Proceedings of the Annual Offshore Technology Conference*, Houston, TX, OTC 4408.

Annevall, T., 1997. Installation and performance of P27 Stevmanta-VLA anchors, In *Proceedings of the 2nd Annual Conference on Mooring & Anchoring,* Aberdeen.

API, 2000. Recommended practice for planning, designing, and constructing fixed offshore platforms—Working stress design, *API RP 2A-WSD,* American Petroleum Institute, Washington, D.C.

Aubeny, C. P., Han, S. W., and Murff, J. D., 2003a. Inclined load capacity of suction caisson anchors, *International Journal for Numerical and Analytical Methods in Geomechanics* 27:1235-1254.

———, 2003b. Refined model for inclined load capacity of suction caissons, *22nd International Conference on Offshore and Arctic Engineering,* Cancun, OMAE 2003-37502.

Aubeny, C. P., Kim, B. M., and Murff, J. D., 2005. Proposed upper bound analysis for drag embedment anchors in soft clay, In *Proceedings of the 2nd International Symposium On Frontiers in Offshore Geotechnics,* Perth.

Aubeny, C. P., Moon, S. K., and Murff, J. D., 2001. Lateral undrained resistance of suction caisson anchors, *International Journal Offshore and Polar Engineering* 11(3): 211-219.

Aubeny, C. P., Murff, J. D., and Roesset, J. M., 2001. Geotechnical issues in deep and ultra deep water, 10th International Conference on Computer Methods and Advances in Geomechanics, Tucson, AZ.

Aubeny, C. D. and Murff, J. D., 2005. Suction caissons and vertically loaded anchors: Design analysis methods, Final report to the Mineral Management Service, MMS project No. 362, Texas A&M University, College Station.

Beck, R. D., Jackson, C. W., and Hamilton, T. K., 1991. Reliable deepwater structural casing installation using controlled jetting, *Society of Petroleum Engineers* 22542, New Orleans.

Bogard, D., Matlock, H., and Chan, J. H. C., 2000. Comparisons of probe and pile tests, In *Proceedings of the Sessions of Geo—New Technological and Design Developments in Deep Foundations (GSP 100),* eds. N. D. Dennis, R. Castelli, and M. W. O'Neill.

Bransby, M. F. and O'Neill, M. P., 1999. Drag anchor fluke-soil interaction in clays, In *Proceedings of the International Symposium on Numerical Models in Geomechanics VII,* 489-494.

Brown, G. A., Nacci, V. A., and Demars, K. R., 1971. Performance of the hydrostatic anchor in sand, In *Proceedings of the Annual Offshore Technology Conference,* Houston, TX, OTC 1472.

Bruce, 2010. *Bruce Anchor Group* web site: www.bruceanchor.com, undated data sheets on various anchors.

Chen, J. Y., Matarek, B., Carpenter, J., Gilbert, R. B., Verret, S., and Puskar, F. J., 2009. Analysis of potential conservatism in foundation design for offshore platform assessment, Report to the Minerals Management Service & API, September.

Chen, W. F., 1975. *Limit analysis and soil plasticity,* Elsevier, Amsterdam.

Chen, W. F. and Liu, X. L., 1990. *Limit analysis in soil mechanics,* Elsevier, Amsterdam.

Chen, W. F. and McCarron, W. O., 1991. Bearing capacity of shallow foundations, In *Foundation Engineering Handbook,* 2nd ed., eds. H. Y. Fang, 144-161, Van Nostrand, New York.

Chen, W. F. and Randolph, M. F., 2005. Centrifuge tests on axial capacity of suction caisson in clay, In *Proceedings of the 2nd International Symposium on Frontiers in Offshore Geotechnics,* Perth.

Clausen, C. J. F., Aas, P. M., and Karlsrud, K., 2005. Bearing capacity of driven piles in sand, the NGI approach, In *Proceedings of the 2nd International Symposium on Frontiers in Offshore Geotechnics,* 677-682, Perth.

Clukey, E. C., Aubeny, C. P., and Murff, J. D., 2003. Comparison of analytical results and centrifuge model tests for suction caissons subjected to combined loads, In *Proceedings of the 22nd International Conference on Offshore Mechanics and Arctic Engineering,* Cancun.

Clukey, E. C., Morrison, M. J., Garnier, J., and Corté, J. F., 1995. The response of suction caissons in normally consolidated clays to cyclic TLP loading conditions, In *Proceedings of the Annual Offshore Technology Conference,* Houston, TX, OTC 7796.

Clukey, E. C. and Phillips, R., 2002. Centrifuge model tests to verify suction caisson capacities for taut and semi-taut legged mooring systems, In *Proceedings of the Deep Offshore Technology Conference,* New Orleans.

Clukey, E. C., Templeton, J. S., Randolph, M. F., and Phillips, R., 2004. Suction caisson response under sustained loop current loads, In *Proceedings of the Annual Offshore Technology Conference,* Houston, TX, OTC 16843.

Coyle, H. M. and Reese, L. C., 1966. Load transfer for axially loaded piles in clay, In *Proceedings of the American Society of Civil Engineers* 92(SM2): 1-26.

Dahlberg, R., 1998. Design procedures for deepwater anchors in clay, In *Proceedings of the Annual Offshore Technology Conference,* Houston, TX, OTC 8837.

Davis, E. H. and Booker, J. R., 1973. The effect of increasing strength with depth on the bearing capacity of clays, *Geotechnique* 23(4): 551-563.

Det Norske Veritas, 2000. Design and installation of drag-in plate anchors in clay, *Recommended Practice RP-E302*, Hovik, Norway.

DNV, 1977. *Rules for the design, construction, and inspection of offshore structures*—Appendix F-Foundations, Det Norske Veritas, Oslo.

Dove, P., Treu, H., and Wilde, B., 1998. Suction embedded plate anchor (SEPLA): A new anchoring solution for ultra-deep water mooring, In *Proceedings of the DOT Conference*, New Orleans.

Doyle, E. H., 1999. Pile Installation Performance for Four TLPs in the Gulf of Mexico, In *Proceedings of the Annual Offshore Technology Conference*, Houston, TX, OTC 10826.

Drucker, D. C. and Prager, W., 1952. Soil mechanics and plastic analysis or limit design, *Quarterly of Applied Mechanics* X(20): 157-165.

D'Souza, R. B., Dove, P. G. S., and Kelly, P. J., 1993. Taut leg spread moorings: A cost-effective stationkeeping alternative for deepwater platforms, In *Proceedings of the Annual Offshore Technology Conference*, Houston, TX, OTC 7203.

Dupal, K., vonEberstein, B., Loeb, D., Xu, H., Grant, J., and Bergeron, B., 2000. Shell's experience with deepwater mooring systems for MODUs, In *Proceedings of the DOT Offshore Technology Conference*, New Orleans.

Ehlers, C. J. and Bowles, W. R., 1973. Underreamed footings support offshore platforms in the North Sea, In *Proceedings of the Annual Offshore Technology Conference*, Houston, TX, OTC 1895.

Ehlers, C. J., Young, A. G, and Chan, J. H., 2004. Technology assessment of deepwater anchors, In *Proceedings of the Annual Offshore Technology Conference*, Houston, TX, OTC 16840.

Eltaher, A., Rajapaksa, Y., and Chang, K. T., 2003. Industry trends for design of anchoring systems for deepwater offshore structures, In *Proceedings of the Annual Offshore Technology Conference*, Houston, TX, OTC 15265.

Focht, J. A. Jr., 1983. Axial capacity of offshore piles in clay, *Shanghai Symposium on Marine Geotechnology and Nearshore Structures*, Shanghai.

Fugro Engineers, 2004. Axial pile capacity design methods for offshore driven piles in sand, *Report to the American Petroleum Institute* 3, August 2004.

Gault, J. A. and Cox, W. R., 1974. Method for predicting geometry and load distribution in and anchor line chain from a single point mooring buoy to a buried anchorage, In *Proceedings of the Annual Offshore Technology Conference*, Houston, TX, OTC 2062.

Gilbert, R. B., Chen, J-Y., Materek, B., Puskar, F., Verret, S., Carpenter, J., Young, A. G, and Murff, J. D., 2010. Comparison of observed and predicted performance for jacket pile foundations in hurricanes, In *Proceedings of the Annual Offshore Technology Conference*, Houston, TX, OTC 20861.

Gilbert, R. B., Choi, Y. J., Dangayach, S., and Najjar, S. S., 2005. Reliability-based design considerations for deepwater mooring system foundations, In *Proceedings of the 2nd International Symposium On Frontiers in Offshore Geotechnics*, Perth.

Goodman, L. J., Lee, C. N., and F. J. Walker, 1961. The feasibility of vacuum anchorage in soil, *Geotechnique* 1(3): 356-359.

Hansen, B., 1970. A revised extended formula for bearing capacity, *The Danish Geotechnical Institute* 98:5-11, Copenhagen.

Helfrich, S. C., Brazil, R. L., and Richards, A. F., 1976. Pullout characteristics of a suction anchor in sand, In *Proceedings of the Annual Offshore Technology Conference*, Houston, TX, OTC 2469.

Helfrich, S. C., Young, A. G, and Ehlers, C. J., 1980. Temporary seafloor support of jacket structures, In *Proceedings of the Annual Offshore Technology Conference*, Houston, TX, OTC 3750.

Hogervorst, J. R., 1980. Field trials with large diameter suction piles, In *Proceedings of the Annual Offshore Technology Conference*, Houston, TX, OTC 3817.

Hossain, M. K. and Briaud, J. L., 1993. Improved soil characterization for pile in sand in API RP-2A, In *Proceedings of the Annual Offshore Technology Conference*, Houston, TX, OTC 7193.

House, A. R., Randolph, M. F., and Borbas, M. E., 1999. Limiting aspect ratio for suction caisson installation in clay, In *Proceedings of the 9th International Offshore and Polar Engineering Conference*, 676-683, Brest.

ISO, 2007. International standard for the design of fixed steel offshore platforms, *ISO/DIS 19902*, International Standards Office, British Standards Institute, London.

Jeanjean, P., 2002. Innovative design method for deepwater surface casings, *SPE Annual Technical Conference and Exhibition,* San Antonio, SPE 77357.

Jeanjean, P., Andersen, K. H., and Kalsnes, B., 1998. Soil parameters for design of suction caissons for Gulf of Mexico deepwater clays, In *Proceedings of the Annual Offshore Technology Conference,* Houston, TX, OTC 8830.

Jeanjean, P., Watson, P. G., Kolk, H. J., and Lacasse, S., 2010. RP 2 GEO: The new API recommended practice for geotechnical engineering, In *Proceedings of the Annual Offshore Technology Conference,* Houston, TX, OTC 20631.

Jeanjean, P., Znidarcid, D., Phillips, R., Ko, H. Y., Pfister, S., Cinicioglu, O., and Schroeder, K., 2006. Centrifuge testing on suction anchors: Double-wall, over-consolidated clay, and layered soil profile, In *Proceedings of the Annual Offshore Technology Conference,* Houston, TX, OTC 18007.

Kim, B. Y., 2005. Upper bound analysis for drag anchors in soft clay, Ph.D. diss., Texas A&M University, College Station.

Kolk, H. J., Baaijens, A. E., and Senders, M., 2005a. Design criteria for pipe piles in silica sands, In *Proceedings of the 2nd International Symposium on Frontiers in Offshore Geotechnics,* 711-716, Perth.

Kolk, H. J., Baaijens, A. E., and Vergobbi, P., 2005b. Results from axial load test on pipe piles in very dense sands, the EURIPIDES JIP, In *Proceedings of the 2nd International Symposium on Frontiers in Offshore Geotechnics,* 661-667, Perth.

Kraft, L. M. and Lyons, C. G., 1974. State-of-the-art: Ultimate axial capacity of grouted piles, In *Proceedings of the Annual Offshore Technology Conference,* Houston, TX, OTC 2081.

Lehane, B. M., Schneider, J. A., and Xu, X., 2005. The UWA-05 method for prediction of axial capacity of driven piles in sand, In *Proceedings of the 2nd International Symposium on Frontiers in Offshore Geotechnics,* Perth.

Lieng, J. T., Hove, F., and Tjelta, T. I., 1999. Deep penetrating anchor: Subseabed deepwater anchor concept for floaters and other installations, In *Proceedings of the 9th International Offshore and Polar Engineering Conference,* 613-619, Brest.

Liu, J., Hu, Y., and Wu, L., 2005. Pullout capacity of circular plate anchors in double-layered clays, In *Proceedings of the 2nd International Symposium on Frontiers in Offshore Geotechnics,* Perth.

Matlock, H., 1970. Correlations for design of laterally loaded piles in soft clay, In *Proceedings of the Annual Offshore Technology Conference,* Houston, TX, OTC 1203.

McClelland, B., 1974. Design of deep penetration piles for ocean structures, *Journal of the Geotechnical Engineering Division* 100(7): 705-747.

Medeiros, C. J., 2001. Torpedo anchor for deep water, In *Proceedings of the Deepwater Offshore Technology Conference,* Rio de Janeiro.

Meyerhof, G. G., 1951. The ultimate bearing capacity of foundations, *Geotechnique* 2(4): 301-322.

———, 1953. The bearing capacity of foundations under eccentric and inclined loads, In *Proceedings of the 3rd ICSMFE,* 1:440-445, Zurich. International Conference on Soil Mechanics and Foundation Engineering.

Mitchell, J. K., 2009. Challenges of regional geotechnics: Things may not be what you think they are (or should be), *Commentary: Geo-Strata, Publication of Geo-Institute-ASCE* 13(4): 14-15.

Murff, J. D., 1987. Pile capacity in calcareous sands—State of the art, *Journal of Geotechnical Engineering* 113(5): 490-507.

Murff, J. D., Aubeny, C. P., and Yang, M., 2010. The effect of torsion on the sliding resistance of rectangular foundations, In *Proceedings of the 2nd International Symposium on Frontiers in Offshore Geotechnics,* Perth.

Murff, J. D. and Hamilton, J. M., 1993. P-Ultimate for undrained analysis of laterally loaded piles, *Journal of Geotechnical Engineering* 119(1): 91-107.

Murff, J. D. and Miller, T. W., 1977. Stability of offshore gravity structure foundations by the upper-bound method, In *Proceedings of the Annual Offshore Technology Conference,* Houston, TX, OTC 2896.

Murff, J. D. and Young, A. G, 2007. Suction caissons for deepwater moorings, *Geo-Strata, Publication of Geo-Institute-ASCE* 8(3): 12-16.

Neubecker, S. R. and Randolph, M. F., 1995. Profile and frictional capacity of embedded anchor chains, *Journal of Geotechnical Engineering* 121(11): 797-803.

O'Loughlin, C. D., Randolph, M. F., and Richardson, M., 2004. Experimental and theoretical studies of deep penetrating anchors, In *Proceedings of the Annual Offshore Technology Conference,* Houston, TX, OTC 16841.

O'Neill, M. W. and Hassan, K. M., 1994. Drilled shafts: Effects of construction on performance and design criteria, In *Proceedings of the International Conference on Design and Construction of Deep Foundations*, U.S. Federal Highway Administration 1:137-187.

O'Neill, M. W. and Murchinson, J. M., 1983. An evaluation of p-y relationships in sands, Report to American Petroleum Institute, University of Houston, Houston, TX.

Paulling, J. R. and Horton, E. E., 1970. Analysis of the tension leg platform, In *Proceedings of the Annual Offshore Technology Conference*, Houston, TX, OTC 1263.

Pelletier, J. H., Murff, J. D., and Young, A. G, 1993. Historical development and assessment of the API-design method for axially loaded piles, In *Proceedings of the Annual Offshore Technology Conference*, Houston, TX, OTC 7157.

Prandtl, L., 1921. Eindringungsfestigkeit und festigkeit von schneiden, Angew, Math. U. Mech., 1(15).

Randolph, M. F., 2003. 43rd Rankine Lecture: Science and empiricism in pile foundation design, *Geotechnique* 53(10): 847-875.

Randolph, M. F., Cassidy, M. J., Gourvenec, S. M., and Erbrich, C., 2005. Challenges of offshore geotechnical engineering, In *Proceedings of the 16th International Conference Soil Mechanics and Geotechnical Engineering* 1:123-176.

Randolph, M. F. and House, A. R., 2002. Analysis of suction caisson capacity in clay, In *Proceedings of the Annual Offshore Technology Conference*, Houston, TX, OTC 14236.

Randolph, M. F. and Murphy, B. S., 1985. Shaft capacity of driven piles in clay, In *Proceedings of the Annual Offshore Technology Conference*, Houston, TX, OTC 4883.

Randolph, M. F., O'Neill, M. P., Stewart, D. P., and Erbrich, C., 1998. Performance of suction anchors in fine-grained calcareous soils, In *Proceedings of the Annual Offshore Technology Conference*, Houston, TX, OTC 8831.

Reese, L. C., 1973. A Design method for an anchor pile in a mooring system, In *Proceedings of the Annual Offshore Technology Conference*, Houston, TX, OTC 1745.

Reese, L. C., Cox, W. R., and Grubbs, B. R., 1974. Field testing of laterally loaded piles in sand, In *Proceedings of the Annual Offshore Technology Conference*, Houston, TX, OTC 2079.

Reese, L. C., Cox, W. R., and Koop, F. D., 1975. Field testing and analysis of laterally loaded piles in stiff clay, In *Proceedings of the Annual Offshore Technology Conference*, Houston, TX, OTC 2312.

Ruinen, R., 2004. Stevmanta VLA installation, A case history, International Mooring Seminar, New Orleans.

Semple, R. M. and Rigden, W. J., 1984. Shaft capacity of driven piles in clay, In *Proceedings of the Symposium on Analysis & Design of Pile Foundations*, ASCE, San Francisco.

Senpere, D. and Auvergne, G. A., 1982. Suction anchor piles—A proven alternative to driving or drilling, In *Proceedings of the Annual Offshore Technology Conference*, Houston, TX, OTC 4206.

Skempton, A. W., 1951. The bearing capacity of clays, In *Proceedings of the Building and Research Congress* 1:180-189, London.

Skempton, A. W. and Northey, R. D., 1952. The sensitivity of clays, *Geotechnique*, vol. 3, no. 1, pp. 40-51.

Templeton, J. S., 2002. The role of finite element in suction foundation design analysis, In *Proceedings of the Annual Offshore Technology Conference*, Houston, TX, OTC 14235.

Terzaghi, K., 1943. *Theoretical soil mechanics*, Wiley, New York.

Thorne, C. P., 1998. Penetration and load capacity of marine drag anchors in soft clay, *Journal of Geotechnical and Geoenvironmental Engineering* 124(10): 945-953.

Toolan, F. E. and Horsnell, M. R., 1992. The evolution of offshore pile design codes and future developments, In *Proceedings of the Conference on Offshore Site Investigation and Foundation Behavior*, Society for Underwater Technology, London.

True, D. G., 1976. Undrained vertical penetration into ocean bottom soils, PhD thesis, University of California, Berkeley.

VanNoort, R., Murray, R., Wise, J., Williamson, M., Wilde, B., and Riggs, J., 2009. Conductor pre-installation, deepwater Brazil, In *Proceedings of the Annual Offshore Technology Conference*, Houston, TX, OTC 20005.

Vesic, A. S., 1975. Bearing capacity of shallow foundations, *Foundation Engineering Handbook*, ed. H. F. Winterkorn and H. Y. Fang, 121-147, Van Nostrand, New York.

Vivitrat, V., Valent, P. J., and Ponterio, A. A., 1982. The influence of chain friction on anchor pile design, In *Proceedings of the Annual Offshore Technology Conference*, Houston, TX, OTC 4178.

Vryhof Anchors, 2005. *Anchor Manual*, krimpen a/d yssel, The Netherlands.

Vryhof Anchors B.V., 2006. *Vryhof anchor manual.*

Wang, M. C., Nacci, V. A., and Demars, K. R., 1975. Behavior of the underwater suction anchor in soil, *Journal of Ocean Engineering* 3(1): 47-62.

Whittle, A. J., Sutabutr, T., Germaine, J. T., and Verney, A., 2001. Prediction and measurements of pore pressure dissipation for a tapered piezoprobe, In *Proceedings of the Annual Offshore Technology Conference*, Houston, TX, OTC 13155.

Wilde, B., Treu, H., and Fulton, T., 2001. Field testing of suction embedded plate anchors, In *Proceedings of the 11th International Offshore and Polar Engineering Conference*, 544-551.

Wroth, C. P., 1972. Discussion on the design and performance of deep foundations, In *Proceedings of the Conference on the Performance of Earth and Earth-Supported Structures* 3:231-234.

Yang, M., 2009. Undrained behavior of plate anchors subjected to general loading, Ph.D. diss., Texas A&M University, College Station.

Young, C. W., 1967. Development of empirical equations for predicting depth of earth penetrating projectiles, Report no. SC-DR-67-60, Sandia National Laboratories.

Young, A. G, Kraft, L. M., and Focht, J. A., 1975. Geotechnical considerations in foundation design of offshore gravity structure, In *Proceedings of the Annual Offshore Technology Conference*, Houston, TX, OTC 2371.

Young, A. G, Quirós, G. W., and Ehlers, C. J., 1983. Effects of offshore sampling and testing on undrained shear strength, In *Proceedings of the Annual Offshore Technology Conference*, Houston, TX, OTC 4465.

Young, A. G, Sullivan, R. A., and Rybicki, C. A., 1975b. Pile design and installation features of the Thistle Platform, In *Proceedings of the SPE European Petroleum Conference*, 8050, London.

Yun, G. J., Maconochie, A., Oliphant, J. and Bransby, F., 2009. Undrained capacity of surface footings subjected to combined V-H-T loading, In *Proceedings of the 19th International Offshore and Polar Engineering Conference*, Osaka.

Zdravkovic, L. and Potts, D. M., 2005. Parametric finite element analyses of suction anchors, In *Proceedings of the 2nd International Symposium on Frontiers in Offshore Geotechnics*, Perth.

Zimmerman, E. H., Smith, M. W., and Young, A. G, 2005. Patent for marine mooring, U.S. Patent No. 6,941,885, filed September 13, 2005.

4

Driven Pile Design for Tension Leg Platforms

Earl H. Doyle, PE, F.ASCE
Geotechnical Consultant

4.1 Introduction

The American Petroleum Institute (API) RP2T (2008) states in the Recommended Practice for the Design of a Tension Leg Platform (TLP), that "A TLP is a vertically moored, buoyant, compliant structural system wherein the excess buoyancy of the platform maintains tension in the mooring system. A TLP may be designed to serve a number of functional roles associated with offshore oil and gas exploitation. It is considered particularly suitable for deepwater applications." The buoyant platform is connected to the seafloor foundation through legs called tendons. The tendons are tensioned by the excess buoyancy effect of the platform by essentially being pulled into the water. Thus, the foundation must be designed to resist the tension load of the tendons. This chapter will discuss the design of the foundation elements typically used to anchor a TLP to the seafloor.

Early TLPs transferred the tension loads applied by the tendons to piling driven below the seafloor through steel templates placed on the seafloor. The most recent structure to be designed this way was Shell Oil Company's Auger TLP, which was installed in the Gulf of Mexico in 2860 ft of water in 1994 and was awarded the ASCE Outstanding Civil Engineering award in 1995. The next TLP to be installed in the Gulf of Mexico was Shell's Mars TLP in 1996 in a water depth of 2933 ft. This structure was the first TLP to use a novel foundation design where the driven piles were directly connected to the tendons, as illustrated in Figure 4.1.

Subsequent to this successful installation, all TLPs have been designed using this method and therefore capture a significant cost savings through reduced fabrication and installation schedule. The pile is fabricated in one piece and is composed of three sections—the receptacle, which the tendon bottom connector is latched into; the main pile section; and the pile shoe. Following driving of the piles to the desired penetration, a separate section containing anodes (Figure 4.2) is placed over the receptacle for corrosion protection and to assist in capturing the tendon bottom during its installation.

4.2 Pile Installation

Each pile is typically lifted off the transport barge and placed on the derrick barge by a dynamically positioned derrick barge using a tandem lifting method. The pile is then lifted off the derrick barge using the same tandem lifting technique and lowered to the seafloor. Four prepositioned buoys are located at each pile site on the seafloor. Two remotely operated vehicles (ROVs) observe

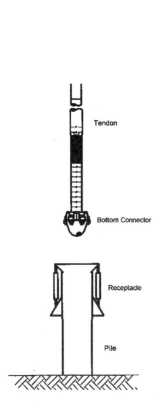

Figure 4.1 Components for directly connecting the pile to a tendon.

Figure 4.2 Anode package and tendon guide cone that is placed over the pile after driving.

the top of the pile and seafloor during the lowering process. Positioning instrumentation is also located on both the ROVs and the pile top to observe the pile's location relative to the intended final seafloor location. A slope indicator is also located at the pile top to ensure that the pile tilt is within specifications prior to lowering the pile to self-weight penetration. A one-diameter *scour* depth is usually allowed in the design to account for the possible deleterious effects of dragging the pile through the soil to stabilize the pile prior to lowering the pile into the soil. The pile is then lowered several meters below the seafloor, and the lateral offsets and vertical tilt are checked to make sure the pile is within dimensional specifications before lowering the pile to its self-weight penetration. Once the pile is lowered to self-weight penetration, it is not possible to pull the pile out using the cranes on the derrick barge. Even though water depths are often in the range of 2500 to 4000 ft, installations have been performed with remarkable precision. The installation performance of three TLPs was reported by Doyle (1999). These installations involved lowering a single pile to the sea floor and accurately installing each pile within a closely confined target circle with a minimum pile tilt. Off-target piles will cause excessive fatigue in the bottom tendon connector and tendons, as well as potentially affecting the tendon axial loads during lateral offset of the TLP during storm conditions. In addition, pile tilt will affect the column buckling characteristics when the hammer is placed on the pile following self-weight penetration. Pile tilt will also increase the bending moments due to the design loads. A 1° installation tilt is usually assumed, and its effect is added to the design loading. Thus, it is important to minimize the off-target range and tilt and to install the piles with as small a variation as possible so that a reasonable tilt and range criterion can be developed. The final tilt for the three TLPs ranged from 0.21 degrees to 0.41 degrees, and the average final range-to-target distances varied from 1.00 ft to 1.08 ft. These remarkably small deviations were achieved by careful planning and attention to detail by the installation contractor and the positioning subcontractor.

It is normal practice to batch set all the piles before starting driving operations. Piles may remain freestanding at their self-weight penetration for several days. The installation crew will typically experience a steep learning curve during stabbing of the first few piles and achieve remarkable time improvements after that.

The piles are typically driven to grade using an underwater Menck hydraulic hammer of the type shown in Figure 4.3. Menck underwater hammers are currently available for water depths up to 8500 ft. One particular issue with the tension piles used in TLPs is fatigue damage during driving. A method to compute driving damage was reported by Hunt et al. (1999). To reduce damage, the hammer energy levels are reduced. Since damage is linearly related to the number of blows and the stress level is a power function, it is better to reduce the energy of the hammer during driving even though the number of blows is increased. For the three TLPs reported by Doyle (1999), final penetrations were within one-fourth inch of the expected penetration.

4.3 Design of Piles

Design practice is guided by two API guidelines. The primary design guideline is RP2T although it defers to API RP2A (2007) whenever a particular design issue is not covered by RP2T. The soil conditions for TLP piles are almost always clay soil sites. No sand sites have been encountered in deepwater. Thus, this chapter will focus only on pile design in clay soils. The elements of pile

Figure 4.3 Menck MHU-500T underwater hydraulic hammer.

design do not differ from pile design for offshore platforms in shallow water, with respect to axial design criterion. Lateral design criterion is basically the same in that the soft clay criteria given in RP2A (also called the Matlock criteria) is used within a beam-column program except that the direct connection method generally produces significantly more lateral deflection then is normally encountered in fixed platform designs where the jacket framing contributes considerable rotational restraint at the pile top. The Matlock criteria given in RP2A is based on pile tests where the maximum lateral deflection was less than about 25% of the pile diameter. If during the design of the piles, the Matlock criterion shows lateral deflection in excess of 25% of the pile diameter, Doyle et al. (2005) suggested different criteria be used. These criteria are based on large deflection centrifuge tests performed on clay soils similar to Gulf of Mexico deepwater soils. One suggestion, for example, is that Matlock's static criteria be used on a remolded shear strength profile.

An innovation of the direct-connect method is to drive the piles, connect the tendons to the piles, and finally to connect the tendons to the hull in one continuous installation. However, piles driven into clay soils do not reach full axial capacity until driving induced pore water pressures have dissipated. Thus, the axial capacity of the piles must be designed to accommodate the timing of the applied loads. In addition, RP2T requires that sufficient axial capacity be reached at first hydrocarbon production. To meet this requirement and to accommodate the timing of the applied loads, the long-term axial safety often is higher than that required by RP2T. An empirical method based on Gulf of Mexico data has been suggested by Bogard and Matlock (1990), Bogard et al. (2000), and Bogard (2001). Analytical techniques using finite element methods and complex soil models have been applied to this problem (e.g., Whittle and Sutabutr 1999, 2005). Pile self-weight penetration and setup data have also been used by Dutt and Ehlers (2009) to show that setup is faster than that predicted by the Bogard empirical method.

4.4 Preliminary Design Considerations

Selection of the pile diameter, wall thickness and length requires knowledge of the static soil design profile, cyclic soil properties, receptacle size, design loads and direction, timing of the applied loads, installation tilt tolerance, and factors of safety. Minimum factors of safety are given by RP2T. The designer must also show that the pile will penetrate during installation to a self-weight penetration that is sufficient to not buckle under the weight of the pile driving hammer; that the piles can be driven to final penetration (including considerations of hammer breakdowns); and that the blows delivered during driving will not cause an excess of driving-induced fatigue damage.

4.5 Soil Properties

Normally, a single site boring is all that is required. The primary reason is that deepwater sites do not normally show significant geotechnical property variations across the seabed footprint of the TLP due to their depositional environment. In addition, one usually does not want the pile behavior at the corners of the TLP to be different from the other corners. Because of the cost and complexity of a geotechnical investigation in deepwater, high-resolution geophysical data is acquired to understand the geology and potential geotechnical complexity of a site. Such studies are termed *integrated geoscience studies*. An accepted definition of an integrated geoscience study is the integration of geophysical data, geologic modeling, geohazard evaluations, and geotechnical engineering to develop geotechnical parameters for design and to assess, if necessary, geological hazards and constraints given that geotechnical knowledge. Several papers have been published that describe integrated geoscience studies (e.g., Doyle 1998; Jeanjean et al. 1998; Young et al. 2009) and detailed descriptions are included in Chapter 2 of this book. Deepwater geotechnical investigations and the tests performed have been described by Pelletier et al. (1997) and Dutt et al. (1992).

A boring usually consists of retrieved push samples, in situ vane shear measurements, and occasionally piezocone data. The piezocone investigation is usually conducted in a separate hole and also may include piezoprobe measurements. A piezoprobe is a cone-type device that is pushed into the bottom of the borehole, and the dissipation of excess pore water pressure is measured over time to obtain a measure of the *in situ* pore pressure (Whittle et al. 2001). Undisturbed undrained shear tests usually performed in the field are: a miniature vane shear (MV), unconsolidated undrained triaxial (UU), and Torvane, as well as an in situ vane shear device called the remote vane (RV). Remolded MV and UU tests are also essential to the pile design to determine self-weight pile penetration and drivability. In the laboratory, advanced static tests usually consist of consolidation tests and direct simple shear (DSS) tests using the normalized soil behavior and the stress history and normalized soil engineering properties (SHANSEP) approach of Ladd and Foott (1974). In general, the shear strength calculated using the DSS and consolidation test results correlate well with the MV and UU data such that the DSS strength is the basis for the interpreted shear strength for deepwater clays. One example is shown in Figure 4.4 for one site in the Gulf of Mexico (1994). The RV data typically show the strongest shear strength, as shown in Figure 4.4.

An extensive laboratory test program was conducted for the Auger TLP design. As a result of that test program, the only cyclic tests usually performed today are stress-controlled DSS tests to

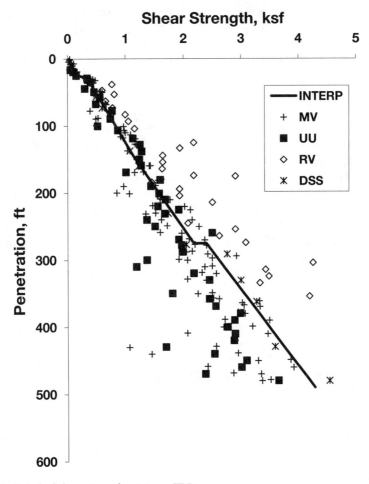

Figure 4.4   Undisturbed shear strengths at Auger TLP.

determine the "threshold cyclic shear strength ratio" (Dutt et al. 1992). The threshold cyclic shear strength ratio is similar to the endurance limit seen in steels in that cyclic loads below the threshold cyclic shear strength ratio do not cause soil failure. A strategy of design therefore is to make sure the safety factor is high enough such that the cyclic loading is always below the threshold cyclic shear strength ratio. If an analytical analysis to determine excess pore water pressure dissipation after driving is conducted (Whittle and Sutabutr, 1999, 2005), additional laboratory tests are required to determine the parameters input to that analysis.

4.6 Axial Capacity

The axial tension capacity is determined using API RP2A guidelines for fixed platforms. Safety factors are given in API RP2T and are higher than that given in API RP2A. Given the range of spacing to diameter ratios for TLPs and based on finite element studies, group effects are not considered for axial capacity calculations. The weight of the soil plug cannot be used in the design of tension piles. The basis for this is the two-way cyclic pile tests conducted by Doyle and Pelletier (1985) that showed the plug weight to disappear under two-way cyclic loading. If other investigations indicate otherwise, the weight of the soil plug may be used if these data are appropriate to the diameter and loading conditions of the pile being designed.

Although conservative, end bearing is not to be counted upon unless appropriately related data indicates otherwise. Thus, the required long-term ultimate axial pile capacity is usually determined from the following equation:

Required Ultimate Capacity Per Pile = (Safety factor) × (Pile Load − Submerged Pile Weight).

4.7 Sustained (Creep) Loading

A sustained load over a long period of time can cause axial pile failure. While little pile data exists, some long-duration model pile test data indicate that at loads above about 30% of the ultimate axial failure load, pile displacement tends to increase with time and long after consolidation should have finished (Edil and Mochtar 1988). Full-scale tests to induce creep rupture of piling have not been reported in the literature. However, Terzaghi and Peck (1964) report in a discussion of remolded clays that, "As soon as the shearing stress in a clay becomes greater than about one-half the peak value, the clay is likely to *creep* at constant shearing stress." Laboratory creep tests show that sustained loads of about 80% of the undrained strength are sufficient to cause failure a few days after the load is applied. However, a conservative approach is to not allow sustained loads to be greater than about 30% of the undrained strength unless truly long-term creep tests show that a larger sustained load would not cause failure.

4.8 Cyclic Stresses

Cyclic stress can lead to pile failure. Laboratory stress-controlled DSS tests show that the *threshold cyclic shear strength ratio* is usually in the range of 0.5 to 0.6. This means that the threshold stress level where cyclic loading is important is when the cyclic component of the load exceeds 50-60% of the ultimate pile capacity. If the entire load was cyclic, the safety factor must be above 1/0.50 to 1/0.60 or 2.00 to 1.67 for cyclic loads to have a deleterious effect on performance. The lowest long-term factor of safety for storm loading is a safety factor of 2.25. Thus, even if the entire load were cyclic, that load is usually less than that required to initiate cyclic failure.

4.9 Installation Setup

As mentioned previously, two methods are usually considered to calculate pile setup as a function of time after installation, namely a Gulf of Mexico based empirical method and an analytical method developed at MIT that is based on site-specific soil data. The Gulf of Mexico setup calculation is based on an empirical method developed by Bogard and Matlock (1990). Subsequent papers (Bogard et al. 2000; Bogard 2001) further clarified the test data used to develop setup relationships. The test results performed at Harvey and Empire resulted in the following equations that Bogard refers to as the lower bound curve (personal communication). The applicable equations for those tests are:

$$F/F_{ult} = (0.2 + 0.84U) \tag{4.1}$$

where

F = pile capacity
F_{ult} = ultimate pile capacity
$U = (t/t_{50})/(1 + t/t_{50})$
t = time, days
$t_{50} = (100 - 2D/tw)D2$—upper bound of data which results in a lower bound for F/Fult, or
$t_{50} = (70 - 1.4D/tw)D2$—lower bound of data, which results in an upper bound for F/Fult
D = Pile diameter, m
t_w = pile wall thickness at bottom of pile, m

Tests were also conducted at West Delta Block 58A, and additional setup analyses were based on a full-scale pile test that resulted in the following modification to F/F_{ult} as:

$$F/F_{ult} = (0.3 + 0.7U) \tag{4.2}$$

Bogard also recommended that the pile *diameter* to *wall thickness at the pile base* ratio (i.e., D/t_w) be kept under 40 because there was no data on higher D/t_w ratios. However, Doyle (1999) evaluated the experience gained from the installation of four Gulf of Mexico TLPs. The pile shoe in all cases was standardized to have a wall thickness about 1.5 times the wall thickness of the section above the shoe and also to have a length of about one pile diameter. Driving setup data was reported for the Mars TLP pile, which had a D/t_w ratio of 48. This data indicated that the initial capacity was at least 30% of the long-term capacity and consistent with a driving resistance based on the remolded shear strength of the soil. The Mars data indicate that the Bogard derived curves are conservative with respect to the D/t_w and soil conditions at the Mars location. Bogard recommends the use of the lower bound curve to be conservative (personal communication). Dutt and Ehlers (2009) stated that, according to hindcast pile driving analyses, about 60 to 80% of the ultimate pile capacity is mobilized in about 7 days, and the extrapolation of their setup model suggests that the setup is almost complete in about 60 days. They also stated that the lower and upper bound estimates of setup presented in their paper for 84-inch and 108-inch diameter piles are applicable for piles with diameters as small as 72 inches. However, they recommend that the guidelines proposed by Bogard et al. (2000) should be used for evaluating pile setup for small diameter piles with an outside diameter less than 48 inch and a D/t_w ratio less than 40, as their correlations are based on static pile load tests.

4.10 Lateral Capacity

Lateral and axial loads are applied at the center of rotation of the tendon bottom connector. A moment of about 200-ft kips is usually applied due to the rotation of the flex element at the same location. The direction of the moment is such that the moment increases the magnitude of compressive outer fiber stress in the pile that results from lateral load. A beam-column program is usually used to analyze the pile response under loading conditions.

Normally two or three piles per corner are used for the direct connect method. As such, lateral group effects may be important if the piles are close together. One method suggested by M. F. Randolph (personal communication) is to calculate the lateral group efficiency as:

$$Z = 0.5 + s/(9D) \leq 1.0$$

where

Z = lateral group efficiency
s = pile spacing
D = pile diameter

For this simple method, the shear strength is reduced by Z for the trailing pile.

One pile diameter of local scour is also assumed in the analysis to account for possible deleterious effects of pile installation.

The soft clay *p-y* criteria in API RP2A was developed for lateral pile tests where the maximum deflections were less than about 25% of the pile diameter. Thus, pile deflections that exceed 25% of the pile diameter will require a different lateral pile criterion. As discussed by Doyle et al. (2004), centrifuge tests were conducted at Cambridge University, and criteria was developed for large diameter piles with lateral pile deflections that exceeded the criteria given in API RP2A. The pile modeled was 96 inches in diameter, and the soils were typical of the deepwater Gulf of Mexico. Doyle et al. (2005) discussed criteria based on the centrifuge tests. They suggested using the criteria developed specifically from the Cambridge tests. Their second suggestion was to develop *p-y* curves using the remolded soil strength and use the *static* criteria (rather than the *cyclic* criteria). They suggested this criterion be used if the maximum pile deflections exceeded those developed during the centrifuge tests, as additional remolding of the soil due to the increased lateral pile deflections would tend to remold the soil to deeper penetrations.

4.11 Installation Clearances

The pile must provide sufficient clearances for the hammer. Menck (personal communication) suggests that the following be considered:

The nominal length of pile in the tendon guide cone (Figure 4.2) is 3.4 to 3.6 meters.
Add a tolerance of 0.1 meters.
Clearance above the mudline should be 0.0 to 0.3 meters.
This determines the height of the pile top above the mudline.

Menck also suggests that holes be included in the pile, which will represent the distance over which air will be in the pile top during driving. This hole must be below the receptacle so that the tendon lugs remain in air during driving. (Otherwise, a water hammer effect will exist on the lugs, which may cause considerable fatigue damage during driving.) Menck also suggests adding a tolerance of 0.6 meters so that the holes are not covered by the possibility of any rising soil plug. This results in the air hole being at least 0.9 meters (3 feet) above the mudline. Typically, the holes are ellipses to reduce driving damage. Another important check is to ensure that the tendons do not come in contact with the pile top during the maximum lateral excursion of the TLP.

4.12 Installation Stability

The pile must be stable when the underwater weight of the hammer is placed on the pile. The first step in determining this is to calculate the self-weight penetration of the pile. Using procedures reported by Doyle (1999), the self-weight penetration is calculated using the remolded shear strength and neglecting the underwater weight of any internal lifting tool (ILT). End bearing of the steel area is usually incorporated into the self-weight calculations. Given that an ILT weighs about 30 kips and is ignored, the self-weight penetration is probably conservative. A typical hammer is a Menck MHU-500T that has an underwater weight of about 161 kips and a center of gravity above the pile top equal to 16 ft. Experience has shown that the hammer placement does not cause any further settling of the pile. This is probably because some time elapses between self-weight penetration and when the hammer is initially placed onto the pile top. The pile is then analyzed at self-weight penetration with a pile tilt of 1 degree and a lateral load at the top of the drive head equal to 5% of the hammer weight and any additional lateral loads against the hammer and pile due to bottom water currents.

It should be noted that the ability to leave a freestanding pile on the seafloor is a result of sufficient self-weight penetration. Gulf of Mexico deepwater normally consolidated strength profiles (Figure 4.3) are amenable to this, but overconsolidated profiles may constrain this option.

4.13 Drivability and Driving Induced Fatigue Calculations

Drivability analyses are usually conducted using a commercially available program called GRL-WEAP (Pile Dynamics 2005). Hammer files, describing their driving characteristics, are built into the program. As discussed previously, starting the hammer at a reduced energy level and incrementally increasing the energy as the pile penetrates into the soil can minimize driving fatigue damage. In order to limit the fatigue damage, several energy combinations are tried and fatigue analyses are conducted before settling on a final driving plan using the procedures presented by Hunt et al. (1999).

Another consideration in the design is to assume that during driving, a hammer breakdown occurs and driving is stopped for a time. The pile must be designed so that the pile can be restarted and driven to final penetration. The amount of time between stopping driving and restarting the pile is usually an owner decision, which is based on the availability of a spare hammer, and/or the time expected to repair the existing hammer. For these calculations, a nonconservative setup calculation is usually assumed.

The tendon receptacle at the top of the pile requires some internal pile head features, either protrusions or recesses, in order to lock the tendon bottom connector securely into the receptacle. The mass of these lugs result in significant local loading and stress concentrations during pile driving. The receptacle is usually a thick wall casting or forging designed to accommodate fatigue damage during the driving operation.

4.14 References

America Petroleum Institute, 2007. Recommended practice for planning, designing and constructing fixed offshore platforms, *Working Stress Design*, 21st ed., API RP 2A-WSD.

America Petroleum Institute, 2008. Planning, designing, and constructing tension leg, *API Recommended Practice 2T*, 3rd ed.

Bogard, J. D. and Matlock, H., 1990. Application of model tests to axial pile design, In *Proceedings of the 22nd Annual Offshore Technology Conference*, Houston, TX, OTC 6376.

Bogard, J. D., 2001. Effective stress and axial pile capacity: Lessons from Empire, In *Proceedings of the 33rd Annual Offshore Technology Conference*, Houston, TX, OTC 13059.

Bogard, J. D., Matlock, H., and Chan, J. H., 2000. Comparison of probe and pile tests in normally consolidated clay, ASCE Geotechnical Special Publication No. 100, In *Proceedings, GeoDenver 2000 Conference*, Denver, CO.

Doyle, E. H., 1994. Geotechnical considerations for foundation design of the Auger and Mars TLPs, BOSS 1994, *7th Int'l Conference Behavior of Offshore Structures*, MIT.

Doyle, E. H., 1998. The Integration of deepwater geohazard evaluations and geotechnical studies, In *Proceedings of the 30th Annual Offshore Technology Conference*, Houston, TX, OTC 8590.

Doyle, E. H.,1999. Pile installation performance for four TLPs in the Gulf of Mexico, In *Proceedings of the 31st Annual Offshore Technology Conference*, Houston, TX, OTC 10826.

Doyle, E. H., Dean, E. T. R., Sharma, J. S., Bolton, M. D., Valsangkar, A. J., and Newlin. J. A., 2004. Centrifuge model tests on anchor piles for tension leg platforms, In *Proceedings of the 36th Annual Offshore Technology Conference*, Houston, TX, OTC 16845.

Doyle, E. H., Dean, E. T. R., and Newlin, J. A., 2005. Lateral design of the ursa tension leg platform, In *Proceedings of the International Symposium on Frontiers in Offshore Geotechnics*, Perth, Australia.

Doyle, E. H. and Pelletier, J. H., 1985. Behavior of a large scale pile in silty clay, In *Proceedings of the XI Int'l Conference on Soil Mechanics and Foundation Engineering*, San Francisco, CA.

Dutt, R. N., Doyle, E. H., and Ladd, R. S., 1992. Cyclic behavior of a deepwater normally consolidated clay, *ASCE Civil Engineering in the Oceans V*, Texas A&M University.

Dutt, R. N. and Ehlers, C. J., 2009. Setup of large diameter pipe piles in deepwater normally consolidated high plasticity clays, In *Proceedings of OMAE 2009 28th International Conference on Offshore Mechanics and Arctic Engineering*, Honolulu, HI.

Edil, T. B. and Mochtar, I. B., 1988. Creep response of model piles in clay, *Journal Geotechnical Engineering* 114(11): 1245-1260.

Hunt, R. J., Chan, J. H.-C., and Doyle, E. H., 1999. Driving fatigue damage estimation for Ursa TLP 96, OD Piles, In *Proceedings of the 9th Int'l Offshore and Polar Engineering Conf & Exhibition*, Brest.

Jeanjean, P., Andersen, K., and Kalsnes, B., 1998. Soil parameters for design of suction caissons for Gulf of Mexico deepwater clays, In *Proceedings of the 30th Annual Offshore Technology Conference*, Houston, TX, OTC 8830.

Ladd, C. C. and Foott, R., 1974. New design procedures for stability of soft clays, *Journal of Geotechnical Engineering* 100:GT7.

Pelletier, J. H., Doyle, E. H., and Dutt, R. N., 1997. Deepwater geotechnical investigations in the Gulf of Mexico, *Underwater Technology* 22 (2): 63-73.

Pile Dynamics, Inc., 2005. GRLWEAP—Wave equation analysis of pile driving, *GRLWEAP Program, Procedures and Models*, Goble Rausche and Likins Associates.

Terzaghi, K. and Peck, R. B., 1964. *Soil mechanics in engineering practice*, John Wiley & Sons, New York.

Whittle, A. J. and Sutabutr, T., 1999. Prediction of pile setup in clay, *Transportation Research Record* 1663:33-41.

Whittle, A. J. and Sutabutr, T., 2005. Parameters for average gulf clay and prediction of pile setup in the Gulf of Mexico, *ASCE Geo-Frontiers*, Austin, TX.

Whittle, A. J., Sutabutr, T., Germaine, J. T., and Varney, A., 2001. Prediction and measurement of pore pressure dissipation for a tapered piezoprobe, In *Proceedings of the 33rd Annual Offshore Technology Conference*, Houston, TX, OTC 13155.

Young A. G, Phu, D. R, Spikula, D. R., Rivette, J. A., Lanier, D. L., and Murff, J. D., 2009. An approach for using integrated geoscience data to avoid deepwater anchoring problems, In *Proceedings of the 40th Annual Offshore Technology Conference*, Houston, TX, OTC 20073.

5

Pipeline Geohazards for Arctic Conditions

Andrew C. Palmer
National University of Singapore

Ken Been
Golder Associates

Any displacement or movement of the soils below the seafloor will have an effect on offshore pipelines, whether they lie on or below the seafloor. This then provides a reasonable definition for a pipeline geohazard—seabed displacements that may cause damage to a pipeline. In this chapter, we will consider several geohazards, but by no means all the geohazards, that might affect pipelines. We will not talk about some of the more common pipeline geohazards such as slope failures, fault movements, earthquakes, and erosion by waves or currents, which are well covered in other texts (for example, Norwegian Geotechnical Institute 2005) and discussed in the broader context of overall project development in Chapter 2 of this text.

Ice gouges and stamukha pits are geohazards resulting from interaction of ice with the seabed. They occur only in relatively shallow waters in cold regions where the sea freezes over in the winter months. Strudel scour is another ice-related geohazard considered in this chapter. Offshore pipelines in ice environments are buried to mitigate the effects of these hazards, and the engineering challenge is to optimize the burial depth—deeper is safer but also more costly. There is a renewed focus by the industry on Arctic oil and gas exploration, fuelled partly by the United States geological survey assessment that up to 25% of the planet's undiscovered hydrocarbon reserves may be found there, partly by the perception that global warming will reduce the severity of the ice regime, and partly because the United States, Canada, and Scandinavian countries that control the reserves allow access by international oil companies. (Russia has allocated its offshore Arctic oil resources to Rosneft and gas resources to Gazprom, but there may still be opportunities for other companies to participate in development of fields, Shtokmann being a good example.) In addition, one of the largest oil fields currently under development is the Kashagan field in the northern Caspian Sea. Despite being at approximately latitude 45°, that part of the Caspian does freeze in winter, and ice action is a major challenge for design and construction of the Kashagan project pipelines. For these reasons, this chapter focuses on ice gouging, strudel scour, and stamukha pits.

5.1 Ice Gouging

5.1.1 Introduction

Sea ice is mobile unless it is landfast. Movement within the ice creates broken fragments of different sizes. Some of the fragments form rubble, and some freeze together to form ice ridges and ice rubble features within the floating ice pack. Single ice features, such as icebergs, can also be transported by wind and currents to locations where the sea itself does not freeze. In shallow or shoaling water, these ice masses run aground and drag on the sea floor. As the grounded ice continues to move, it digs into the seabed and cuts gouges (in Canada called scours). The mechanism is shown schematically in Figure 5.1 below.

Ice gouges have been mapped on the seabed in the Beaufort Sea (offshore Canada and Alaska), offshore Sakhalin Island in the far east of Russia, and in the North Caspian sea, among other regions. Scour marks have in many cases been tracked for several kilometers with sidescan or multibeam sonar. Maximum scour depths up to 2 or 3 m are not uncommon in soft soils in the Beaufort Sea, although most observed scours tend to be less than 1 m deep. Widths of scours range from a few meters for single keels, up to tens of meters for multi-keel features. Figure 5.2 shows scour marks observed from a helicopter through very clear shallow water in the North Caspian Sea (Been et al. 2008). Relict ice gouges are also visible in glaciolacustrine sediments in Norway, Scotland, and Canada (Woodworth-Lynas and Guigne 1990; Thomas and Connell 1985; and Longva and Thoresen 1990) and show evidence of deformation of the soils below and adjacent to the gouges. Ice gouging is considered to be widespread, and areas of the world where it is of concern to offshore developments are illustrated in Figure 5.3.

Ice gouges create hazards for offshore pipelines. It is easy to demonstrate that pipes should be buried below the deepest gouges, as the ice gouging forces might be in the range of 10 to 100 MN (Palmer et al. 1990). It is not obvious, however, how much deeper pipelines should be buried to avoid excessive displacements of the soil immediately below the gouging ice keel. Figure 5.4 illustrates the ice gouging problem, in which the key components are:

1. What is the depth of gouge?
2. What are the soil displacements below the gouge depth?
3. What are the strains in a pipeline buried at depth z below the gouge depth?

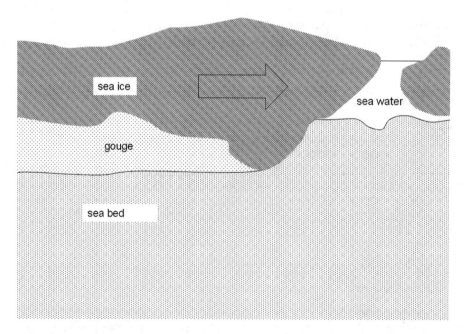

Figure 5.1   Ice gouging schematic.

Figure 5.2 Ice gouges observed from a helicopter in shallow, clear water in the Caspian Sea (Been et al. 2008, photo courtesy K. R. Croasdale).

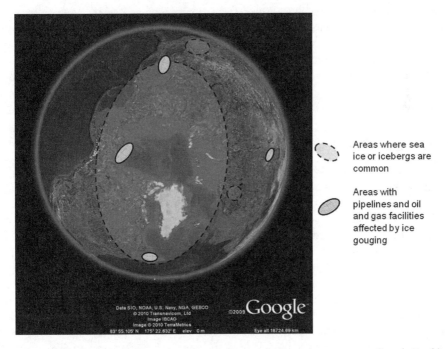

Figure 5.3 Areas of seasonal sea ice where ice gouging might occur (map courtesy Google Earth).

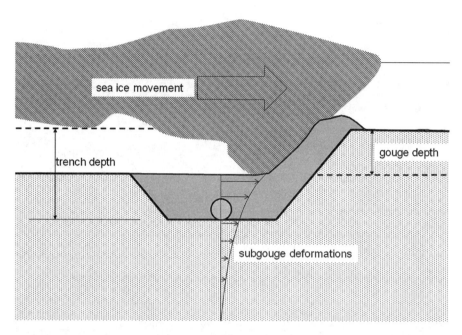

Figure 5.4   The ice gouging problem for offshore pipelines. Pipeline must be designed against soil displacements below gouging ice keels.

To keep the problem tractable, we can ignore the fact that the buried pipeline is an inclusion in the soil that will affect the displacement of the soil. Under this idealization, the ice drives the deformation of the soil, and the soil drives the deformation of the pipeline, but the presence of the pipeline has no effect on the soil or the ice. We consider the so-called free-field soil displacements in the absence of a pipeline and then apply these soil displacements to the pipeline in a structural model that is uncoupled from the model to determine the soil displacements. However, there is evidence from finite-element calculations that the presence of a large-diameter pipeline significantly modifies the movement of the soil (Konuk et al. 2006) and perhaps also the movement of the ice. Small-scale experiments confirm this, but more work is needed.

5.1.2 Ice Gouge Surveys

Information about ice gouges is all indirect, in that no one has ever observed an ice gouge happening (and unless someone is brave enough to go diving under the ice during a storm that is sufficiently severe to move the ice, this is unlikely to happen soon). This indirect information comes from seabed mapping surveys. Ice gouges show up on sidescan sonar images and multibeam or single beam echosounder records. The Canadian and United States geological surveys have both been actively mapping gouges in the Beaufort Sea since the 1970s. In particular, the Geological Survey of Canada has maintained a repetitive mapping program in which the transects and features are surveyed over several years, so that new scours can be distinguished from old scours (Blasco et al. 2007); however, we do not know which fragment of ice causes what gouge.

     The gouge survey database is the starting point for determining the pipeline burial depth. From the database, we are able to determine the shape of the gouge mark in the seabed and the frequency of occurrence of scours. Subject to the important caveat below, we can then assume that the gouge mark has the same shape as an apparently rigid ice keel that caused the gouge, but there are several problems with this assumption:

- Ice keels are not rigid and infinitely strong but are usually agglomerations of fragments of ice (Croasdale et al. 2005). The ice fragments may be only marginally stronger than the seabed they are gouging, and the fragments may be held together only by buoyancy or by some combination of buoyancy and weak ice bonds created by later freezing. There is plenty of evidence from gouge marks that ice keels do in fact break and change shape during gouging.
- The side walls of a gouge can collapse back into the gouge after the ice keel has passed, thus changing the apparent shape of the gouge.
- Active seabed sedimentation tends to fill in gouges and, depending on how long between the gouge formation and the gouge survey (Palmer and Niedoroda 2005), can significantly alter the depth, breadth, and apparent shape of the gouge.
- The gouging process requires a large horizontal force between the ice and the seabed, but it also generates a large vertical force. The vertical force lifts the ice and may also cause it to pitch and roll. Significant lifting has been observed in the field, and simple calculations confirm that the soil resistance is large enough to result in a lifting mechanism. As a result, the bottom of an observed gouge may be significantly higher than the depth to the lowest point of the ice when the ice is floating freely.

Most deep gouges are formed in the late winter. Most surveys of gouge depth are carried out several months later, in the brief open-water period of the Arctic summer. Simple calculations show that the seabed is easily disturbed by storms and that the wave height required to initiate sediment transport is surprisingly small. Sediment that is mobilized by a storm moves across the seabed and is carried into any gouge it encounters. A gouge forms a sediment trap. Once in the trap, the sediment stays there. The gouge depth measured in a subsequent gouge survey is therefore smaller than the depth of the original gouge.

The significance of this mechanism of gouge infill was pointed out a long time ago (Niedoroda and Palmer 1986) and later quantified by more detailed calculations (Palmer and Niedoroda 2005). Many gouge surveys have neglected to recognize its significance, and the results need to be treated cautiously for that reason. It is widely thought that gouge intensity is smaller in very shallow water, 5 to 10 m deep, and larger in somewhat deeper water, so that the deepest gouges are in about 20 m water depth. This may be seriously misleading and not a true guide to the reality of gouging: it may simply reflect the fact that gouges infill more rapidly in shallow water, because the waves required to start the seabed sediment moving are smaller in shallow water.

There may be different ways to determine extreme gouge depths. Gouging remolds the soil so that the soil is frequently remolded down to the maximum gouge depth and left undisturbed below. There is an analogy with the phenomenon called *plough pan* in agriculture. If a field is repeatedly plowed to the same depth, the undisturbed soil below that depth forms a hard layer that can obstruct the drainage of water and block the downward growth of roots. Farmers break up that layer by occasionally plowing with chisel plows that reach into the hard layer. Similarly, we might expect a break in soil properties at the maximum gouge depth and that the break could be detected by sounding with cone penetrometers and vane shear devices. That option would have the great advantage that it would pick up the depth of large subgouge deformations, which is more important to the pipeline designer than the depth of the gouges themselves (Palmer 1997).

5.1.3 Ice Gouging and Subgouge Displacements

The next step in pipeline design is to determine the soil displacements caused by a given scour shape. While this might sound easy, it is not. Much of conventional soil mechanics is based on limiting strains and loads on the soil to ensure stability of foundations, retaining walls, and other structures. Consider the problem in terms of conventional soil mechanics (Figure 5.5). If the ice keel were vertical, the problem would resemble that of a retaining wall (Figure 5.5a), and we could use passive earth pressure theory as a starting point. It is only a starting point, however, since we normally design soil retention systems for limited displacement, not the many meters of displacement associated with ice gouging. At the other extreme, we would have vertical loads on almost level ground, and bearing capacity theory would be the starting point (Figure 5.5b). Gouging is quite different. Neither passive pressure nor bearing capacity theory account for the kinematics associated with the rupture surface propagation and continuous motion of a gouging ice keel. In gouging, the strains are very much larger, typically of the order of 0.1 or 1, rather than 0.001. We are concerned with displacements (rather than forces) in and around the shear zone. There is limited precedent for this type of problem.

The state of practice for subgouge displacements that developed during the 1990s was based on the Pressure Ridge ice scour experiment (PRISE) joint industry project carried out by C-CORE at Memorial University of Newfoundland. During PRISE, physical model tests of scouring in sands and clays were carried out in a geotechnical centrifuge and empirical relationships between subgouge displacement, and gouge width and depth were developed (Woodworth-Lynas et al. 1996). Since then, additional centrifuge and 1g scale tests have been carried out to investigate scaling effects using different materials and a wider range of test conditions (Allersma and Schoonbeek

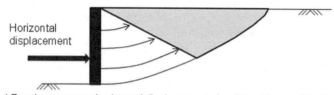

a) Passive pressure: horizontal displacement of wall, load is calculated

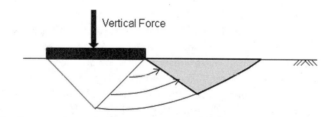

b) Bearing capacity: vertical loading on horizontal foundation

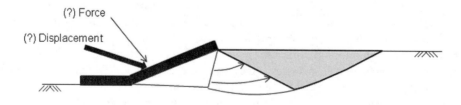

c) Ice scour: displacement and force depend on ice motion, and are both unknown

Figure 5.5   Comparison of ice gouging problem to conventional soil mechanics (passive pressure and bearing capacity).

2005; Liferov et al. 2004; Been et al. 2008). Two joint industry projects are also currently active in Canada. There is much controversy about how far below the ice keel the subgouge deformations extend.

We can examine the mechanism of ice gouging through a set of intermediate scale physical tests on clay reported in Been et al. (2008). We pushed rigid steel indenters horizontally through a bed of medium stiff clay. The front face, or attack, angles of the indenters were 15°, 30°, and 45°. The tests performed with 45° and 30° keels produced a mound and a subgouge displacement pattern that was significantly different from the mound and subgouge displacements produced with the 15° angle keels.

Figures 5.6 and 5.7 illustrate the scour mechanisms observed in terms of the soil particle trajectories in each case. Under the advancing 15° keel, the soil in front of the keel is compressed vertically but also moves outward from the center line toward the edge of the keel. Beyond the edge, the soil displacement is upward into the mound. By contrast, for 30° and 45° keels, the soil in front of the keel and within the scour was initially pushed in the direction of movement so that as the keel advances, the direction of soil movement relative to the keel is upward into the mound. Thereafter, the movement is lateral as the mound clears to the side of the keel. A similar difference between steep and low angle keels was obtained qualitatively in numerical simulations and has been observed in tests on sands.

Keel angle is therefore an important consideration for subgouge displacements, but there is very little known about keel attack angles. Measurements in the range of 30° to 60° from horizontal have been made with upward-looking sonar and by drilling through grounded ice features, but neither of these conditions can be considered representative of a gouging ice keel. During gouging, the forces between the soil and the ice fragments in the keel will change the shape of the keel through abrasion, compression, breakage, and maybe plucking blocks out of the keel. Also,

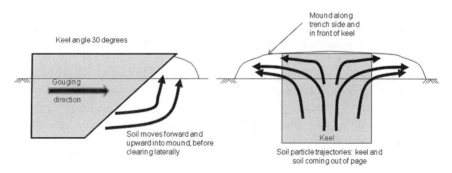

Figure 5.6 Observed soil displacement patterns; ice gouge by model 30° indenter in clay.

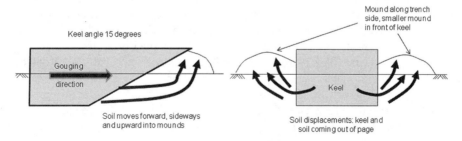

Figure 5.7 Observed soil displacement patterns; ice gouge by model 15° indenter in clay.

the keel may pitch or tilt forward as the bottom of the keel comes into contact with the seabed, thereby reducing the apparent attack angle. Nor is there any reason why the advancing face of a keel should be a plane so that the keel angle in any given keel is not a constant. Attempts have also been made to examine gouge records at locations where the gouging direction has changed. The concept here is that the impression left in the seabed at the end of a gouge, just before it turns, represents the shape of the gouging keel. While these observations tend to support keel angles in the range of 30° to 40°, there are uncertainties associated with stability of the imprint on the seabed and infill by sedimentation, as discussed earlier, for gouge surveys in general. (The natural angle of repose of many soils is about 30°, and so caution is needed when interpreting anything from observations of 30° slopes.)

Typical measured subgouge displacements below a rigid, horizontal moving, indenter gouging in clay are shown on Figure 5.8, from Been et al. (2008). This was a 45° keel, 1.5 m wide, gouging to a depth of 0.15 m. Both vertical and horizontal displacements were measured. The step change in horizontal displacement at a depth of about 300 mm occurred on a plane of weakness that was thought to exist between the layers of clay blocks that were used to prepare the soil bed, which can be seen on the photograph inset on Figure 5.8. Allersma and Schoonbeek (2005) as well as Woodworth-Lynas et al. (1996) show similar results from smaller scale centrifuge testing. (Centrifuge testing constitutes the largest nonproprietary database of subgouge displacements.) Great care is needed in using all of this data and the empirical relationships developed from them, because subgouge displacements in any one test are a function of many parameters, including:

- Undrained shear strength of the clay (larger displacements apparently occur in softer clays)
- Density of sand (larger displacements occur in looser sands)
- Indenter (keel) attack angle (as discussed, the gouge mechanism can change)
- Roughness of the indenter/keel face (affects the mechanism and forces transferred to the soil)
- Whether the indenter is constrained to move horizontally or whether it can move vertically as well (vertical movement affects the forces and likely reduces the depth of subgouge displacements)
- The shape of the underside of the indenter (most, but not all, indenters in tests have a horizontal base behind the front face and the length of this base matters)
- Rate of gouging, and whether the soil is undrained, drained, or partially drained (most actual gouging is undrained, but slow, small scale, and centrifuge scale tests in sands may be drained)
- Gouge depth, gouge width, and the ratio between gouge depth and width
- Test scale, or acceleration, in the case of centrifuge testing

Numerical models are an obvious alternative to physical models to better understand and quantify subgouge displacements, but the problem is fiendishly difficult to model realistically. Since the subgouge displacements are so dependent on soil behavior, a realistic model needs to be 3-D and needs to consider evolving dilatancy and very large strains, as well as coupled pore water drainage effects. Nevertheless, finite element analyses have been attempted to determine subgouge displacements (e.g., Lach and Clark 1996; Konuk and Gracie 2004; Konuk et al. 2005; Konuk and Yu 2007; Fredj and Comfort 2008). Large strain finite element formulations using adaptive meshing techniques (e.g., an ALE or CEL formulation in LS Dyna or Abaqus) have shown promise as a predictive tool (Palmer et al. 2005), but they still lack a generally available adequate constitutive relationship and coupled flow. Quality physical data for calibration and verification of the numerical models for subgouge displacements is also very limited.

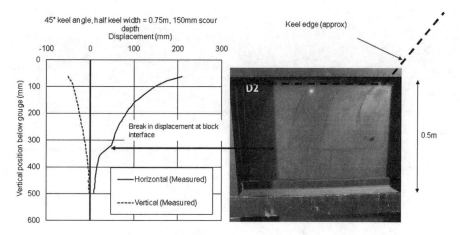

Figure 5.8 Typical subgouge displacements in indenter test (from Been et al. 2008) (Vertical scale on photograph is not same as vertical axis on graph)

5.1.4 Pipeline Strain Calculations

The final part of the problem, determining the pipeline strains for a given set of soil conditions and subgouge displacements, is less controversial and is commonly carried out using the Winkler spring model for the soil-pipe interaction. This approach has been applied for many years for onshore pipelines that are subject to ground displacements, such as occurs when pipelines cross active faults and landslide zones. Another especially interesting case where soil displacements can cause distress to pipelines occurs in northern climates where frost heave and/or permafrost thaw takes place after pipeline construction and operations.

In addition to recommended procedures and standards (e.g., DNV OS F101; ASCE 2001), there are several structural analysis programs that are suitable for Winkler spring analysis of soil-pipe interactions, and they may include different levels of complexity for both the pipe (e.g., beam or shell elements) and soil springs (e.g., linear or nonlinear). Recent studies (e.g., Fredj and Comfort 2008; Konuk et al. 2006) have shown that the Winkler model may be conservative, because it does not allow for horizontal transmission of shear stress within the soil. 3-D finite element modeling of the full ice-soil-pipeline interaction is recommended to reduce or eliminate uncertainties.

A reason for this is that the presence of a large-diameter pipeline significantly modifies the movement of the soil and perhaps also the movement of the ice. The idealization that has been adopted to separate the ice-soil interaction from the soil-pipeline interaction, which is needed to make the problem more tractable, is at the root of the problem. More work is needed on the ice gouging problem.

5.2 Strudel Scour

A strudel is a hole in the sea ice though which water flows downward: the word is the German for whirl, as in apple strudel (Reimnitz and Bruder 1972; Reimnitz et al. 1974). Strudels occur in the Arctic spring, when the rivers thaw but the sea ice is still frozen. River melt water floods across the sea ice. If there is a hole in the ice, the river water flows downward through the hole (Figure 5.9). Why the water flows downward can be seen by thinking of a hole in a horizontally continuous layer of ice with water both above and below it. The ice is lighter than water, and therefore tends to float upward, which it can only do if water flows downward through the hole. The driving force

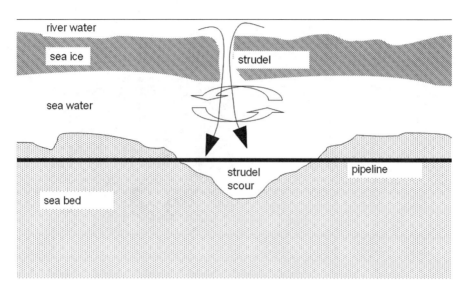

Figure 5.9   Strudel scour schematic.

is the density difference between ice and water, not the density difference between fresh and salt water; almost the same thing would happen if the salt water were above the ice and the fresh water were below it. Within the hole, the flow tends to become a rotating vortex by the same instability mechanism that leads to rotation in the water flowing out of a washbasin (but that has nothing to do with Coriolis acceleration or the rotation of the Earth, as a popular misconception has it). The downward-flowing water creates a jet in the water beneath. Solomon (2007) reports an average depth of 0.76 m and width of 14 m for strudel scours in the Alaskan Beaufort Sea, with maximum depths of 1.8 m and widths of 40 m.

Strudel formation is a possible hazard to an Arctic pipeline. The pipeline will almost certainly be trenched, and probably buried, to protect it from the ice. The powerful downward-directed jet below a strudel reaches down to the seabed and can move the seabed sediment. The pipeline can be exposed, and the jet can excavate a scour hole under the pipeline. The pipeline is less protected against ice because it is exposed. A combination of a high water velocity in a hole and a long span across the scour hole might be enough to create vortex-excited oscillations, which could lead to fatigue failure. If the hole is wide, the unsupported pipeline span might bend downward to the extent that it deforms plastically and buckles.

The probability of intersection between a pipeline and a strudel scour hole can be estimated from the strudel density. Imagine an area A traversed by a pipeline, and suppose there to be N randomly and independently distributed strudels, each of which might create damage within a radius r from the center of the strudel. The probability that the pipeline will lie within the damage radius of one of the strudels is then approximately $2r(N/A)$ per unit length of pipeline. If there are 6 strudels per square kilometer, and the damage radius is 5 m, the probability is $2(0.005)(6) = 0.06$/km.

However, this calculation assumes that the strudels are uniformly distributed. It has been suggested that the presence of a pipeline might provoke the formation of strudels. A seabed pipeline will almost certainly be warmer than the seawater above it. Heat transfer from the pipeline might form a convection cell, in which warmer water above the pipeline rises and cooler water flows in to replace it. The rising warm water will reach the underside of the sea ice, already very close to its melting point, and lead to additional melting. This would thin the ice along a line above the pipeline and create a line of weakness in the ice. That in turn will make it more likely that a strudel will form above the pipeline. Factors that might increase the probability of a strudel include:

1. Thin ice
2. The absence of current beneath the ice (because a current will transfer heat away from the pipeline and prevent stable convection cells from being formed)
3. A hot pipeline without thermal insulation
4. Ice that has already been fractured, by ice movement, by ridging, by proximity to a shear zone or a platform, or by the passage of icebreaking ships

The risk of vortex-excited oscillation of a pipeline intersected by a strudel was examined by Palmer (2000). Vortex-excited oscillations depend on the reduced velocity *VR* defined by

$$V_R = \frac{U}{ND} \tag{5.1}$$

where

U = the velocity of the water flowing across the pipeline, resolved in the direction perpendicular to the pipeline axis,
N = the natural frequency of flexural oscillations (and almost invariably it is only the lowest natural frequency that is relevant), and
D = the outside diameter of the pipeline.

If the free span of a pipeline across a scour hole is L (not necessarily larger than the strudel scour hole diameter) the lowest natural frequency N of bending oscillations of the span is (Palmer and King 2008)

$$N = \frac{C_1}{L^2} \sqrt{\frac{F}{m}} \sqrt{1 - \frac{P}{P_E}} \tag{5.2}$$

where in addition

C_1 = a constant that reflects the end conditions and can be taken as 3.5 if the ends are constrained against rotation and as 1.6 if the ends are free to rotate (the most conservative assumption);
F = the flexural rigidity (units of force × length^2);
m = the mass per unit length of the pipeline, allowing for the added mass of the water surrounding the pipeline;
P = the axial force in the pipeline and its contents, compressive positive; and
P_E = the Euler buckling force $C_2 F/L^2$, where C_2 is 4π for constrained ends and π for ends free to rotate.

The axial force depends on whether or not the pipeline is free to move longitudinally.
If it is not free to move, P is given by

$$P = \pi R^2 p (1 - 2v) + 2\pi R t E \alpha \theta \tag{5.3}$$

where

R = the mean radius of the pipeline (measured to half-way through the wall thickness),
p = the internal pressure,
v = Poisson's ratio,
t = the wall thickness,
E = the elastic modulus,
α = the linear thermal expansion coefficient,
θ = the temperature increase, operating temperature minus installation temperature.

In-line oscillations (in the direction of the pipeline axis) are found to begin when VR reaches about 1.5 but are usually so small that the stresses they induce will not create fatigue damage. Cross-flow oscillations (perpendicular to the pipeline axis) can be much larger: they begin when VR is about 3, reach a peak when VR is about 5, and then slowly decline if VR is increased further. The amplitude of the oscillations depends on the level of damping, and that damping includes both structural damping with the pipeline steel and weight coating and geotechnical damping from the seabed soil at and beyond the ends of the span. Most published measurements of vortex-induced vibrations refer to systems with unusually low damping and may not be a reliable guide to vibrations of a pipeline span in a strudel.

Calculations to quantify the level of risk (Palmer 2000) suggest that damaging oscillations of a large-diameter pipeline intersected by a strudel are unlikely, because they would require an unusual combination of a large strudel centered close to the pipeline, a long span, low damping, and high current within the strudel scour pit. A small-diameter pipeline or an umbilical might be more vulnerable.

Most instances of vortex-excited oscillation in offshore engineering occur when the water velocity across the pipeline or structure is uniform or nearly uniform along the length. A strudel is different, because if the pipeline forms a span across a strudel scour, and if the flow is rotating within the scour pit, the relative velocity acts in one direction on one-half of the span and in the opposite direction on the other half. This would appear to make oscillations even less likely, but it is possible that it might excite a second mode with a node close to the center of the span; as far as is known, that possibility has not been investigated.

A strudel might be so large that a pipeline might bend downward under its own weight, to the extent that the pipe wall would buckle and rupture. This is again possible but unlikely, because very large movements would be required before the wall would rupture.

In many instances, it will be judged unnecessary to deploy special measures to protect against strudel scour. Sometimes the risk of damage might be thought large enough to justify protection. A pipeline can then be protected by excavating a trench deep enough to safeguard the pipeline against the more serious threat of ice gouging, then pulling or lowering the pipeline into the trench, and then backfilling with rock or with mattresses. The rock would need to be in pieces large enough to be stable against the current under a strudel and that could be assessed by the methods used for rubble mound breakwaters and conventional scour protection. Alternatively, the trench might be backfilled with natural seabed soil, and the soil then covered with mattresses.

5.3 Stamukha Pits

Stamukhi are heavily grounded ice rubble features (Figure 5.10). Pits are created under stamukhi by the weight of the over-lying rubble and/or milling action of ice blocks during stamukha formation. If a stamukha forms over a pipeline, there are two interactions to be considered, as illustrated in Figure 5.11.

If the pit depth, or stamukha penetration depth, is greater than the burial depth to top of pipe, then direct contact of the ice rubble on the pipeline is possible (see Figure 5.11A). Even if the pit depth is less than the pipeline burial depth, or the pit forms adjacent to the pipelines (see Figure 5.11B), the lateral and vertical soil displacements that occur during pit formation will cause loads on the pipeline that need to be considered in pipe design.

Figure 5.10 Stamukha (heavily grounded ice rubble) in the Caspian Sea (photo courtesy K.R. Croasdale).

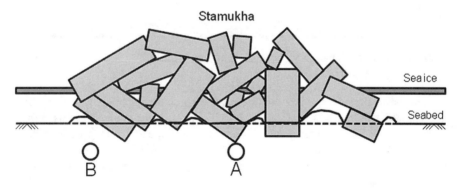

Figure 5.11 Stamukha interaction with pipelines: A) direct contact, or B) pipe affected by soil displacements.

5.3.1 Stamukha—Pipeline Interaction Frequency

If we start with an areal density of stamukha, (number per km^2 per year), and an average stamukha diameter of D, then the annual number of stamukha that overlap with a pipeline corridor of length L and width w is (Figure 5.12)

$$N = \rho (D + w) L \qquad (5.4)$$

Considering direct contact between the stamukha and the pipeline, the width w is the pipeline diameter, but since soil displacements due to pit formation must also be considered, the corridor width might be in the order of 3 m.

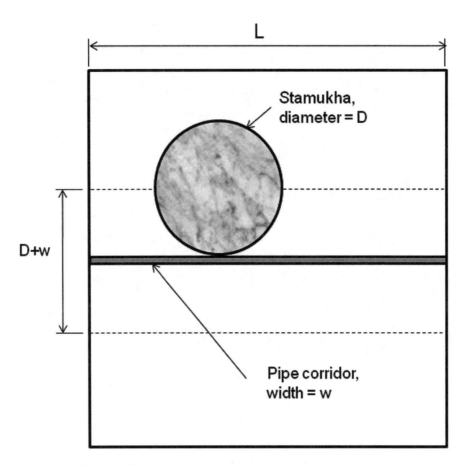

Figure 5.12   Stamukha—pipeline interaction geometry.

Next, we need to consider the fact that there are a large number of contact points or pits between the stamukha and the seabed. First, assume that the stamukha contact points are spaced on average at $s\ m^{-1}$. Then consider that the average length of the pipe under a stamukha of diameter D is the average chord length or 0.785 D. We can now calculate the number of pits over the pipeline in a given year as the number of stamukha over the pipeline times the number of pits that interact with the pipeline—that is:

$$R = N \cdot (0.785D) \cdot s \qquad\qquad (5.5)$$

Pits in the seabed have been observed from seabed surveys, using instruments such as sidescan sonar and multibeam echo sounder. In addition, stamukha have been drilled from the surface with thermal drills to determine the ice profile and seabed contact depth. (A thermal drill is a probe that penetrates ice by circulating warm fluid or steam through the probe.) Pit depths will be highly dependent on the strength of the seabed and the height of the stamukha, which determines the average pressure on the seabed so that regional specific surveys around any particular site would be needed to develop a distribution of pit depths. However, it is common to consider pit depths as having an exponential distribution. Although there is no apparent physical reason for an exponential distribution, it generally provides a reasonable fit to the data and simplifies the mathematics to determine a design pit depth with a given return period.

The probability of exceedance for pit depths (p_{ex}) is:

$$P_{ex} = \frac{1}{T \cdot R} \tag{5.6}$$

where T is the return period and R is the rate of pits over the pipeline per year. Equating the exceedance probability for pit depths with the exceedance probability for the exponential distribution and solving for the depth yields the expression for the return period depth:

$$d_{T} = \mu \cdot \ln(T \cdot R) \tag{5.7}$$

For illustrational purposes, we can now calculate the design pit depth for a 50 km pipeline ($L = 50$ km), which is affected by a pit within 2 m of the pipe ($w = 4$ m). Assume also that stamukha and pit surveys have been carried out to yield the following parameters:

- Average stamukha diameter $D = 200$ m
- Annual density of stamukha (number/km²/yr), $\rho = 0.02$ km^{-2}yr^{-1}
- Mean pit depth, $\mu = 0.3$ m
- Spacing between pits is 5 m, that is, $s = 0.2$ m^{-1}

Using (5.4) and length units of km, the number of stamukha in the pipeline corridor per year N is:

$$N = 0.02 \cdot (0.200 + 0.004) \cdot 50 = 0.204 \text{ stamukha per year}$$

The number of pits over the pipeline per year R is then (5.5):

$$R = 0.204 \cdot (200 \times 0.785) \cdot 0.2 = 6.4 \text{ pits per year}$$

The 100-year return period depth can be calculated using (5.7):

$$d_{100} = 0.3 \cdot \ln(100 \cdot 6.4) = 1.94m$$

5.3.2 Effect of Pits on Pipelines

Another key parameter that has not been mentioned so far is the diameter of the stamukha pit determined from the stamukha surveys, which may typically be in the order of 1 to 3 m. Larger pits are unlikely since the stamukha is made up of ice rubble derived from relatively thin ice (in the order of a meter or less), and large vertical forces would be required to form the pits (i.e., the stamukha height would be excessive). However, larger pits in deep, soft clays cannot be ruled out altogether.

Two load cases for the pipeline need to be considered, as shown conceptually in Figure 5.11, a vertical and a lateral loading case. The vertical load case (Figure 5.13) considers that the ice rubble can come into contact with the pipeline, which is a more extreme case than if the pit is shallower than the pipeline burial depth and the ice does not contact the pipe. For bending, the pipe can be analyzed as a beam supported by soil springs that are generally assumed to be nonlinear. In addition to bending, local denting of the pipeline must also be considered, especially for thin-walled pipe or in the absence of a concrete weight coating for protection.

The pressure from the ice will not be uniform, and the analysis needs to consider a range of pressures. The highest ice-pipe contact stresses might occur over a relatively small length of pipe under an individual block of ice, and this provides the design case for local denting or buckling of the pipe. As the length of pipe increases, the average contact stress will diminish, and bending of the pipe becomes the design case. The relationship with contact pressure and area for local ice loads is well documented (e.g., ISO 19906:2010), however, the applicability of this data from surface ice

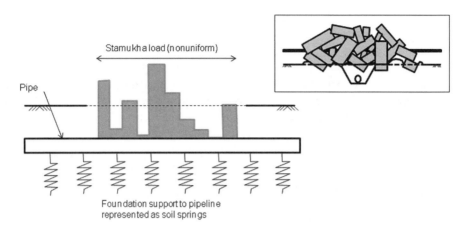

Figure 5.13   Analysis of vertical stamukha loading on pipe.

to stamukha loading is uncertain. The ice at the base of a stamukha is submerged in water and is therefore close to its melting point. In this regard, it is relatively weak ice. In addition, the load will be limited by the mechanics of the interaction between the ice blocks in the stamukha and the height of the stamukha itself.

The average vertical pressure under a stamukha can be calculated from the height of the stamukha as follows:

$$P_{ice} = (1 - n_{ice}) \cdot (h \cdot \gamma_i + d \cdot (\gamma_i - \gamma_w)) \tag{5.8}$$

where n_{ice} is the porosity of the ice (typically equal to 0.25); h is the sail height of the stamukha; d is the water depth; γ_i is the unit weight of the ice, which is approximately equal to 8.8 kN/m^3; and γ_w is the unit weight of the water. The loading on any section of a pipeline will not be the average; it may be higher as loads from overlying ice blocks may become concentrated in smaller areas. A concentration factor as high as five times has been used in some projects, with pressures over several meters of pipeline in the order of 100 to 300 kPa.

Pipeline bending is very sensitive to the loaded length, so an analysis should include a range of loaded pipeline lengths in the order of 1 to 20 m. Although Figure 5.13 only illustrates a short length of pipeline, the analysis model needs to consider a much longer section of pipeline and to include axial forces in the pipe developed from soil-pipe interaction, as well as operating temperatures and pressures.

Lateral loading of a pipeline as a result of a stamukha pit forming immediately adjacent to the pipe can be modeled with a 3-D finite element approach, but it is not clear that such an approach is justified given all of the other uncertainties in the behavior of the stamukha and pit formation. An approximate analytical approach is to consider undrained or constant volume penetration and the strain path for *simple pile penetration* after Baligh (1986). The equation of a particle originally located at a radial distance ro from the pile (or pit) centerline is

$$\left(\frac{r}{R}\right)^2 = \left(\frac{r_o}{R}\right)^2 + \frac{1}{2}(1 + \cos \phi) \tag{5.9a}$$

$$\phi = \arctan\left(\frac{r}{z}\right) \tag{5.9b}$$

where R is the radius of the pile shaft (or pit), and r and z are the radial and vertical (cylindrical) coordinates of the particle. The maximum displacement r is given when $\cos(\phi) = 1$, and the radial

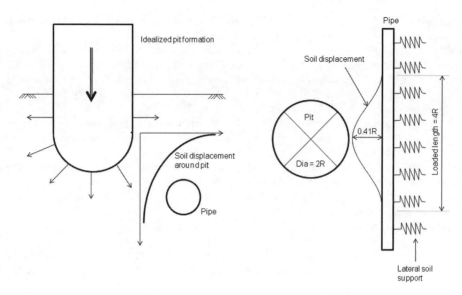

Figure 5.14   Lateral soil displacements to pipeline from cylindrical pit formation.

coordinate for a point starting at the edge of the pile $r_o = R$ will be $r = \sqrt{2}R$, giving a displacement due to pile penetration of $(\sqrt{2} - 1)R$, or something less than 0.25 diameters.

This is the maximum displacement adjacent to the center of the pit. The displacement will reduce with depth, giving less displacement for pipe burial below the pit depth, and will also reduce away from the center of the pit, as illustrated on Figure 5.14. A reasonable approximation is that the displacement decays to zero (in the direction of the pipe) at a distance of one radius from the edge of the pit, given a total loaded length of pipe of four radii or two diameters. Similarly, the displacement can be considered to decay with depth to zero at a depth of one diameter below the pit depth.

5.4 References

ASCE, 2001. Guidelines for the design of buried steel pipe, American Lifelines Alliance.

Allersma, H. G. B. and Schoonbeek, I. S. S., 2005. Centrifuge modeling of scouring ice keels in clay, Paper presented at the *International Conference on Offshore and Polar Engineering*, ISOPE2005, 2005-JSC-427.

Baligh, M. M., 1986. Undrained deep penetration, I: Shear stresses, *Geotechnique* 36(4): 471-485.

Been, K., Sancio, R. B., Ahrabian, D., van Kesteren, W., Croasdale, K. R., and Palmer, A. C., 2008. Subscour displacement in clays from physical model tests, *International Pipelines Conference*, Calgary, Alberta, Canada, IPC2008-64186.

Blasco, S., Bennett, R., and MacKillop, K., 2007. Seabed ice scour research in the Beaufort Sea, Presentation to PERD/CCTII workshop on Engineering Issues for Offshore Beaufort Sea Development, Calgary, Alberta.

Croasdale, K. R., Comfort, G., and Been, K., 2005. Investigation of ice limits to ice gouging, In *Proceedings of the 18th International Conference on Port and Ocean Engineering under Arctic Conditions* 1:23-32.

DNV, 2007. *Offshore Standard DNV-OS-F101, Submarine Pipeline Systems.*

Fredj, A., Comfort, G., and Dinovitzer, A., 2008. A case study of high pressure/high temperature pipeline for ice scour design using 3-D continuum modeling, In *Proceedings of the ASME 27th International Conference on Offshore Mechanics and Arctic Engineering*, Estoril, Portugal, OMAE2008-57702.

ISO 19906: 2010. Petroleum and natural gas industries—Arctic offshore structures.

Konuk, I. and Gracie, R., 2004. A 3-dimensional Eulerian finite element model for ice scour, In *Proceedings, International Pipelines Conference*, Calgary, Canada, IPC 04-0075, 8p.

Konuk, I. and Yu, S., 2007. A pipeline case study for ice scour design, In *Proceedings of the Offshore Mechanics and Arctic Engineering Conference*, San Diego, California, OMAE2007-29375.

Konuk, I, Yu, S. and Fredj, A., 2006. Do Winkler models work: A case study for ice scour problem, In *Proceedings of the 25th International Conference on Offshore Mechanics and Arctic Engineering*, Hamburg, Germany, OMAE2006-92335.

Konuk, I., Yu, S. and Gracie, R., 2005. An ALE FEM model of ice scour, In *Proceedings of the 11th International Conference of the International Association of Computer Methods and Advances in Geomechanics*, Turin, Italy, 8p.

Lach, P. R. and Clark, J. I., 1996. Numerical simulation of large soil deformation due to ice scour, In *Proceedings of the Canadian Geotechnical Conference* 1:189-198.

Liferov, P. and Hoyland, K. V., 2004. In situ ice ridge scour tests: Experimental setup and basic results, *Cold Regions and Ice Technology* 40:97-110.

Longva, O. and Thoresen, M. K., 1990. Iceberg deformation and erosion in soft sediments, southeast Norway, *Marine Geology* 92:87-104.

Norwegian Geotechnical Institute, 2005. Offshore geohazards, summary report, research institution-based strategic project, 2002-2005, NGI-Report 20021023-2, December 8, 2005, http://www.ngi.no/static-pages/sip8/default.asp.

Niedoroda, A. W. and Palmer, A. C., 1986. Subsea trench infill, In *Proceedings of the 18th Annual Offshore Technology Conference*, Houston, TX, OTC 5340, 4:445-452.

Palmer, A. C., 1997. Geotechnical evidence of ice scour as a guide to pipeline burial depth, *Canadian Geotechnical Journal* 34:1002-1003.

Palmer, A. C., 2000. Are we ready to construct submarine pipelines in the Arctic? In *Proceedings of the 32nd Annual Offshore Technology Conference*, Houston, TX, OTC12183.

Palmer, A. C. and Kaye, D., 1991. Rational assessment criteria for pipeline spans, In *Proceedings of the Offshore Pipeline Technology Seminar*, Copenhagen.

Palmer, A. C. and King, R. A., 2008. *Subsea Pipeline Engineering*, 2nd ed., Pennwell, Tulsa, OK.

Palmer, A. C., Konuk, I, Comfort, G., and Been, K., 1990. Ice gouging and the safety of marine pipelines, In *Proceedings of the 22nd Offshore Technology Conference*, Houston, TX, 3:235-244, OTC6371.

Palmer, A. C., Konuk, I., Niedoroda, A. W., Been, K., and Croasdale K. R., 2005. Arctic seabed ice gouging and large sub-gouge deformations, In *Proceedings of the International Symposium on Frontiers in Offshore Geotechnics*, 645-650, Perth.

Palmer, A. C. and Niedoroda, A. W., 2005. Ice gouging and pipelines: Unresolved questions, In *Proceedings 18th International Conference on Port and Ocean Engineering under Arctic Conditions*, 1:11-21, Potsdam, NY.

Reimnitz, E. and Bruder, K. F., 1972. River discharge into an ice-covered ocean and related sediment dispersal, Beaufort Sea coast of Alaska, *Geological Society of America Bulletin* 83:861-866.

Reimnitz, E., Rodeick, C. A., and Wolf, S. C., 1974. Strudel scour: a unique Arctic marine geologic phenomenon, *Journal of Sedimentary Petrology* 44(2): 409-420.

Solomon, S. M., 2007. Shallow water geohazards in the Mackenzie Delta region, Presentation to *PERD/CCTII Workshop on Engineering Issues for Offshore Beaufort Sea Development*, Calgary, Alberta.

Thomas, G. S. P. and Connell, R. J., 1985. Iceberg drop, dump, and grounding structures from pleistocene glacio-lacustrine sediments, Scotland, *Journal of Sedimentology Petrology* 55(2): 243-249.

Woodworth-Lynas, C. M. T. and Guigne, J. Y., 1990. Iceberg scours in the geological record: Examples from glacial Lake Agassiz, In *Glaciomarine Environments: Processes and Sediments*, eds. J. A. Dowdeswell and J. D. Scourse, *Geological Society Special Publication* 53:217-233.

Woodworth-Lynas, C. M. L., Nixon, J. D., Phillips, R., and Palmer, A. C., 1996. Subgouge deformations and the security of Arctic marine pipelines, In *Proceedings of the 28th Annual Offshore Technology Conference* 4:657-664, Houston, TX, OTC8222.

6

The Application of Centrifuge Model Testing to Deepwater Geotechnical Problems

Edward C. Clukey
Geotechnical Advisor
BP America, Houston

Mark F. Randolph
Professor of Civil Engineering
Centre for Offshore Foundation Systems, University of Western Australia, Perth

6.1 Introduction

Over the course of the past 20 years, centrifuge testing has evolved from primarily a research tool to a tool capable of resolving critical design issues for foundations. This transition has become no more obvious than in the offshore industry where, because of the size of the foundations, it is often difficult to perform reasonably scaled 1-g model tests to verify designs. Some of the earlier centrifuge successes were described by Murff (1996) in a paper that mostly focused on shallow water applications. This chapter expands on those applications into deeper water and further demonstrates the effectiveness of the centrifuge tool beyond the research realm.

The offshore industry and centrifuge testing have evolved, with much of the attention now directed at deepwater applications. As the industry expanded into deepwater, new types of foundations were required. One of the most significant changes once water depths reached about 1300 to 1400 m was the switch from driven piles to suction caissons for both exploration and production facilities. Centrifuge testing, carried out at various testing facilities throughout the world, provided critical and timely information to verify design approaches for suction caissons. Initial tests were aimed at simply verifying the holding capacities for typically sized suction caissons in typical deepwater soil conditions. As the testing evolved, the results started to provide a more fundamental understanding of foundation behavior and helped to resolve some important design issues regarding foundation performance. These more recent results should provide a basis for updates to future design codes.

In addition to suction caissons, the deepwater expansion provided the motivation for innovation into other anchoring systems. Two other foundation anchorages (suction embedded plate anchors [SEPLA] and torpedo anchors) have likewise benefited from the ability of centrifuge testing to provide high quality technical data in a timely and cost effective way. Although these types of foundations have primarily been used for exploration, mobile offshore drilling units (MODUs), they have now been used in some limited production applications because of their reduced costs. Although, in some cases, field tests were performed to validate design approaches for these foundations, the additional data gained by centrifuge tests helped provide the confidence needed to extend the range of potential applications.

Beyond foundation anchorages, the move to deepwater has enhanced the use of subsea production systems and the need to understand how mudmats and flowlines respond to thermally induced expansion. Thermal considerations and the accompanying flowline expansion can create very complex loadings on subsea structures. These loads can include combinations of sliding, overturning moments, and torsion. Once again, centrifuge testing has provided a useful and convenient means for understanding how mudmats respond under these complex load combinations. However, in order to properly understand the magnitudes of these loads, the designer must first understand the flowline kinematics and soil restraints from thermal effects. Centrifuge testing combined with other 1-*g* types of tests have provided the information required to better understand the pipe-soil interaction, including the embedment process and responses under both axial and lateral movements. These tests were particularly useful because of difficulties in determining soil characteristics for analytical models due to the very low stresses applied to the soil. The results also showed that the soil restraint was, at times, significantly more than predicted by existing analytical models. The enhanced restraints help reduce or eliminate costly subsea equipment often used to mitigate expansion effects.

The chapter will begin with a review of some of the basic principles that underlie centrifuge model testing. This review will then be followed with a description of the use of centrifuge model testing in helping develop suction caisson technology, now one of the primary foundations for deepwater applications. Centrifuge model tests used to verify holding capacities for other innovative foundations, such as SEPLAs and torpedo anchors, are then described. Insight gained from centrifuge model tests for fatigue problems for a drilling conductor is discussed, and a final application is presented that addresses recent work on modeling pipeline-soil interaction. Following these various example applications, a brief summary is provided of current geotechnical centrifuge facilities around the world that are active in deepwater modeling, and some of the logistics and rationale for centrifuge modeling, particularly with respect to alternatives such as numerical modeling, are discussed. A detailed review of centrifuge modeling in offshore geotechnics is also provided by Gaudin et al. (2010).

6.2 Principles of Centrifuge Model Testing

The underlying principle of centrifuge model testing is to maintain 1:1 scaling of stress and related quantities of similar dimension, such as soil strength and modulus, by subjecting a model with linear dimensions scaled as 1:*n* relative to the prototype to an acceleration field of *n* times gravity *g* (Schofield 1980; Taylor 1995). The use of similar soil types in model and prototype, with the same density and friction angle, allows other scaling ratios to be determined—for example with forces (being the product as stress times area) scaling as 1:n^2. A full summary of scaling relationships has been presented by Garnier et al. (2007).

Two particular scaling issues are worth noting. The first is in respect to the (incorrect) scaling of particle size, which is usually maintained at approximately 1:1, rather than the true scaling of 1:*n* for linear dimensions. For fine-grained soils, this is not a significant issue as the particles are still (typically) many hundreds of times smaller than critical dimensions of any structure, and the soil will respond essentially as a continuum. However, for coarser grained material such as sand, it is necessary to ensure that the mean grain size is sufficiently small (e.g., less than about 3 to 5%) compared with key dimensions of the model, such as the diameter of a pipeline or foundation

(Ovesen 1979). Even then, particle size effects may lead to some error in the scaling of pre-failure displacements, since the ratio of *elastic* displacement to particle size may be quite small (Palmer et al. 2003).

The other problem area of scaling is in respect to time, since different scaling ratios apply to different types of time-dependent processes. From a dynamics viewpoint, time scales as the square root of length/acceleration, hence $1:n$. Therefore dynamic phenomena such as earthquakes must be scaled with a frequency that is n times that of the prototype event. From a consolidation perspective, however, the time scale of consolidation scales with the square of the linear dimension, hence as $1:n^2$, assuming similar soil and pore fluid in model and prototype. In order to match the time scales for dynamic events and consolidation, it is necessary to scale the viscosity of the pore fluid by $n:1$, thus changing the time scale for consolidation to $1:n$. A third consideration is that of strain rate, which may be viewed as the ratio of velocity to linear dimension. During a dynamic event, where the velocity scales as $1:1$, strain rates will therefore be n times higher than in the prototype event, leading to some enhancement of the shear strength, at least for typical geomaterials.

The balance between consolidation rate and strain rate leads to some difficulties when modeling continuous processes or cyclic loading events. During a continuous process such as a penetrometer test, the degree of consolidation is controlled by the normalized velocity $V = vd/c_v$, where v is the velocity, d the penetrometer diameter, and c_v the consolidation coefficient (Finnie and Randolph 1994). Assuming that the consolidation coefficients in model and prototype soils are similar, it would be necessary to scale the velocity by n in order to match the degree of consolidation. In fine-grained soils, however, it is usually sufficient to demonstrate that V exceeds the threshold at which partial consolidation starts to occur so that the penetrometer test is fully undrained. The actual velocity may be chosen so as to achieve similar strain rates in model and prototype, which suggests that the velocity in a continuous event should be scaled as $1:n$ (maintaining the ratio v/d constant, thus effectively scaling time as $1:1$).

Similar considerations apply to modeling cyclic loading, where a critical issue is the degree of consolidation that occurs over the time to apply a given number of loading cycles. Since the time scale of consolidation is dictated by $c_v t/d^2$ (where d is the relevant linear dimension), ideally, the time for the loading should be scaled by $1:n^2$. However, that would typically introduce unwanted inertial effects (for example a test carried out at 100 g, with linear dimensions scaled as $1:100$, would require cyclic loading at a frequency of 10^4 greater than in the prototype event, so $\sim 10^3$ Hz to model typical wave-induced cyclic loading). In practice, a compromise is usually followed, with the cyclic loading carried out at 1 to 10 Hz (so typically between $n^{0.5}$ and n times the prototype loading), and a check made that the critical number of cycles can be applied without significant consolidation. Extended sequences of cyclic loading, with 100 to 1000 cycles, will generally allow significant consolidation to occur at model scale and so will not simulate the correct buildup of excess pore pressure or loss of strength. The elevated frequency of cyclic loading will also lead to higher strain rates in the model and hence enhanced strength of the soil.

Since the time scale for viscous effects scales as $1:1$, centrifuge model tests cannot usefully simulate shear creep—the test would need to be run for the same length of the time as relevant for the prototype. However, the same may not be true for volumetric creep or secondary compression, depending on what assumption is made regarding the underlying time scale of the secondary compression. The common assumption that *isotach* compression curves are approximately evenly spaced for each tenfold increase in time *relative to the time for primary consolidation* (Mitchell 1993) implies that the timescale for secondary consolidation in a centrifuge test scales in a similar way to primary consolidation, namely at $1:n^2$.

6.3 Previous Centrifuge Model Testing

Centrifuge model testing has been used previously on a variety of offshore geotechnical problems. As described by Murff (1996), these included a variety of applications from the impact of jack-up spudcans on the behavior of piles on steel jackets, the lateral performance of piles in calcareous and clay soils, the foundation response for large gravity-based structures, and the use of suction caissons for tension leg platform (TLP) applications. More recently, Doyle et al. (2004) have described the use of centrifuge modeling to assess the interaction of closely spaced piles for a TLP application. However, as discussed by Murff, the offshore oil and gas industry has been somewhat slow in adopting this modeling technique, a surprising situation given the size of offshore foundations and other components and the associated high costs involved with performing large 1-g tests whose physical size is still often an order of magnitude or more or less than the prototype.

As Murff noted, many of the early programs of centrifuge modeling were carried out as part of general research within Exxon Production Research and were not specifically focused on a direct application for a particular project. The one exception was a program to investigate the lateral response for piles driven in calcareous soils and used to anchor a strut support system for an offshore jacket (Wesselink et al. 1988). However, nearly all of these programs demonstrated the effectiveness of techniques used subsequently to address specific project issues.

One notable exception where the technique was used with potentially disappointing results was a program to investigate the liquefaction potential of a Moliqpak arctic oil rig sitting on dredged sand (Jeffries and Been 2006). Although the test results appeared to suggest that liquefaction would not occur during ice-generated dynamic loading, the soil did, in fact, liquefy during such a loading event. However, uncertainties with respect to the actual loading may have contributed to differences between the centrifuge results and the observed field behavior. For similar types of dynamic problems, centrifuge testing has been used extensively to investigate earthquake related geotechnical issues (Kutter and Balakrishnan 2000) and is one of the major components in a national effort to better understand earthquakes by U.S. researchers (Network Earthquake Engineering Simulation, NEES, https://www.nees.org/).

Much of the work cited by Murff was performed by Exxon (now ExxonMobil) and did not represent a more universal acceptance of the technique across the industry. The initial focus of using the technique for research and the potential perceived complexity of the technique are possible reasons for the reluctance of the offshore community to fully utilize the advantages of this type of testing. Despite these reservations, the technique has continued to evolve since the time that Murff first published his assessment on its usage in the offshore industry.

Programs both at ExxonMobil and at BP were initiated to investigate the response of large gravity-based structures under earthquake loading. Perhaps though, the greatest usage occurred with assisting in the development of suction caisson technology. The need for suction caisson technology was required due to increased difficulties encountered with pile driving in deepwater (greater than about 1250 m). Suction caissons offered an alternative approach to these installation challenges. However, load test data to provide equivalent levels of confidence for predicting foundation performance under field loading conditions were not available. Centrifuge testing was, therefore, used to provide this much-needed information in a timely manner and to improve the overall foundation reliability. The following section highlights some of the important contributions made through centrifuge testing for suction caisson technology.

6.4 Suction Caissons

Suction caissons were initially used for *deepwater* applications in the North Sea (Støve et al. 1992) for TLPs in about 300 to 400 m water depths. These initial applications used clustered arrangements of caisson cells for each corner foundation. The soils for these initial applications were overconsolidated while the caisson embedment/diameter aspect ratios, using an equivalent diameter for the cluster, were relatively small (about 0.5). A series of about 10:1 1-*g* model tests were performed in the field to verify the Snorre foundation design (Dyvik et al. 1993).

Because of the North Sea success and the lower cost of these foundations in the North Sea, Exxon production research initiated a series of centrifuge tests in the early 1990s to determine if this relatively new concept could be economically adopted for the Gulf of Mexico (Clukey and Morrison 1993; Morrison et al. 1994; Clukey et al. 1995). Once again, the work was focused on using suction caissons for TLP applications where the loads were primarily uplift loads with small offsets from the vertical. Several of the tests performed were used to simulate the 1-*g* model test performed to confirm the Snorre foundation design (Morrison et al. 1994). A clustered arrangement of cells was still considered appropriate since it provided flexibility during installation by providing the potential for separately controlled suction within each cell. However, to better understand the fundamental behavior of suction caissons, most of the tests were performed on single cell caissons. The design loads were again considered to be mostly vertical uplift.

The results of these tests confirmed the feasibility of using suction caissons in the Gulf of Mexico. However, piles continued to be the most viable option as water depths for TLPs extended to about 1250 m, due mainly to the design, fabrication, and installation efficiencies associated with the more mature pile technology. As water depths continued to increase to even deeper water in the late 1990s, other types of facilities such as SPARs, production semi-submersibles, and floating, production storage, and offloading (FPSO) facilities became the preferred facility option. At that time, underwater pile driving capabilities were not able to extend to these depths, and suction caissons began to be used in the Gulf of Mexico. The first application was ExxonMobil and BP's Hoover-Diana deep draft caisson vessel in 1999. As additional projects evolved in greater than 1250 m water depths, suction caissons continued to be the preferred foundation type until about 2007 when pile driving capabilities were extended to these deeper water depths.

The use of suction caissons extended to most parts of the offshore world during the 1990s, with applications off the coasts of West Africa (Colliat et al. 1995, 1996), Brazil (Mello et al. 1998), and Australia (Randolph et al. 1998; Erbrich and Hefer 2002). A detailed summary up to 2004 is given in Andersen et al. (2005). Centrifuge model tests formed part of the design validation process for some of these applications (Randolph et al. 1998).

The geotechnical community became increasingly interested in advancing the knowledge base and fundamental understanding of the effects of suction on the caisson response, particularly for cases where the suction caissons were required to sustain uplift loading. Important differences between piles and suction caissons were identified. These included:

- For driven piles, most of the capacity is derived from external skin friction, whereas suction caissons provide more of a balance between external skin friction and the suction or reverse end bearing at the bottom of the caisson. The magnitude of the reverse end bearing under various loading conditions needed to be established.
- For driven piles, there is little practical interaction between the lateral capacity, which is governed by the near surface soil-pile interaction, and the axial capacity, governed by the soil at a substantial depth; much greater interaction occurs for suction caissons.
- The wall thickness to diameter ratios (t/D) for suction caissons are significantly smaller than piles. This is an important consideration in estimating set-up times and may potentially reduce the external skin friction compared with driven piles.
- The application of a reduced internal pressure (referred to as suction) could also cause relatively more soil to enter the caisson than enters a pile during installation, potentially causing a reduction in external skin friction and lower friction coefficients (α values) on the outside of the caisson (Andersen and Jostad 2002).

- With the advent of deeper water and the trend toward facilities with taut or semi-taut legged facilities, the applied load on the foundations has a significantly greater lateral component. To more effectively resist this component, the mooring line is attached on the side of the caisson below the mudline. The impact of this raised additional uncertainties regarding the interaction of the vertical and horizontal loads. In addition, there were concerns about whether gapping occurs on the backside of the caisson and the influence that might have on capacity.
- In the Gulf of Mexico, for facilities in deepwater (greater than 1250 m), a significant number of designs are driven by loop currents instead of shorter duration hurricane loads. Peak or near peak loop current loads will last over a significantly longer period of time than storm or hurricane loads, potentially causing a reduction in the ultimate holding capacity.
- As for piles, cyclic loading is also a consideration for suction caissons. However, for suction caissons in deeper water, cyclic loading is generally all in the same direction and potentially less severe than cyclic loading on piles used for traditional jacket foundations. The less severe cyclic loading could result in greater operational holding capacity in comparison with the monotonic capacity, although, as noted above, a gap could also form on the backside of the caisson, thereby reducing capacity.
- There are two major installation issues related to suction caissons. If the suction pressure required for installation is too high, then either the steel caisson could buckle, as has occurred, or the soil plug could get pulled up inside the caisson, potentially limiting the final penetration depth or the holding capacity of the caisson.

One of the first sets of centrifuge tests on suction piles addressed a number of these issues (Clukey and Phillips 2002; Clukey et al. 2003). These tests were performed to simulate suction caissons similar to those intended to be installed on several deepwater BP projects in the Gulf of Mexico. These tests focused on measuring the total holding capacity for suction caissons with similar geometries to those planned for field installations. Tests were performed by both pushing the caissons into the soil and by using a combination of pushing and suction. The tests were corrected for rate effects in the centrifuge where the tests were run to failure in less than 30 seconds. The shear strength of the clay (kaolin) was determined with a cone penetrometer calibrated to the direct simple shear strength measurements made on samples extracted from the centrifuge.

Inclined loading tests were performed with loading angles ranging from about 25° to 42° from the horizontal with the loading point located about two-thirds down the side of the caisson. Test results from two tests with loading angles of 40° and 42° are shown in Figure 6.1. A few tests were performed with an outward bevel at the bottom of the caisson to investigate if the external shin friction could be enhanced by forcing more displaced soil to the outside of the caisson.

These tests helped confirm design methodologies and demonstrated:

- For the tests performed by pushing the caissons into the soil, the reverse end bearing capacity factor Nc was 9.4 if the American Petroleum Institute (API) procedures to estimate the external skin friction on the sides of the caisson were followed (Clukey & Phillips 2002).
- A comparison of two tests, one pushed into place and the other installed with a combination of pushing and suction, showed a reduction of about 25% in external skin friction in the latter test, although this difference has not been substantiated in later testing.
- The tests with various loading angles generally confirmed analytical solutions for the interaction between the vertical and horizontal loads (Figure 6.2, Clukey et al. 2003).
- The tests with the bevel were determined to be inconclusive in terms of increasing the capacity by enhancing the external skin friction.

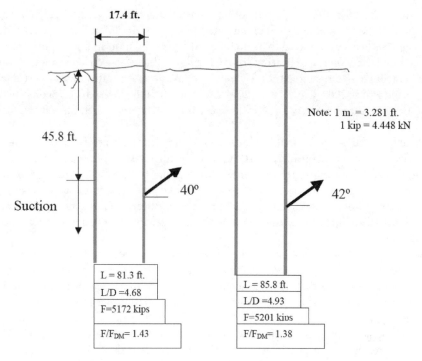

Figure 6.1 Suction caisson test results for caissons installed with suction (left) and by pushing (right). The ratio F/F_{DM} is the measured capacity to design methodology used when tests were performed. Dimensions are in prototype units (Clukey and Phillips 2002).

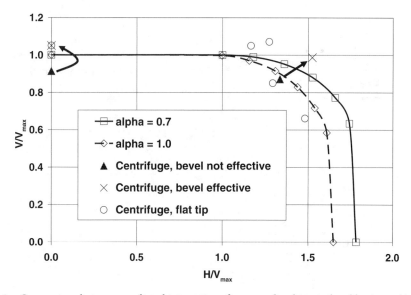

Figure 6.2 Comparison between predicted interactions for normalized vertical and horizontal load components and centrifuge test data. Predictions were made using two assumed values of caisson/soil interface resistance factor alpha (Clukey et al. 2003, with permission from ASME).

Subsequent testing addressed the impact of long-term loading and cyclic loading effects on the capacity of the caissons. These tests demonstrated that the degradation from cyclic loading was minimal, and the loads would need to be maintained for several months before pore pressure dissipation reduced the reverse end bearing. The reduction in the reverse end bearing was confirmed by other centrifuge investigators at the University of Western Australia and numerical analyses (Randolph and House 2002; Clukey et al. 2004; Chen and Randolph 2007b). The combined results showing the normalized capacity Q/Q_{ult}, where Q_{ult} is the static undrained capacity, vs. hold time is shown on Figure 6.3.

It should be noted, however, that the reverse end bearing will also be reduced for longer term loads because of rate effect considerations. While the centrifuge technique can give some insight into the effects of strain rate in a relative sense—for example, by varying the time to failure by two or three orders of magnitude—it may not be a particularly effective tool for investigating shear creep at very low absolute rates of strain. The time scale of shear creep is generally assumed to be the same at model and prototype scales, and performing centrifuge tests for very long periods of time is not practical.

Jeanjean et al. (2006) performed a series of tests that separated the internal side resistance, external skin friction, and reverse end bearing. The results from these tests showed there was no difference between caissons that were pushed into the soil and those installed with a combination of pushing and suction. However, the average external skin friction was found to be less than API predictions for piles, with an average mobilized α value of 0.85 for fully set-up conditions. The reverse end bearing had a peak bearing capacity factor (N_c) of 12 at large displacements. However, this value was reached at large displacements after the external skin friction had softened. The bearing capacity factor at the peak load was found to be close to 9. Similar magnitudes of design parameters were obtained from research on suction caisson response carried out at the University of Western Australia (Chen & Randolph 2007a), again with no difference in capacity observed between caisson installed by jacking or using suction.

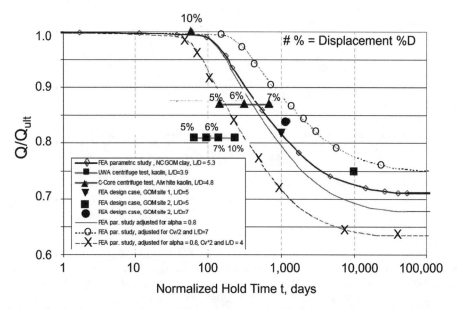

Figure 6.3 Comparison of centrifuge and finite element results showing reduction in normalized holding capacity with holding time (Clukey et al. 2004) © 2004 OTC. Reproduced with permission of the copyright owner. Further reproduction prohibited without permission.

Figure 6.4   Photograph showing larger diameter suction caissons being towed for installation in overconsolidated soils.

The dependence of the bearing capacity factor on displacement also suggests that the correct bearing capacity factor for design may depend on the caisson geometry. The caissons tested had a length to diameter ratio (L/D) of 4. If the L/D ratio was less, the end bearing would be more dominant (assuming caisson sufficiently deep to neglect surface effects), and the bearing capacity factor at the peak load could increase some. For greater L/D ratios, the opposite could be true.

Most deepwater locations consist of normally consolidated to slightly overconsolidated soil conditions. Therefore, most of the centrifuge testing performed to date has focused on performing tests in these types of soils. However, in regions where massive submarine slides have occurred, the soils can be significantly overconsolidated. BP had a Gulf of Mexico project where the two suction caissons were placed on a large 700 m escarpment with highly overconsolidated soils. To verify the design of these caissons (Figure 6.4), centrifuge tests were performed at the University of Colorado (Jeanjean et al. 2006). The three tests that were performed had capacities that were 4 to 16% greater than the design predictions. The design predictions were therefore considered to be slightly conservative.

The potential for a gap (Figure 6.5) to form behind a suction caisson was illustrated in model tests carried out in a lightly overconsolidated carbonate mud, simulating conditions for a development by Woodside in the Timor Sea off the north coast of Australia (Randolph et al. 1998). Numerical analysis has indicated that this can reduce the lateral capacity by up to about 20% (Supachawarote et al. 2005).

6.5 Alternate Deepwater Foundation Solution

6.5.1 Suction Embedded Plate Anchors

Another use for suction caissons is to install plate anchors. These SEPLAs have been used extensively for anchoring MODUs that are used for exploration or production drilling (Dove et al. 1998; Wilde et al. 2001). The SEPLA anchor and placement at the end of the suction caisson is

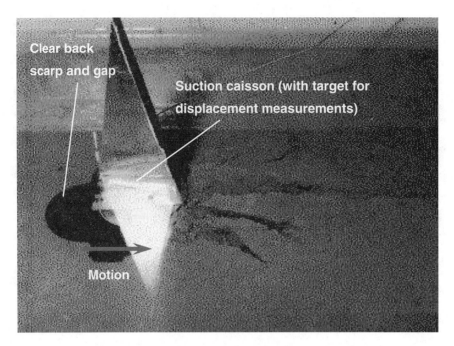

FIGURE 6.5a   Observation of gap formation in centrifuge model test.

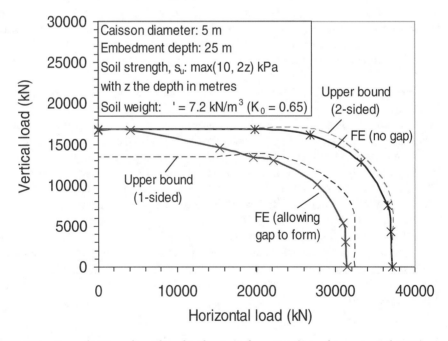

FIGURE 6.5b   Example FE results with and without gap formation (Supacharawote et al. 2005).

shown on Figure 6.6, while a centrifuge model used for testing is shown on Figure 6.7. Key questions concerning the behavior of SEPLAs include:

- The amount of vertical movement (loss of embedment) required to key the anchor in order to mobilize the ultimate pullout capacity (UPC)
- Determination of the appropriate bearing capacity factor to estimate the ultimate holding capacity
- The impact of cyclic loading and creep on the UPC
- Determination of any tendency for the anchor to roll in the direction perpendicular to axis of loading or pull out in the direction in which it was installed

Figure 6.6   SEPLA anchor and placement at end of suction caisson.

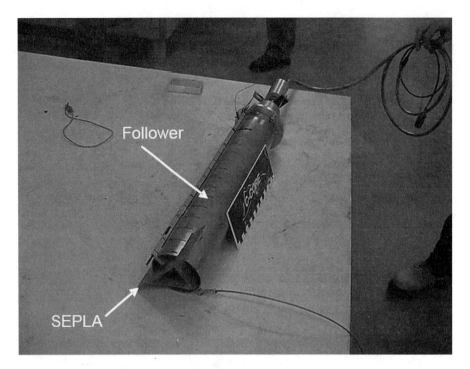

Figure 6.7 Centrifuge model of SEPLA and installation suction caisson.

Model tests reported by Gaudin et al. (2006) showed loss of embedment of between 1 and 1.5 times the anchor height. These tests were on square anchors, with a centrally positioned padeye offset from the plane of the anchor by 0.66 times the anchor dimension. A loading angle of between 50° and 60° was adopted during keying and pullout, which is somewhat higher than typical in field conditions.

An interesting observation from the model tests of Gaudin et al. (2006) was the effect of partial remolding of the soil due to suction caisson installation and retrieval. This led to a 20% reduction in the SEPLA capacity if the anchor was tested within an equivalent prototype time t of about 7 months (nondimensional time factor, $T = c_v t/D^2$ of 0.25, where c_v is the consolidation coefficient consistent with that from a piezocone dissipation test and D the caisson diameter). The full capacity was recovered for post-installation consolidation times of 1.5 years ($T > 0.5$).

There is still a degree of uncertainty with respect to embedment loss during keying and the influence of post-installation setup time on the anchor capacity. Now that SEPLA use for temporary moorings (e.g., for MODUs) is established, there is motivation to extend application to more permanent moorings. The higher level of reliability required for permanent moorings is still a driver for further studies, through both physical modeling and advanced numerical analysis. Initial results from 3-D large deformation finite element analysis (Wang and Randolph 2009) have shown the importance of the anchor weight and eccentricity of the keying load relative to the centroid of the anchor. Both quantities affect the relative magnitudes of moment and (net) uplift load, and hence the loss of embedment during keying.

6.5.2 Torpedo Anchors

Another form of anchor that has been used with success in waters offshore Brazil are dynamically penetrated anchors (Medeiros 2001, 2002; Brandão et al. 2006). These are torpedo shaped, typically 12 to 15 m long, 0.8 to 1.2 m in diameter and weighing up to 1000 kN in air (Figure 6.8a). A similar concept is undergoing development in European waters (Lieng et al. 1999, Figure 6.8b). A related type of anchor has also been developed recently for use in the Gulf of Mexico (Zimmerman et al. 2009).

Torpedo anchors are designed to be released from a height of 50 to 100 m above the seabed, free falling to impact soft seabed sediments at velocities of 10 to 30 m/s, finally embedding by 2 to 3 times their length (Figure 6.9). The *efficiency* of such anchors is low, with holding capacities of 3 to 6 times the anchor weight in air reported from field experience (Medeiros 2001). However, the low cost of the anchor and simple installation procedure render them a cost-effective alternative to other deepwater anchoring systems.

In principle, the capacity of torpedo anchors is somewhat insensitive to the precise strength profile, since greater embedment will occur for lower strength gradients. However, prediction

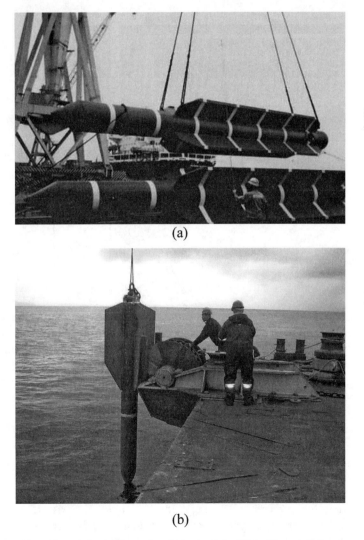

(a)

(b)

Figure 6.8 Dynamic anchors (a) torpedo anchor (Araujo et al., 2004 with permission from ASME); (b) deep penetrating anchor at 1:3 scale.

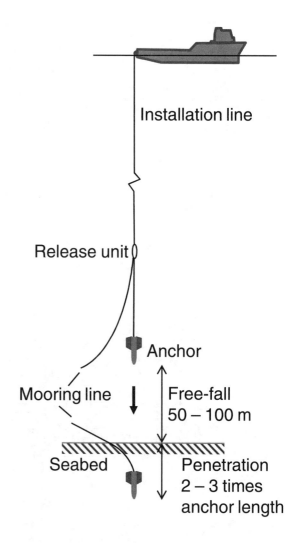

Figure 6.9 Dynamic anchor installation procedure.

of the embedment depth for a given strength profile is complicated by the very high strain rates involved during penetration (normalized velocities, v/d, where d is the anchor diameter, of up to 25 s^{-1}) and by hydrodynamic aspects. The latter may involve entrainment of a boundary layer of water immediately adjacent to the anchor, but analysis also suggests that it is important to consider a hydrodynamic drag resistance (in additional to conventional geotechnical friction and bearing resistance) in the soft sediments near the mudline.

Centrifuge model tests have played a significant role in developing a database of performance for torpedo anchors and in establishing calculation techniques for predicting embedment and ultimate holding capacity. The high g field in the centrifuge allows relevant impact velocities to be reached from only modest drop heights. A typical setup is shown in Figure 6.10 (Richardson 2008). Anchors are modeled at a scale of 1:200, with length up to 75 mm and diameter of 6 mm. They are released from heights of up to 300 mm but at 200 g. A guide provides the necessary resistance to Coriolis (circumferential) accelerations that would otherwise occur, and the impact velocity is measured by a series of photo emitter receiver pairs. Tests have been conducted covering a variety of tip geometries and with varying numbers of anchor flukes (0, 3, or 4) at the rear of the anchor (O'Loughlin et al. 2004, 2009).

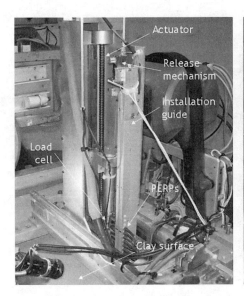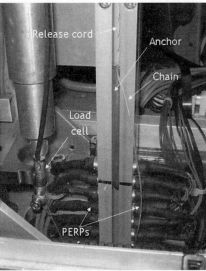

Figure 6.10 Experimental arrangement for dynamic anchor installation in beam centrifuge (courtesy of Mark Richardson).

Prediction of the anchor embedment follows a conventional approach (Audibert et al. 2006; O'Loughlin et al. 2009), with the net embedment force determined from the anchor weight less the frictional and tip bearing resistance, and a drag term expressed in terms of a drag coefficient C_d. This term is important at shallow depths, where the velocity is high and soil resistance is low; a drag coefficient of 0.24 has been suggested.

The frictional and tip resistance are computed in terms of similar α and N_c values as used for pile installation but incorporating a rate parameter R_f expressed as:

$$R_f = 1 + \lambda \log\left(\frac{\dot{\gamma}}{\dot{\gamma}_{ref}}\right) \text{ or } R_f = \left(\frac{\dot{\gamma}}{\dot{\gamma}_{ref}}\right)^{\beta}$$

for logarithmic or power law respectively. The strain rate $\dot{\gamma}$ may be taken (to a first approximation) as v/d, which may be several orders of magnitude greater than the reference strain rate at which the shear strength is determined.

A comparison of predicted and measured velocity profiles for an anchor of length 75 mm (15 m prototype) is shown in Figure 6.11, with the best fit values of the rate parameters λ and β indicated. These values are significantly higher than the nominal 10% increase in shear strength per decade of strain rate typically found at laboratory strain rates, and there is a trend for higher values for the higher impact velocities. One of the limitations of reduced scale modeling, for both laboratory and centrifuge conditions, is that strain rates are much higher than in the field. For the model torpedo anchor, the maximum strain rate for an impact velocity of 20 m/s is ~30 × 10³ s^{-1}, in stark contrast to a standard simple shear test with a strain rate of perhaps 15 × 10⁻⁶ s^{-1}, so 9 orders of magnitude lower.

Following installation, the soil surrounding the anchor will re-consolidate, and the eventual anchor holding capacity can be estimated using standard approaches as used for pile foundations, although an α value of 0.8 for normally consolidated clay was found to give a better fit to data than taking α = 1 (O'Loughlin et al. 2004). However, because the anchor is a full-displacement device (inducing plastic strains over a larger distance into the soil domain), consolidation times can be substantial.

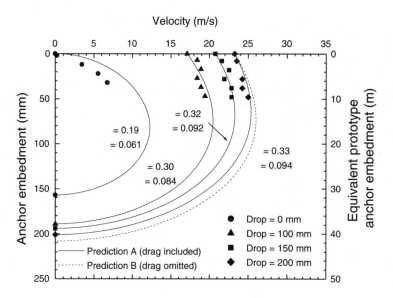

Figure 6.11 Predicted and measured velocity profiles (O'Loughlin et al. 2009 with permission from ASME).

Figure 6.12 shows data from centrifuge model tests on torpedo anchors that were load tested to failure at different times following installation (Richardson et al. 2009). The anchors were either released directly at the soil surface (so-called quasi-static or QS tests) or installed from the full drop height (dynamic or *DY*). The data are presented in terms of a nondimensional time factor $T = c_h t/d^2$, where c_h is the coefficient of consolidation estimated from piezocone dissipation tests (around 5.5 m²/yr for the kaolin used in the model tests). From these results, assuming values of c_h

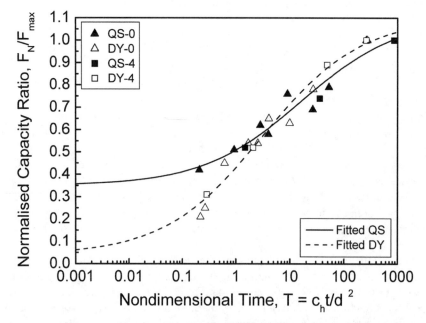

Figure 6.12 Dependence of normalized capacity ratio on consolidation time and extrapolated short term normalized capacity ratios (Richardson et al. 2009 with permission from ASCE).

in the range 10 to 30 m²/yr, the time required for 50% consolidation of a prototype torpedo anchor with a shaft diameter of 1.2 m would be 35 to 100 days. Similarly, the time for 90% consolidation would be about 2.5 to 7 years. These are much longer times than the 90% consolidation periods of 20 to 60 days suggested for 2.1-m-diameter open ended piles (Dutt and Ehlers 2009) and less than 90 days for larger diameter suction caissons (Jeanjean 2006). This longer setup time may be a result of water entrainment during the installation process, coupled with the larger *equivalent diameter* (in respect of the area of steel per unit length) compared with open-ended piles or suction caissons. The results suggest the need for caution in design assumptions regarding torpedo anchor capacity within the first few months following installation.

6.6 Conductor Fatigue

A significant problem for vertical top-tension risers (Figure 6.13) for deepwater facilities concerns the fatigue of the riser, particularly at the threaded connections where they are typically joined about every 9 m. An important part of the fatigue assessment for these risers is the lateral soil response. This response is typically modeled as discrete p-y springs.

As described by Jeanjean (2009), analytical assessments of a threaded connection about 9 m below the mudline, using API recommended values of the p-y springs, initially suggested a potentially short fatigue life. Since the fatigue assessment is very sensitive to the soil stiffness at small displacements, as described by Jeanjean (2009) and Templeton (2009), both centrifuge testing and finite element modeling were performed to investigate the p-y response and determine if the p-y approximations for piles were too conservative.

As shown on Figure 6.14, the agreement between the finite element and centrifuge test results was very good for static monotonic loading, with both methods showing that the API method

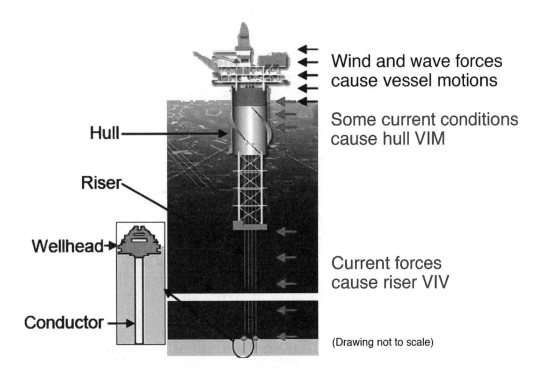

Figure 6.13 Riser/Conductor system for deepwater facility (Jeanjean 2009) © 2009 OTC. Reproduced with permission of the copyright owner. Further reproduction prohibited without permission.

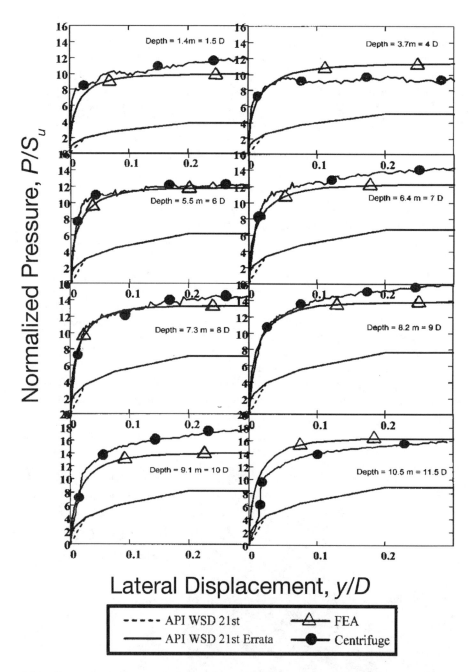

Figure 6.14   Comparison of *p-y* curves derived from centrifuge testing, 3-D finite element analyses vs. API (Jeanjean 2009) © 2009 OTC. Reproduced with permission of the copyright owner. Further reproduction prohibited without permission.

underpredicts both the soil stiffness at small displacements and the ultimate capacity. The changes to the *p-y* curves added significantly to the assessed fatigue life of the threaded connector.

In addition to the static monotonic tests performed in the centrifuge, cyclic tests were also performed in the centrifuge. The initial intent of these tests was to determine if low-level cyclic loading could strain harden the soil and provide enhanced soil stiffness during subsequent loading. As shown in Figure 6.15, the soil stiffness at small displacements did not appear to change for tests subsequent to small amplitude cyclic loading. However, the small amplitude cyclic loading did increase the ultimate limit state (plastic hardening) of the soil response at large displacements. This observation could have important implications on assessments of pile-founded platforms subject to large hurricane loads and may help explain, for instance, why the failure modes for platforms during some of the recent hurricanes in the Gulf of Mexico appeared more in the jacket legs than in the foundations themselves.

The other benefit derived from the cyclic tests was their use in determining the appropriate soil response during cyclic loading relative to the methods used in the analytical programs. Currently, most riser analysis programs used in design are capable of implementing linear springs. A few can now incorporate nonlinear work hardening springs. At this time, none can directly account for the effects of cyclic loading. It is also important to note that, in Figure 6.15, the cyclic unload-reload soil stiffness was never less than the tangent stiffness used in design.

Overall, the results of the centrifuge program to investigate the response of conductors used with top tension risers demonstrated that:

- API procedures for laterally loaded pile underpredicts both the soil stiffness at small displacements and the ultimate capacity under lateral loading.
- There was excellent agreement between finite element simulations of the conductor problem and the centrifuge results.
- Small amplitude cyclic loading did not result in increased soil stiffness at small displacements.

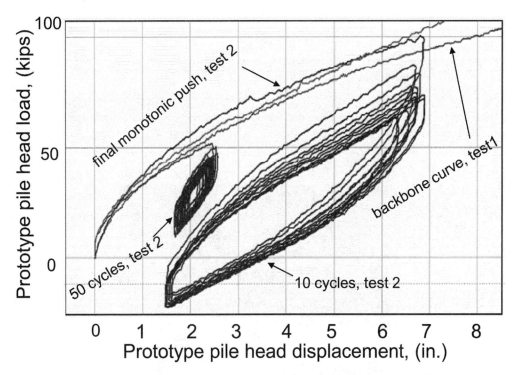

Figure 6.15   Cyclic load test data obtained for conductor fatigue problem (Jeanjean 2009) © 2009 OTC. Reproduced with permission of the copyright owner. Further reproduction prohibited without permission.

- Cyclic loading tests demonstrated that using the tangent stiffness derived from static loading is conservative when assessing the fatigue life of a threaded connector when compared to the analytical techniques used to address the problem.
- Prior cyclic loading did increase the ultimate lateral capacity of the conductor. This may have important implications for platform assessments under hurricane loads.

6.7 Flowlines and Pipelines

As mentioned in the introduction to this chapter, the increasing importance of flowlines, pipelines, and steel catenary risers (SCRs) in deepwater developments has focused attention on pipe-soil interactions (Bruton et al. 2006). Particular challenges include the low effective stress level and low shear strengths relevant for these interactions, which mostly occur within the upper 0.5 m of the seabed, and also the large cyclic motions undergone during thermally induced lateral buckling of flowlines and environmentally induced motions in the touchdown zone of SCRs.

Model testing, both on the laboratory floor and under centrifuge conditions, has been relied upon to develop design parameters covering pipeline embedment, axial friction, and lateral resistance during large displacements. Estimating pipeline embedment is critical since it has a marked effect on the axial and lateral design parameters and also on the thermal transfer response of the pipeline. This requires (Randolph and White 2008):

1. Accurate assessment of the intact and remolded strength profile in the upper meter of the seabed
2. Theoretical vertical bearing capacity solutions, including allowance for soil heave and buoyancy of the partially embedded pipeline
3. Determination of the amplified pipeline contact stress at the touchdown zone during installation
4. Allowance for the dynamic motions in the touchdown zone experienced by each segment of the pipeline during the installation process

Centrifuge model testing has made an essential contribution to evaluation of dynamic lay effects through generic studies and by means of site-specific testing programs for particular offshore developments. A typical model test arrangement is shown in Figure 6.16 with a pipe segment of 20 mm diameter tested at 40 *g*, hence modeling a 0.8 m diameter prototype. Two series of pipe lay

Figure 6.16   Typical centrifuge model test arrangement (courtesy of David White).

simulations were conducted in different soil types—kaolin clay and a high plasticity from offshore Angola. Both samples were reconsolidated on the centrifuge from a slurry to achieve normally consolidated strength profiles with a very small intercept at the mudline. Dynamic lay effects were modeled by applying a series of cyclic horizontal displacements while holding the vertical load constant (Cheuk and White 2011).

The results from a test on kaolin clay representing a pipe imposing a vertical load of 2.2 kN/m (a nominal bearing stress, V/D, of 2.75 kPa) on the seabed are summarized in Figure 6.17, showing the pattern of cyclic horizontal motion, the resulting embedment, and also the corresponding vertical and horizontal loads. Note that these have been normalized by dividing by the pipeline diameter and the shear strength at the pipeline invert, hence as the pipeline embeds further, the value of V/Ds_u decreases from an initial value of 3 (static embedment) to below 2. Reduced vertical capacity due to the combined horizontal and vertical loading, softening of the soil due to remolding, and physical plowing of the soil during the horizontal motion all contribute to the dynamic embedment of the pipe.

The results from four tests, two in each type of clay, are shown in Figure 6.18. These summarize the influence of vertical load on the rate of embedment for the selected sequence of horizontal motion and also highlight a significant difference in the behavior of the two clay types. There is generally greater additional embedment for the tests in high plasticity clay, perhaps a result of water entrainment and hence greater softening of this clay compared with the kaolin, even though the sensitivity observed from cyclic T-bar tests was similar for the two clays.

Once in operation, a flowline may buckle laterally as a result of thermal expansion and internal pressurization. The lateral displacements may be several meters, with soil berms developing at the extreme lateral excursions of the flowline during repeated heating and cooling cycles. A schematic of the resulting lateral resistance and the development of berms at either end of the range of motion is shown in Figure 6.19. A detailed snapshot of a berm from a centrifuge model test is illustrated in Figure 6.20.

The pattern of lateral resistance consists of: an initial brittle peak at the start of each change in direction due to transient suction on the trailing edge of the pipe; a steady state (or sometimes slightly rising) resistance during the main part of the motion; and then an abrupt rise in resistance as the berm is contacted at the extremities of the cyclic displacement range. With incremental

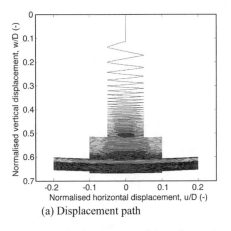

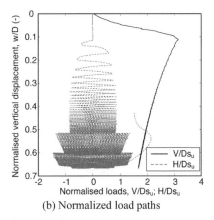

(a) Displacement path   (b) Normalized load paths

Figure 6.17 Results from tests modeling the pipeline lay process (Cheuk and White 2011, with permission from ICE Publishing).

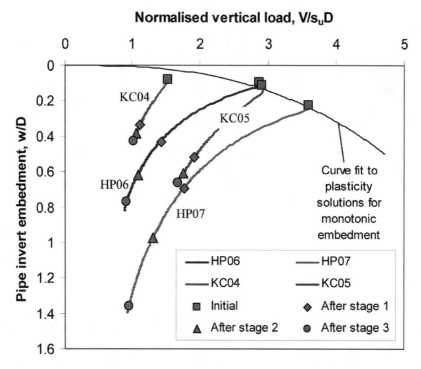

Figure 6.18   Accumulation of pipe embedment due to clay effects (Cheuk and White 2011, with permission from ICE Publishing).

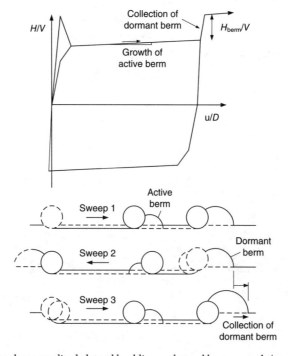

Figure 6.19.   Modeling large amplitude lateral buckling cycles and berm growth (courtesy of David White).

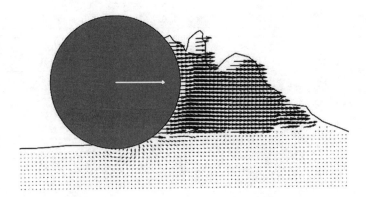

Figure 6.20.   Modeling large amplitude lateral buckling cycles and berm growth (White et al. 2007, with permission from David White).

embedment over many cycles, the range before the increase in resistance will gradually decrease due to accumulation of the berm. Theoretical attempts to evaluate each component of this pattern of lateral resistance are still being developed, and there is currently heavy reliance on empirical fits to data from physical model tests.

6.8 Facilities, Logistics, and Rationale for Centrifuge Modeling

Details of the more than 100 centrifuge facilities in operation worldwide are provided on the website of the Technical Committee on Physical Modeling in Geotechnics of the International Society for Soil Mechanics and Geotechnical Engineering (www.tc2.civil.uwa.edu.au). Only a fraction of these are currently active in modeling deepwater applications for the oil and gas industry, and these are summarized in Table 6.1—note that this is not an inclusive list but reflects those most active over the last decade or so. There is a separate group of centrifuge facilities that have tended to specialize in modeling seismic soil-structure response, details of which may be found on the NEES website.

Sizes of containers range from plan areas of 0.25 m² to 2 m², with sample depths of 0.3 to 1 m, but often the headroom above the soil surface is a critical dimension. Characterization of the sample is achieved by penetrometer testing with cones (in some cases piezocones), typically of 10 mm to 13 mm diameter, used for sand (sometimes also in clay) and predominantly T-bar (Stewart and Randolph 1991, 1994) or ball penetrometers used for clay. The T-bar penetrometer factor relating bearing resistance (as a net pressure) to shear strength is generally taken as 10.5 or 11. The small diameter of the T-bar (commonly 5 mm) allows superior definition of the strength profile in the upper part of the soil sample, although for pipeline modeling, where strength definition in the upper 5 to 20 mm of the sample is critical, a shallow penetration correction to the T-bar factor is required (White et al. 2010).

A successful program of centrifuge model tests requires significant planning by the client who should develop a draft specification for the tests to be undertaken with clear objectives identified. Often a tender document will be prepared and submitted to three or four potential centrifuge facilities. Detailed planning then proceeds with the manager of the preferred facility establishing final scaling levels for the model, sample preparation and characterization strategies, and the testing arrangements in each soil sample. The extent to which interpretation of the test data is undertaken by the centrifuge team, or left for the client, should also be clarified. A minimum report to the client will provide full details of the testing, with plotting and digital files of the data acquired

Table 6.1 Details of centrifuge facilities active in offshore applications

Country	Location	Radius (m)	Max. g level	Payload (kg)	Capacity (g-tonne)
Australia	University of Western Australia,[1] Perth	1.8	200	200	40
Brazil	State University of Norte Fluminense, Rio de Janeiro	3.0	100	1000	100
Canada	C-CORE St John's	5.0	100	2200	220
France	Laboratoire Centrale des Ponts et Chaussées, Nantes	5.5	100	1200	200
Holland	Deltares Delft	6.0	300	5500	1650
Singapore	National University of Singapore, Singapore	1.9	200	400	40
United Kingdom	Cambridge University,[1] Cambridge	4.1	150	2000	150
	Dundee University, Dundee	3.2	150	1500	150
United States	University of Colorado, Boulder	6.0	200	2000	400
	Rensselaer Polytechnic Institute, New York	3.0	200	1000	100

Note: [1]These universities also operate drum centrifuges of various sizes.

and sufficient analysis of the data to establish their integrity and consistency. More detailed inter-pretation, if required, would generally be provided separately.

The full process from tendering to final delivery of data may take 12 months or more (mainly depending on the number of soil samples to be prepared and the preparation technique), although a more typical period would be around six months. For site-specific modeling to resolve an issue that is proving critical in design, a fast-track arrangement may be possible, particularly for facilities operating smaller centrifuges. Even so, a minimum period of about two months would generally be required in order to plan the tests, fabricate suitable models, conduct the tests, and report results.

Costs for a program of centrifuge model tests vary quite widely depending on the size of centrifuge being operated and the complexity of the tests. Often the cost is dictated more by the number of separate *boxes* (or samples) to be prepared rather than the number of tests, at least for applications where a number of tests can be conducted in each sample. Typically, a client should allow costs of $30,000 to $60,000 in 2010 prices per reported (successful) test.

The choice of soil in which to undertake the model tests will depend on whether the objective is to establish the generic response of a particular application, such as suction caissons, torpedo anchors, etc., or whether the modeling is directed at a specific project. In the former case, it has been common to adopt some form of kaolin as a generic clay and a locally available (with respect to the facility) silica or carbonate sand for coarser grained soils. The choice of kaolin, a somewhat silty clay, is partly because of its relatively high coefficient of consolidation, hence it allows mod-erate sample consolidation times. Layered profiles of sand and clay—for example, to examine suction caisson penetration through soft clay into sand—may be prepared with relative ease, even to the extent of using silicon oil as the pore fluid in the sand (to help scale consolidation times or seepage rates) and water in the clay.

Kaolin has a relatively low sensitivity (between 2 and 3) and a relatively high unit weight, so it may be preferable to use natural soil, recovered in bulk from offshore as box cores or gravity piston cores, particularly for the very low unit weight, high plasticity West African clays. These clays have a much lower consolidation coefficient, requiring long consolidation periods, but will provide a better match to prototype conditions.

Most deepwater sites have strength profiles that increase linearly with depth, sometimes with a slight crust at the surface, in the upper 0.5 m or so. The most direct method to achieve a linear strength profile is to consolidate the soil from a slurry *in-flight*, although this may take several days of centrifuge spinning. Any surface crust may be achieved by scraping off some of the surface material or even by temporarily lowering the water table to just below the soil surface for a short period, allowing slight desiccation.

The relative merits of physical (centrifuge) and numerical modeling have been discussed widely—see in particular Gaudin et al. (2010). The principal advantage of any model test is the use of *real* soil, enabling rheological aspects such as softening during cyclic loading or large shear strains to be captured naturally. Centrifuge model testing allows particular insight into applications involving large relative motion between structure and soil (e.g., penetration and keying of anchors, dynamic embedment and lateral sweeping of pipelines, transient uplift, and moment capacity of mudmats), since these are difficult to model with confidence using numerical techniques. The recent development of visualization techniques, such as particle image velocimetry (White et al. 2003), has added a new dimension to the quality of output from model testing.

The main limitations of centrifuge model testing are associated with scale effects, which include: the relative size of the structure and soil grains (e.g., in respect to the tip of a model suction caisson relative to sand grains); the high strain rates relative to prototype situations; and reduced consolidation times (proportional to the square of the linear dimension so n^2 faster for a 1:n model). Additional limitations relate to difficulties in simulating precisely a given soil profile, particularly if there is complex layering, and also in simulating the exact construction sequence that might be used offshore.

Ideally, numerical and physical modeling should be undertaken in parallel, with insight from both allowing development or calibration of a (simplified) calculation model for use in the main design. Although we live increasingly in a virtual world, crosschecking with the reality of soil behavior under complex conditions involving cyclic loading, partial remolding, water entrainment, etc., remains essential.

6.9 Conclusions

This chapter has highlighted recent uses of geotechnical centrifuge testing for deepwater geotechnical challenges. The technique has advanced considerably in the past 10 to 15 years and is now frequently used by both practitioners and researchers in the offshore community. Perhaps its use is no better illustrated than in the contribution it has made to the advancement and development of suction caisson technology. Suction caissons have become the dominant anchoring technology for anchorages of production facilities as well as various other types of subsea systems once water depths extended to greater than 1250 meters. In contrast to driven piles, which relied more on a variety of 1-g model test results to form the empirical data base for design, suction caissons have relied on centrifuge test results to form the basis for determining the holding capacity under a wide range of loading conditions.

In addition to suction caissons, centrifuge testing has helped establish the viability of other types of alternate anchorage systems such as SEPLAs and torpedo anchors. The technique has also been successfully used to determine the response of such diverse deepwater components as conductors, as well as flowlines and pipelines, when subjected to complex loadings from fatigue to thermal expansion.

The use of the centrifuge technique has clearly provided a valuable and cost effective tool for deepwater geotechnical engineering. However, potential users of results obtained with the technique need to be aware of the limitations of the technique. Scaling relationships for all aspects of a problem need to be examined to determine if the technique can be successfully employed and, if so, how the tests should be performed to derive the most benefit from testing. Opportunities to compare centrifuge test results with large-scale 1-*g* tests, field data, or robust analytical and numerical solutions are, of course, always encouraged.

Despite some potential limitations, the technique has proven to be a valuable asset in the toolkit of an offshore geotechnical engineer, augmenting other approaches to solve a complex array of difficult geotechnical problems. It is expected that the centrifuge community will continue to expand and offer further improved modeling capabilities that will enhance our knowledge and reduce limitations in understanding and designing for challenging deepwater conditions.

6.10 References

Andersen, K. H. and Jostad, H. P., 2002. Shear strength along outside wall of suction caisson in clay after installation, In *Proceedings of the 12th ISOPE conference* 2002-PCW-02, Kyushu, Japan.

Andersen, K. H., Jeanjean, P., Luger, D., and Jostad, H. P., 2005. Centrifuge tests on installation of suction anchors in soft clay, *Ocean Engineering.* 32(7): 845-863.

Araujo, J. B., Machado, R. D., and Medeiros, C. J., 2004. High holding power torpedo pile—Results for the first long term application, In *Proceedings of the 23rd International Conference Offshore Mechanics and Arctic Engineering, ASME,* Vancouver, OMAE2004-51201.

Audibert, J. M. E., Movant, M. N., Won, J. Y., and Gilbert, R. B., 2006. Torpedo piles: Laboratory and field research, In *Proceedings of the 16th International Offshore and Polar Engineering Conference,* 462-468, San Francisco, CA.

Brandão, F. E. N., Henriques, C. C. D., Araújo, J. B., Ferreira, O. C. G., and Amaral, C. D. S., 2006. Albacora Leste field development—FPSO P-50 mooring system concept and installation, In *Proceedings of the 38th Offshore Technology Conference,* Houston, TX, OTC 18243.

Bruton, D. A. S., White, D. J., Cheuk, C. Y., Bolton, M. D., and Carr, M. C., 2006. Pipe-soil interaction behaviour during lateral buckling, including large amplitude cyclic displacement tests by the Safebuck JIP, In *Proceedings of the 38th Offshore Technology Conference,* Houston, TX, OTC 17944.

Chen, W. and Randolph, M. F., 2007a. Radial stress changes and axial capacity for suction caissons in soft clay, *Géotechnique* 57(6): 499-511.

Chen, W. and Randolph, M. F., 2007b. Uplift capacity of suction caissons under sustained and cyclic loading in soft clay, *Journal of Geotechnical and GeoEnvironmental Engineering* 133(11): 1352-1363.

Cheuk, C. Y. and White D. J., 2011. Modelling the dynamic embedment of seabed pipelines, *Géotechnique* 61(1): 39-57.

Clukey, E. C. and Morrison, M. J., 1993. A centrifuge and analytical study to evaluate suction caissons for TLP applications in the Gulf of Mexico, In *Proceedings of the ASCE Conference on Foundations, Geotechnical Special Publications* 38:141-156.

Clukey, E. C. and Phillips, R., 2002. Centrifuge model tests to verify suction caisson capacities for taut and semi-taut legged mooring systems, In *Proceedings of the Deep Offshore Technology Conference,* New Orleans.

Clukey, E. C., Aubeny, C. P., and Murff, J. D., 2003. Comparison of analytical and centrifuge model tests for suction caissons subjected to combined loads, In *Proceedings of the 22nd International Conference on Offshore Mechanics and Arctic Engineering,* OMAE-37503.

Clukey, E. C., Morrison, M. J., Garnier, J., and Corté, J. F., 1995. The response of suction caissons in normally consolidated clays to cyclic TLP loading conditions, In *Proceedings of the 27th Offshore Technology Conference,* Houston, TX, OTC 7796.

Clukey, E. C., Templeton, J. S., Randolph, M. F., and Phillips, R. A., 2004. Suction caisson response under sustained loop-current loads, In *Proceedings of the 36th Offshore Technology Conference,* Houston, TX, OTC 16843.

Colliat, J-L., Boissard, P., Andersen, K. H., and Schroeder, K., 1995. Caisson foundations as alternative anchors for permanent mooring of a process barge offshore Congo, In *Proceedings of the 27th Offshore Technology Conference,* Houston, TX, OTC 7797.

Colliat, J-L., Boissard, P., Gramet, J-C., and Sparrevik, P., 1996. Design and installation of suction anchor piles at a soft clay site in the Gulf of Guinea, In *Proceedings of the 28th Offshore Technology Conference*, Houston, TX, OTC 8150.

Dove, P., Treu, H., and Wilde, B., 1998. Suction embedded plate anchor (SEPLA): A new anchoring solution for ultra-deepwater mooring, In *Proceedings of the Deep Offshore Technology Conference*, New Orleans.

Doyle, E. H., Dean, E. T. R., Sharma, J. S., Bolton, M. D., Valsangkar, A. J., and Newlin J. A., 2004. Centrifuge model tests on anchor piles for tension leg platforms, In *Proceedings of the 36th Offshore Technology Conference*, Houston, TX, OTC 16845.

Dutt, R. N. and Ehlers, C. J., 2009. Set-up of large diameter driven pipe piles in deepwater normally consolidated high plasticity clays, In *Proceedings of the 28th International Conference on Offshore Mechanics and Arctic Engineering*, Honolulu, HI, OMAE2009-79012.

Dyvik, R., Andersen, K. H., Hansen, S. B., and Christophersen, H. P., 1993. Field tests of anchors in clay, *Journal of Geotechnical Engineering* 119(10): 1515-1531.

Erbrich, C. and Hefer, P., 2002. Installation of the Laminaria suction piles—A case history, In *Proceedings of the 34th Offshore Technology Conference*, Houston, TX, OTC 14240.

Finnie, I. M. S. and Randolph, M. F., 1994. Punch-through and liquefaction induced failure of shallow foundations on calcareous sediments, In *Proceedings of the International Conference on Behaviour of Offshore Structures*, 217-230, Boston.

Garnier, J., Gaudin, C., Springman, S. M., Culligan, P. J., Goodings, D., Konig, D., Kutter, B., Phillips, R., Randolph, M. F., and Thorel, L., 2007. Catalogue of scaling laws and similitude questions in centrifuge modeling, *International Journal of Physical Modelling in Geotechnics* 7(3): 1-24.

Gaudin, C., Clukey, E. C., Garnier, J., and Phillips, R., 2010. New frontiers for centrifuge modelling in offshore geotechnics, In *Proceedings of the 2nd International Symposium on Frontiers in Offshore Geotechnics*, Perth, ISFOG2010.

Gaudin, C., O'Loughlin, C. D., Randolph, M. F., and Lowmass, A., 2006. Influence of the installation process on the performance of suction embedded plate anchors, *Géotechnique* 56(6): 381-391.

Jeanjean, P., 2006. Set-up characteristics of suction anchors for soft Gulf of Mexico clays: Experience from field installation and retrieval, In *Proceedings of the 38th Offshore Technology Conference*, Houston, TX, OTC 18005.

Jeanjean, P. J., 2009. Development of *p-y* springs for fatigue analyses, In *Proceedings of the 41st Offshore Technology Conference*, Houston, TX, OTC 20158.

Jeanjean, P., Znidarcic, D., Phillips, R., Ko, H. Y., Pfister, S., and Schroeder, K., 2006. Centrifuge testing on suction anchors: Doublewall, stiff clays, and layered soil profile, In *Proceedings of the 38th Offshore Technology Conference*, Houston, TX, OTC 18007.

Jeffries, M. G. and Been, K., 2006. *Soil liquefaction a critical state approach*, Taylor and Francis, New York and London.

Kutter, B. L. and Balakrishnan, A., 2000. Dynamic model test data from electronics to knowledge, Keynote lecture at Centrifuge 98, 2:931-943, Balkema, Rotterdam.

Lieng, J. T., Hove, F., and Tjelta, T. I., 1999. Deep penetrating anchor: Subseabed deepwater anchor concept for floaters and other installations, In *Proceedings of the 9th International Offshore and Polar Engineering Conference*, 613-619, Brest.

Medeiros, C. J., 2001. Torpedo anchor for deep water, In *Proceedings of the Deepwater Offshore Technology Conference*, Rio de Janeiro.

Medeiros, C. J., 2002. Low cost anchor system for flexible risers in deep waters, In *Proceedings of the 34th Offshore Technology Conference*, Houston, TX, OTC 14151.

Mello, J. R. C., Moretti, J., Sparrevik, P., Schroeder, K., and Hansen, S. B., 1998. P19 and P26 moorings at the Marlim Field: The first permanent taut leg mooring with fibre ropes and suction anchors, In *Proceedings of the Conference on Floating Production Systems, FPS'98*, London.

Mitchell, R. J., 1993. *Fundamentals of Soil Behavior*, 2nd ed., John Wiley and Sons, New York.

Morrison, M. J., Clukey, E. C., and Garnier, J., 1994. Behaviour of suction caissons under static uplift loading, In *Proceedings of the International Conference Centrifuge 94*, 823-828, Singapore.

Murff, J. D., 1996. The geotechnical centrifuge in offshore engineering, In *Proceedings of the 28th Offshore Technology Conference*, Houston, TX, OTC 8265.

O'Loughlin, C. D., Randolph, M. F., and Richardson, M., 2004. Experimental and theoretical studies of deep penetrating anchors, In *Proceedings of the 35th Offshore Technical Conference*, Houston, TX, OTC 16841.

O'Loughlin, C. D., Richardson, M. D., and Randolph, M. F., 2009. Centrifuge tests on dynamically installed anchors, In *Proceedings of the International Conference on Offshore Mechanics and Arctic Engineering*, Honolulu, HI, OMAE2009-80238.

Ovesen, N. K., 1979. The scaling law relationship: Panel discussion, In *Proceedings of the 7th European Conference on Soil Mechanics and Foundation Engineering* 4:319-323, Brighton.

Palmer, A. C., White, D. J., Baumgard, A. J., Bolton, M. D., Barefoot, A. J., Finch, M., Powell, T. A., Faranski, A. S., and Baldry, J. A. S., 2003. Uplift resistance of buried submarine pipelines: Comparison between centrifuge modelling and full-scale tests, *Géotechnique* 53(10): 877-883.

Randolph, M. F. and House, A. R., 2002. Analysis of suction caisson capacity in clay, In *Proceedings of the 34th Offshore Technology Conference*, Houston, TX, OTC 14236

Randolph, M. F., O'Neill, M. P., Stewart, D. P., and Erbrich, C. T., 1998. Performance of suction anchors in fine-grained calcareous soils, In *Proceedings of the 30th Offshore Technology Conference*, Houston, TX, OTC 8831.

Randolph, M. F. and White, D. J., 2008. Pipeline embedment in deep water: Processes and quantitative assessment, In *Proceedings 40th Offshore Technology Conference*, Houston, TX, OTC 19128.

Richardson, M. D., 2008. Dynamically installed anchors for floating offshore structures, PhD thesis, University of Western Australia.

Richardson, M. D., O'Loughlin, C. D., Randolph, M. F., and Gaudin, C., 2009. Setup following installation of dynamic anchors in normally consolidated clay, *Journal of Geotechnical and GeoEnvironmental Engineering* 135(4): 487-496.

Schofield, A. N., 1980. Cambridge geotechnical centrifuge operations, *Géotechnique* 30(3): 227-268.

Stewart, D. P. and Randolph, M. F., 1991. A new site investigation tool for the centrifuge, In *Proceedings of the International Conference on Centrifuge Modelling*, 531-538, Boulder, CO.

Stewart, D. P. and Randolph, M. F., 1994. T-Bar penetration testing in soft clay, *Journal of Geotechnical Engineering Division*, ASCE, 120(12): 2230-2235.

Støve, O. J., Bysveen, S., and Christophersen, H. P., 1992. New foundation systems for the Snorre development, In *Proceedings of the 24th Offshore Technology Conference*, Houston, TX, OTC 6882.

Supachawarote, C., Randolph, M. F., and Gourvenec, S., 2005. The effect of crack formation on the inclined pull-out capacity of suction caissons, In *Proceedings of the Conference of the International Association of Computer Methods and Analysis in Geomechanics*, IACMAG-05, 577-584, Torino.

Taylor, R. N., ed., 1995. *Geotechnical centrifuge technology*, Blackie Academic Press, London.

Templeton J. S., 2009. Finite element analysis of conductor/seafloor interaction, In *Proceedings of the 41st Offshore Technology Conference*, Houston, TX, OTC 20197.

Wang, D., Hu, Y., and Randolph, M. F., 2009. Keying of rectangular plate anchors in normally consolidated clays, Paper Submitted for Review.

Wesselink, B. D., Murff, J. D., Randolph, M. F., Nunez, I. L., and Hyden, A. M., 1988. Analysis of centrifuge model test data from laterally loaded piles in calcareous sand, *Engineering for Calcareous Sediments*, 1:261-270, Perth.

White D. J. and Cheuk C. Y., 2008. Modelling the soil resistance on seabed pipelines during large cycles of lateral movement, *Marine Structures* 21(1): 59-79.

White D. J., Dingle H. R. C., and Gaudin C., 2007. SAFEBUCK JIP Phase II: Centrifuge modelling of pipe-soil interaction: Factual report, Report for Boreas Consultants, UWA, GEO 07396.

White, D. J., Gaudin, C., Boylan, N., and Zhou, H., 2010. Interpretation of T-bar penetrometer tests at shallow embedment and in very soft soils, *Canadian Geotechnical Journal* 47(2): 218-229.

White, D. J., Take, W. A., and Bolton, M. D., 2003. Soil deformation measurement using Particle Image Velocimetry (PIV) and photogrammetry, *Géotechnique* 53(7): 619-631.

Wilde, B., Treu, H., and Fulton, T., 2001. Field testing of suction embedded plate anchors, In *Proceedings of the 11th International Offshore and Polar Engineering Conference* 2:544-551.

Zimmerman, E. H., Smith, M. W., and Shelton J. T., 2009. Efficient gravity installed anchor for deepwater mooring, In *Proceedings of the 41st Offshore Technology Conference*, Houston, TX, OTC 20117.

7

Reliability of Offshore Foundations

Robert B. Gilbert, PhD, PE
Brunswick-Abernathy Professor
The University of Texas at Austin

7.1 Introduction

Reliability is an important consideration in planning, designing, constructing, and operating off-shore facilities. The consequences of a failure for these facilities can be severe, and the costs to increase the reliability can be large. An assessment and analysis of the reliability helps to balance the risks of failure with the costs of conservatism.

Reliability analyses for offshore facilities have been conducted since at least the 1960s. These analyses have been used to (1) compare alternative designs, (2) develop designs and design guidance to achieve target levels of reliability, (3) optimize resources for planning and for operation and maintenance, and (4) understand and quantify how uncertainty impacts the performance of components and systems.

Applications include a variety of different types of foundations for structures: driven piles for steel jackets (Marshall and Bea 1976; Anderson et al. 1982; Stahl 1986; Tang et al. 1990; Tang and Gilbert 1993; Horsnell and Toolan 1996; Bea et al. 1999), shallow foundations for gravity-based structures (e.g., Kraft and Murff 1975; Wu et al. 1989), shallow foundations for jack-up drill rigs (e.g., Nadim and Lacasse 1992), tension piles for tension leg platforms (e.g., Ronold 1990) and suction caisson anchors for floating production systems (Clukey and Banon 2000; Gilbert et al. 2005). With facilities moving into deeper water, there has been greater emphasis recently on the reliability of seafloor facilities such as pipelines and flowlines (e.g., Muhlbauer 2004; Sweeney 2005; Brown et al. 2006). Another common application for offshore reliability analyses is for natural hazards such as earthquakes (e.g., Bruschi et al. 1996), submarine slopes (e.g., Bea 1998; Nadim et al. 2003; Nodine et al. 2009) and ice loads (e.g., Wu et al. 1989).

The objective of this chapter is to describe and illustrate practical means for assessing and analyzing the reliability of offshore foundations. The emphasis is on fundamental principles and methods that can be applied to a wide variety of project-specific problems and decisions. Examples based on real-world analyses are presented to demonstrate the implementation of these fundamentals; while the examples focus specifically on foundations for production facilities, the principles and methods being illustrated are general.

7.2 Methods for Reliability Analysis

The objective of a reliability analysis is to assess the probability that the foundation will perform adequately:

$$Reliability = P(Foundation\ Performs\ Adequately) \tag{7.1}$$

where $P(Foundation\ Performs\ Adequately)$ is the probability of the event that the foundation performs adequately. It is important to carefully define the event *Foundation Performs Adequately* at the start:

- What is the intended purpose of the foundation?
- Over what time period is the foundation intended to perform adequately?
- What are the sources of uncertainty in whether or not the foundation will perform adequately?
- Are there different levels of inadequate performance?
- What are the consequences of inadequate performance?
- Can inadequate performance be detected and are there early indications?
- What is the target reliability for the foundation?

Example 1. Performance Considerations for Riser Foundation

Consider a suction caisson foundation that will hold a deepwater riser down to the sea floor. The intended purpose is to hold the buoyant riser in place when it is installed, when it is in operation with product flowing through it, and when it is shut in for maintenance. The tension load on the foundation is greatest during installation of the riser. The expected period of operation is 20 years. When the riser is first installed, it may be left for up to 6 months before it is put into operation. The time between installation of the suction caisson and attachment of the riser is not known but could be as short as 24 hours. There is uncertainty in the load due to variations in weights about the specified values, variations in environmental loads with time, uncertainty in the metocean characteristics, variations in the densities of the fluids in the riser, and the possibility of not setting the buoyancy properly during installation. There is uncertainty in the capacity due to uncertainty in the soil properties, uncertainty in the model used to predict uplift capacity, and uncertainty in the increase of capacity with time after installation. *Inadequate Performance* is that the uplift capacity of the suction caisson, the caisson pulls out of the sea floor, and the riser pops up to the water surface. There is a large tolerance in vertical displacement of the suction caisson if there is not an uncontrolled release of the riser. The consequences of inadequate performance include damage to the riser and the installation vessel, delay in production, and possible death or injury to the installation crew. An estimated equivalent cost of failure is between \$100 million and \$1 billion. It would be possible but very expensive to monitor the displacement of the suction caisson when the riser is attached and to have contingency measures in place to reduce the buoyant load of the riser in the event that an uplift failure of the suction caisson seems imminent. The target annual reliability is a probability of at least $1.0 - 1.0 \times 10^{-4} = 0.9999$ that the foundation will perform adequately.

The event that the *Foundation Performs Adequately* can generally be expressed as the event that the capacity of the foundation is greater than the load:

$$Reliability = P(Capacity > Load) \qquad (7.2)$$

For convenience in maintaining significant figures in making calculations and communicating results, the reliability is typically cast in terms of the probability of failure:

$$Reliability = 1 - P(Capacity \le Load) = 1 - P(Failure\ to\ Perform\ Adequately) = 1 - p_F \quad (7.3)$$

where p_F is the probability that the foundation fails to perform adequately. Methods for calculating the probability of failure are summarized in the following sections.

7.2.1 Simplified Analytical Models

The classic reliability model is for a capacity and load that are each represented as normally distributed, statistically independent random variables (e.g., Benjamin and Cornell 1970; Ang and Tang 1975):

$$p_F = P(\text{Capacity} \le \text{Load}) = P(R \le S) = \Phi\left(-\frac{\mu_R - \mu_S}{\sqrt{\sigma_R^2 + \sigma_S^2}}\right) = \Phi(-\beta) \tag{7.4}$$

where R is the capacity with a mean value of μ_R and a standard deviation of σ_R, S is the load with a mean value of μ_S and a standard deviation of σ_S, β is called the reliability index, and $\Phi(.)$ is the standard normal function, which is widely tabulated in textbooks and in spreadsheet programs. The relationship between the reliability index and the probability of failure is shown in Figure 7.1.

Normal distributions can generally provide a rough representation of uncertainty for capacities and loads because values relatively close to the mean are the most likely, and values more than several standard deviations away from the mean are not likely. However, normal distributions do allow for the possibility of negative values, which may not be realistic. An extension of this reliability model is for a capacity and load that are each represented as lognormally distributed, statistically independent variables; a lognormal distribution is asymmetrical and only allows for the possibility of positive values. In this case, the probability of failure is obtained as follows:

$$p_F = P(\text{Capacity} \le \text{Load}) = P(R \le S) = \Phi\left(-\frac{\ln(r_{\text{median}}/s_{\text{median}})}{\sqrt{\delta_R^2 + \delta_S^2}}\right) = \Phi(-\beta) \tag{7.5}$$

where r_{median} and s_{median} are the median values for the capacity and load, respectively, and δ_R and δ_S are the coefficient of variation values for the capacity and load, respectively. The median and

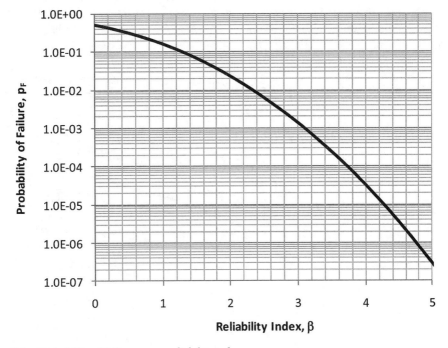

Figure 7.1 Probability of failure versus reliability index.

coefficient of variability value for a lognormally distributed random variable are obtained from the mean value and the standard deviation as follows:

$$x_{median} = e^{\ln(\mu_X) - (1/2)\ln(1 + \delta_X^2)} \tag{7.6}$$

$$\delta_X = \frac{\sigma_X}{\mu_X} \tag{7.7}$$

A convenient simplification for calculating the probability of failure with lognormal variables (Equation 7.5) is to use a median factor of safety and a total coefficient of variation:

$$p_F = \Phi\left(-\frac{\ln(FS_{median})}{\delta_{total}}\right) = \Phi(-\beta) \tag{7.8}$$

where $FS_{median} = r_{median}/s_{median}$ and $\delta_{total} = \sqrt{\delta_R^2 + \delta_S^2}$. The median factor of safety is related to the factor of safety or the load and resistance factors used to check the design in a typical code format as follows:

$$FS_{median} = FS_{design} \frac{(r_{median}/r_{design})}{(s_{median}/s_{design})} \tag{7.9}$$

or

$$FS_{median} = \left(\frac{\psi_S}{\phi_R}\right) \frac{(r_{median}/r_{design})}{(s_{median}/s_{design})} \tag{7.10}$$

where FS_{design} is the factor of safety in an allowable stress design format (e.g., API 2000), ψ_S and ϕ_R are the load and resistance factors in a limit state or load and resistance factor design format (e.g., API 1991), and r_{design} and s_{design} are the nominal values for the foundation capacity and load, respectively, that would be used in the design check.

The ratios of median to design values for the capacity and load indicate any biases between the median or most likely value and the value assumed for design. The total coefficient of variation in Equation 7.8 indicates the overall uncertainty in both the load and the capacity. Figure 7.2 shows

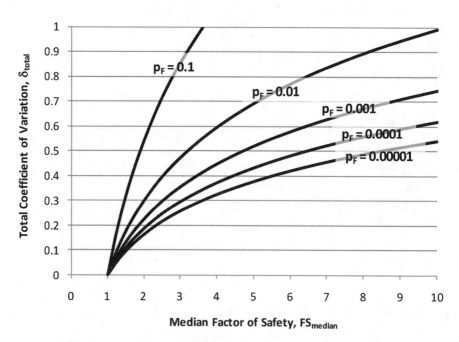

Figure 7.2 Probability of failure for lognormal load and capacity.

how the probability of failure is related to the median factor of safety and the total coefficient of variation for lognormal variables.

Example 2. Resistance Factor for Driven Pile Design with Jacket Structure

Consider a driven steel pipe pile for a fixed jacket structure in 200 m of water. The design life for the structure is 20 years. The event of concern is that the pile will fail axially in tension due to the maximum overturning moment that occurs on the jacket during a hurricane. The target reliability is for a probability of failure in the design life that is less than 0.001. The structure is located in a new geologic setting with clays that have higher void ratios but similar undrained shear strengths compared to typical normally consolidated marine clays, meaning that there is uncertainty about whether to treat them as normally consolidated or overconsolidated in estimating axial pile capacity. The objective is to establish an appropriate resistance factor for a load and resistance factor design check.

The nominal design load corresponds to the 100-year combination of winds, waves, and currents: s_{design} = 18 MN. The median value for the maximum load in a 20-year exposure is approximately 0.7 times the design load: s_{median}/s_{design} = 0.7. Due to variability in the occurrence and intensity of hurricanes, the coefficient of variation in the load is 0.3: δ_S = 0.3. The load factor is 1.5.

The coefficient of variation in the design axial pile capacity is taken to be 0.4, which is relatively large due to the unusual geologic setting: δ_R = 0.4. Also, the ratio of the median capacity to the design capacity is assumed to be 1.0: r_{median}/r_{design} = 1.0.

The total coefficient of variation is $\delta_{total} = \sqrt{\delta_R^2 + \delta_S^2} = \sqrt{0.4^2 + 0.3^2} = 0.5$.

From Figure 7.2, the required value of the median factor of safety for a probability of failure of 0.001 is 4.7. From Equation 7.10, the required resistance factor is then:

$$\phi_R = \left(\frac{\psi_S}{FS_{median}}\right)\frac{(r_{median}/r_{design})}{(s_{median}/s_{design})} = \left(\frac{1.5}{4.7}\right)\frac{1.0}{0.7} = 0.46.$$

In other words, the design value for the axial pile capacity needs to be at least (1.5/0.46) = 3.3 times the design value for the axial load, or $r_{design} \geq 3.3(18.0 \text{ MN})$ = MN. The equivalent factor of safety, 3.3, is relatively large in this case due to the large uncertainty in the axial capacity in this unusual geologic setting.

Example 3. Reliability Assessment for Driven Piles Supporting Tension Leg Platform

Consider a driven pile for a tension leg platform in 1000 m of water. The design life for the structure is 20 years. The event of concern is that the pile will fail axially in tension due to the maximum tendon load that occurs during a swell (a specific design sea state). The target reliability is for a probability of failure in the design life that is less than 0.0001. The steel piles have already been ordered based on a preliminary design check, but refinements to the design loads have raised a concern that longer piles will be needed. The objective is to assess the probability of failure for the existing design to decide whether or not the pile lengths need to be changed.

The nominal design load corresponds to the 100-year combination of winds, waves, and currents: s_{design} = 20 MN. The median value for the maximum load in a 20-year exposure is approximately 0.8 times the design load: s_{median}/s_{design} = 0.8. Most of this bias is due to a compounding of conservative assumptions about coincident directions for waves and winds, variations in pretension, and seafloor subsidence. Since most of the total tendon load is pretension and not from environmental forces, the coefficient of variation in the load is very small δ_S = 0.1.

With the existing piles that have already been ordered, the design value for the axial pile capacity is 56 MN, giving a design factor of safety of 56/20 = 2.8. The intent was to achieve a larger design factor of safety of 3.0, which is the motivation in assessing the reliability. The coefficient of variation in the design axial pile capacity is taken to be 0.25. Also, the ratio of the median capacity to the design capacity is assumed to be 1.0: $r_{median}/r_{design} = 1.0$. The total coefficient of variation is $\delta_{total} = \sqrt{\delta_R^2 + \delta_S^2} = \sqrt{0.25^2 + 0.1^2} = 0.27$. The median factor of safety is obtained from Equation 7.9:

$$FS_{median} = FS_{design}\frac{(r_{median}/r_{design})}{(S_{median}/S_{design})} = (2.8)\frac{1.0}{0.8} = 3.5.$$

From Figure 7.2, probability of failure is well below the target value of 0.0001.

The probability of failure is small in this case because there is relatively little uncertainty in the tendon load for this tension leg platform. This analysis supports a decision to stay with the initially specified lengths for the piles because the reliability is acceptable even though the factor of safety is not as large as intended.

7.2.2 Analytical Approximation—First Order Reliability Method

In more general problems, the probability of failure can be expressed as follows:

$$p_F = P[g(\overline{X}) \leq 0] \tag{7.11}$$

where $g(\overline{X})$ is a function that indicates the performance of the foundation, taking on values less than zero when the foundation fails, and $\overline{X}$ is a vector of n random variables upon which the performance depends (the input variables). The First Order Reliability Method, or FORM, provides a useful technique to approximate the probability of failure analytically for complex cases where the performance function can be a nonlinear function of the input variables and the input variables can be correlated and can have marginal probability distributions of any type. With this method, the probability of failure is approximated as follows (e.g., Ang and Tang 1984; Madsen et al. 1986; Thoft-Christensen and Murotsu 1986; and Melchers 1987):

$$p_F \cong \Phi(-\beta) \tag{7.12}$$

where the reliability index β is given by

$$\beta = -\frac{\frac{\partial g}{\partial s}\Big|_{s^*}^T \overrightarrow{s^*}}{\left(\frac{\partial g}{\partial s}\Big|_{s^*}^T [C_s]\frac{\partial g}{\partial s}\Big|_{s^*}\right)^{1/2}} \tag{7.13}$$

and

- $\overrightarrow{s^*}$ is a vector with standard normal transformed values from a particular realization of the input variables $\overrightarrow{x^*}$

$$s^*_i = \Phi^{-1}[F_{X_i}(x^*_i)] \tag{7.14}$$

where $Fx_i(x^*_i)$ is the cumulative distribution function for input variable i;

- C_S is a matrix with correlation coefficients, $\rho x_i\, x_j$, between all pairs of the input variables X_i and X_j,

$$C_S = \begin{bmatrix} \rho_{X_1, X_1} & \cdots & \rho_{X_1, X_N} \\ \vdots & \ddots & \vdots \\ \rho_{X_N, X_1} & \cdots & \rho_{X_N, X_N} \end{bmatrix} \tag{7.15}$$

- and $\overline{\dfrac{\partial g}{\partial s}\Big|_{s^*}}$ is a vector with partial derivatives of the performance function $g(\overrightarrow{X})$ with respect to the standard normal transformed values of the input variables evaluated at $\overrightarrow{s^*}$:

$$\frac{\partial g}{\partial s_i}\bigg|_{s^*} = \left(\frac{\partial g}{\partial x_i}\bigg|_{x^*}\right)\frac{\phi\{\Phi^{-1}[F_{X_i}(x^*_i)]\}}{f_{X_i}(x^*_i)} \qquad (7.16)$$

where $\dfrac{\partial g}{\partial x_i}\Big|_{x^*}$ is the partial derivative of the performance function evaluated at the set of input values in $\overline{x^*}$, $\phi(.)$ is the standard normal probability density function, and $f_{X_i}(x^*_i)$ is the probability density function for input variable i.

The key to this method is finding the realization of input variables $\overline{x^*}$ such that Equation 7.12 provides a reasonable approximation. This point in the n-dimensional space of $\overline{X}$ is found by minimizing β with respect to $\overline{x}$ subject to $g(\overline{x}) = 0$. The set of values in $\overline{x^*}$ is referred to as the Most Probable Point, and it physically represents the most likely combination of input variables that will lead to the event of failure.

Example 4. FORM Analysis for Suction Caissons Anchors in Spar Mooring System

Consider the suction caisson for the mooring system of a spar in 1000 m of water. The design life for the structure is 20 years. The event of concern is that the caisson will fail to hold the line load during a hurricane. The characterization of possible sea states due to hurricanes at this location was updated in 2006 based on new data from recent hurricanes. The objective is to evaluate the probability of failure for a design that was based on the older, pre-2006 characterization of sea states.

The load on the suction caisson is described by the tension and the angle in the mooring line at the seafloor, T and θ (Figure 7.3). The maximum tension on the most heavily loaded mooring line during a 3-hour sea state in a hurricane is related to the significant wave height for that sea state (Figure 7.4). The angle of the line at the seafloor is related to the tension in the line (Figure 7.5). The relationships in Figures 7.4 and 7.5 for this spar were developed using a numerical model (OTRC 2006) that couples the hydrodynamic loading on the hull with dynamics of the mooring and riser system.

The updated characterization of possible sea states due to a hurricane is shown in Figure 7.6; this information has been included in updated design guidance for this

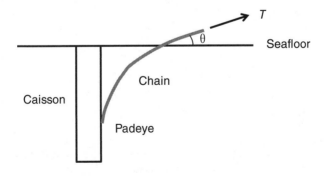

Figure 7.3 Mooring line load on suction caisson.

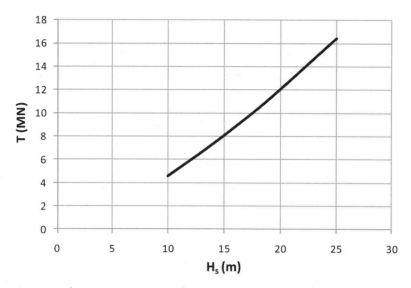

Figure 7.4   Maximum line tension versus significant wave height in 3-hour sea state.

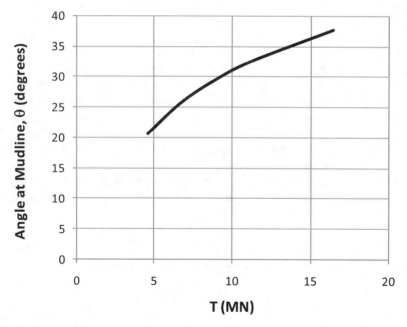

Figure 7.5   Angle at seafloor versus tension in line.

location. The probability distribution for the maximum significant wave height in a 20-year design life is obtained from the following relationship since events of exceedance are independent between years:

$$P(\text{maximum } H_s \text{ in 20 years} > h_s) = 1 - e^{-v h_s t} \tag{7.17}$$

where v_{h_s} is the annual frequency of exceedance for that value of h_s (from Figure 7.6) and t is the design life of 20 years. The resulting probability distribution for the maximum

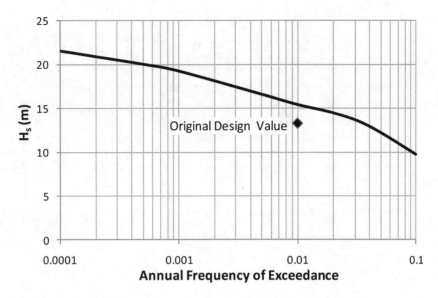

Figure 7.6 Annual frequency that a given value for the significant wave height will be exceeded

significant wave height in 20 years is shown in Figure 7.7; this distribution is reasonably modeled using a Gumbel Type I Maximum Value distribution, $F_{H_s}(h_s) = e^{-e^{-\alpha_H(h_s-u_H)}}$ with $u_H =$ 12.3 m and $\alpha_H = 0.611$ m^{-1} (Figure 7.7).

In order to account for variability of wave heights within a sea state and also for uncertainty in the relationship between the maximum line tension and the significant wave height, the following model is used:

$$T = N_T \mu_{T(H_s)} \tag{7.18}$$

where N_T is a normally distributed correction factor with a mean of 1.0 and a standard deviation of 0.06, and $\mu_{T(H_s)}$ is the expected relationship between the maximum line tension and the significant wave height in a 3-hour sea state (Figure 7.4).

In order to account for dynamic variations in the relationship between the line tension and the angle at the seafloor, the angle when the maximum tension occurs is modeled as $\theta = \mu_{\theta(T)} + \Delta\theta$ where $\mu_{\theta(T)}$ is the expected relationship between the angle and tension shown in Figure 7.5, and $\Delta\theta$ is a normally distributed variable with a mean value of 0 and a standard deviation of 2 degrees.

The caisson was designed based on the pre-2006 characterization of sea states, which corresponded to a design value for the significant wave height of 13 m and a design tension of 8.65 MN at an angle of 28.5° to the horizontal at the seafloor for the most heavily loaded line. The soil is a normally consolidated marine clay with a design profile of undrained shear strength versus depth shown in Figure 7.8. The design capacity of the suction caisson, including the segment of mooring line chain that extends below the seafloor to the padeye, was estimated using a numerical model for the caisson combined with an analytical model for the chain (OTRC 2006). The following input were used in estimating the design capacity: the side shear correction factor α was 0.8, and the reverse end bearing factor N was 9. The resulting design was a caisson with a diameter of 4 m, a length of 24 m, a load attachment or padeye at 16 m below the seafloor, and a chain with 0.133-m thick links. The design capacity of the caisson is compared with the design load in Figure 7.9. The capacity of the caisson is governed by its vertical uplift capacity; the design factor of safety for vertical loading is 2.3.

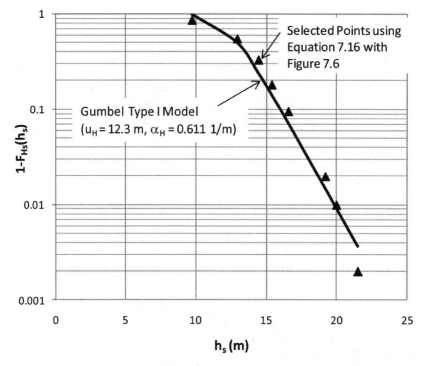

Figure 7.7   Probability distribution for maximum value of significant wave height in 20 years.

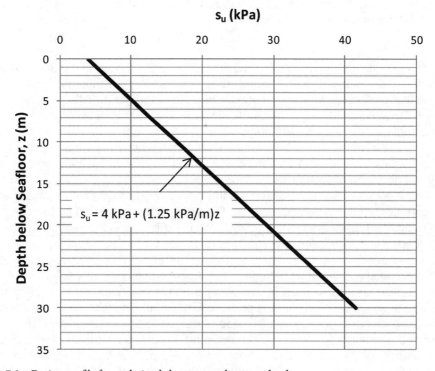

Figure 7.8   Design profile for undrained shear strength versus depth.

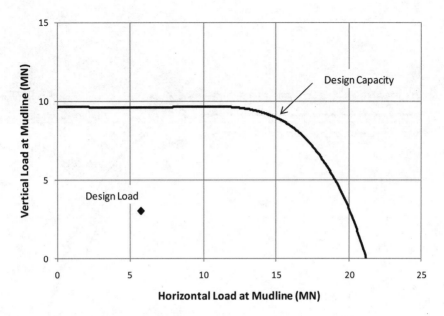

Figure 7.9 Comparison of design load and design capacity

In order to assess the reliability of the suction caisson with the updated metocean description, the following input variables for the capacity are modeled as random variables:

- The average gradient of increasing undrained shear strength with depth s_{uz} is modeled as a lognormal distribution with a mean of 1.25 kPa/m and a coefficient of variation of 0.2. This random variable accounts for uncertainty in the undrained shear strength due to systematic test variations as well as spatial variability in the soil properties.
- The input variables N and α are modeled with a bivariate lognormal distribution with the following parameters: $\mu_N = 11.0$, $\delta_N = 0.25$, $\mu_\alpha = 0.8$, $\delta_\alpha = 0.2$ and $\rho_{N,\alpha} = -0.8$. This joint random variable accounts for uncertainty in the actual capacity compared to the design capacity, and is based on comparisons of measured and predicted capacities (e.g., Najjar 2005). The values for α and N are negatively correlated because axial capacity measurements can be explained by different combinations of α and N since the axial capacity increases with both variables. The sensitivity of the caisson capacity to α and N is shown in Figure 7.10.

The input for the FORM analysis is shown in Figure 7.11. The performance function is formulated as follows:

$$g(\vec{X}) = \frac{T_{v,capacity}}{T_{v,load}} - 1. \tag{7.19}$$

where $T_{v,capacity}$ is the vertical component of the capacity evaluated from the capacity interaction curve at the horizontal component of the applied load $T\sin\theta$, and $T_{v,load}$ is the vertical component of load $T\sin\theta$.

An iterative procedure is required to find the most probable point and the corresponding reliability index. The results for the first iteration are shown in Figure 7.12. A starting point is selected for $\vec{x}^*$ by guessing a combination of values that is reasonably close to

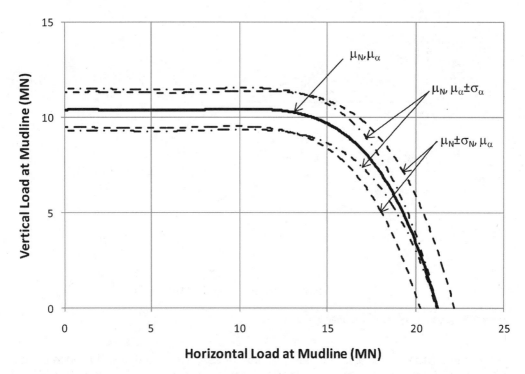

Figure 7.10   Sensitivity of caisson capacity to input variables α and *N*.

the most probable point. The input variables for the capacity model N (x_2^*) and α (x_3^*), were set to their mean value minus one standard deviation while the main input for the load model H_s (x_4^*) and N_T (x_5^*) were set to their mean value plus one standard deviation to reflect that a failure will most likely involve a lower than expected capacity and a higher than expected load. The value for $\Delta\theta$ (x_6^*) was set to its mean value since the event of failure is relatively insensitive to this variable. Finally, the constraint that $g(\overline{x^*}) = 0$ is satisfied by finding the value of s_{uz} (x_1^*) such that $T_{v,capacity} = T_{v,load}$; this step requires an iterative procedure due to the complexity of the model for capacity. The partial derivatives $\left.\dfrac{\partial g}{\partial s_i}\right|_{s^*}$ are found from Equation 7.16, where the partial derivatives $\left(\left.\dfrac{\partial g}{\partial x_i}\right|_{\overline{x^*}}\right)$ are approximated numerically using finite differences and the cumulative distribution function values $F_{x_i}(x^*_i)$ and probability density function values $f_{x_i}(x^*_i)$ are calculated from the probability distributions for the input variables. The reliability index for the first iteration β is then calculated from Equation 7.13 (Figure 7.12). The next iteration proceeds by using the information from the first iteration to establish a new set of values for $\overline{x^*}$ that will minimize β subject to $g(\overline{x^*}) = 0$. In this example, the method of Lagrange's multiplier is used to adjust x_2^* to x_5^* by approximating $g(\overline{x^*})$ as a linear function, and then x_1^* is found such that $g(\overline{x^*}) = 0$. This procedure is continued until there is convergence in β.

The results from the FORM analysis are shown in Figure 7.13. Convergence was achieved relatively rapidly, within five iterations, indicating that the assumed starting point is reasonably close to the Most Probable Point. The most likely combination of the input variables at failure, the Most Probable Point, provides insight into the physical problem. The Most Probable Point is depicted in Figure 7.14. Failure is most likely governed by the

Caisson Dimensions								
D (m)	4							
A (m²)	12.57							
L (m)	24							
	Variable	Units		Probability Distribution				
X_1	S_{uz}	kPa/m	Lognormal	λ [ln(kPa/m)]	0.204	ζ	0.198	
X_2	N		Lognormal	λ	2.368	ζ	0.246	
X_3	α		Lognormal	λ	-0.243	ζ	0.198	
X_4	H_s	m	Type I	u_H (m)	12.355	α_H (1/m)	0.611	
X_5	N_T		Normal	μ	1.000	σ	0.060	
X_6	$\Delta\theta$	degrees	Normal	μ (deg)	0.000	σ (deg)	2.000	
μ_X		1.25	kPa/m					
		11						
		0.8						
		13.3	m					
		1						
		0	deg					
σ_X		0.25	kPa/m					
		2.75						
		0.16						
		2.1	m					
		0.06						
		2	deg					
$C_S = \rho_{Xi,Xj}$	1	0	0	0	0	0		
	0	1	-0.8	0	0	0		
	0	-0.8	1	0	0	0		
	0	0	0	1	0	0		
	0	0	0	0	1	0		
	0	0	0	0	0	1		

Figure 7.11 Input to FORM analysis of suction caisson.

axial capacity of the caisson, although there is a small component of interaction with the lateral capacity (Figure 7.14). Therefore, increasing the reliability would best be accomplished by increasing the axial versus the lateral capacity of the caisson, although the lateral capacity cannot be ignored to the interaction. The most probable load at failure is relatively farther above the design load than the most probable capacity at failure is below the design capacity (Figure 7.14). This relationship is also indicated by the percentile values of the input variables at failure ($F_{X_i}(x^*_i)$ in Figure 7.13): the most probable load is a 99.6th percentile value, while the most probable capacity, as represented by the gradient of increasing undrained shear strength with depth, is an 11th percentile value. Failure is not very sensitive to the capacity model factors, N and α, since the most likely values at failure for these variables have percentile values close to 50% (Figure 7.13). While there is significant uncertainty in these model factors, the negative correlation between them dampens their influence on the axial capacity.

Iteration	1					
x^*	0.61944	kPa/m		T	8.9069	MN
	8.2500			θ	29.4193	deg
	0.6400			$T_{H,load}$	7.758	MN
	15.4000	m		$T_{V,load}$	4.375	MN
	1.0600			S_{utip}	18.866	kPa
	0.0000	degrees		$T_{H,capacity}$	7.758	MN
				$T_{V,capacity}$	4.375	MN
				g	0.000	
				TH'	36.3960	
				TV'	24.016	
dg/dx^*	2.21E+00					
	4.82E-02					
	9.75E-01					
	-1.86E-01					
	-1.96E+00					
	-2.33E-02					
	$F_X(x^*)$	$f_X(x^*)$				
	2.84E-04	8.58E-03				
	1.48E-01	1.14E-01				
	1.52E-01	1.86E+00				
	8.56E-01	8.14E-02				
	8.41E-01	4.03E+00				
	5.00E-01	1.99E-01				
s^*	-3.4461		dg/ds^*	0.272		
	-1.0453			0.098		
	-1.0277			0.124		
	1.0617			-0.520		
	1.0000			-0.118		
	0.0000			-0.047		
β	3.03440					

Figure 7.12   Results from first iteration of form analysis

The probability of failure in the 20-year design life from the FORM analysis is $p_F \cong \Phi(-2.945) = 0.0016$. This probability is relatively high compared to what is typically targeted in practice—for example, Goodwin et al. (2000) suggest a target probability of failure for a mooring system in a 20-year design life that is below 0.001. The probability is relatively high because the caisson was designed using less severe metocean conditions

Iteration	5						
x*	0.96525	kPa/m			T	13.4828	MN
	10.5198				θ	35.3428	deg
	0.7617				$T_{H,load}$	10.998	MN
	21.2171	m			$T_{V,load}$	7.799	MN
	1.0292				S_{utip}	27.166	kPa
	0.3460	degrees			$T_{H,capacity}$	10.998	MN
					$T_{V,capacity}$	7.801	MN
					g	0.000	
					TH'	34.6475	
					TV'	24.705	
dg/dx*	1.33E+00						
	4.00E-02						
	7.57E-01						
	-1.14E-01						
	-1.71E+00						
	-1.82E-02						
	$F_X(x^*)$	$f_X(x^*)$					
	1.14E-01	1.01E+00					
	4.77E-01	1.54E-01					
	4.41E-01	2.62E+00					
	9.96E-01	2.71E-03					
	6.87E-01	5.91E+00					
	5.69E-01	1.97E-01					
s*	-1.2063		dg/ds*	0.254			
	-0.0582			0.104			
	-0.1490			0.114			
	2.6158			-0.550			
	0.4869			-0.102			
	0.1730			-0.036			
β	2.94521						

Figure 7.13   Results from FORM analysis.

than those now expected, based on recent storm data. The relationship between the probability of failure and the mean value for the maximum significant wave height in 20 years is shown in Figure 7.15. When the foundation was designed, the mean value for the maximum significant wave height in 20 years was close to 10 m, and the estimated probability of failure with this characterization of the metocean conditions is about ten times smaller than the updated estimate with a mean significant wave height of 13.3 m (Figure 7.15). The decision about whether to improve the reliability of the foundation based on the new

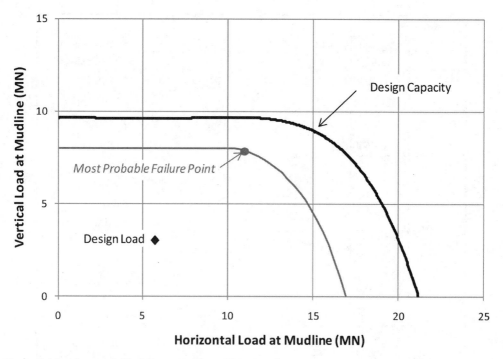

Figure 7.14   Most probable failure point from FORM analysis.

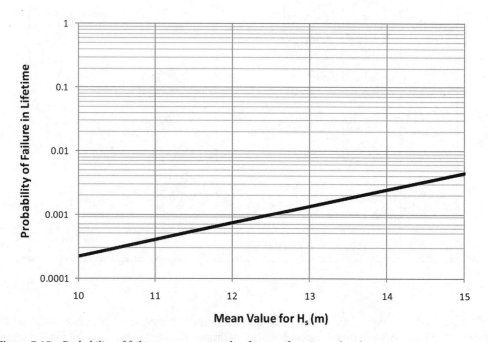

Figure 7.15   Probability of failure versus mean value for significant wave height.

metocean characterization will depend on a comparison of the consequences of a failure and the costs of possible alternatives to increase the reliability.

7.2.3 Numerical Approximation—Monte Carlo Simulation

The probability of failure in Equation 7.11 can also be expressed as follows:

$$p_F = \int_{-\infty}^{+\infty} \cdots \int_{-\infty}^{+\infty} I[g(\vec{x}) \le 0] f_{X_2 \ldots X_n}(x_1, \ldots, x_n)\, dx_1 \ldots dx_n \tag{7.20}$$

where

$$I[g(\vec{x}) \le 0] = \begin{array}{l} 0 \text{ if } g(\vec{x}) > 0 \\ 1 \text{ if } g(\vec{x}) \le 0 \end{array}.$$

This integral can then be approximated with numerical integration as follows:

$$p_F \cong \frac{1}{N} \sum_{k=1}^{N} I[g(\vec{x}_k) \le 0] \tag{7.21}$$

where $\vec{x}_k$ is sampled randomly from $f_{X_2 \ldots X_n}(x_1, \ldots, x_n)$ N times, which is known as Monte Carlo simulation (e.g., Ang and Tang 1984; Madsen et al. 1986; and Melchers 1987). While it is not necessarily a very efficient method of numerical integration, it is generally easy to implement and is becoming more popular as computers become faster. More importantly, it provides an unbiased estimate for P_F, meaning that the estimate approaches the correct value as N approaches infinity.

The precision of the estimate for the probability of failure depends on the number of realizations N and the magnitude of the probability of failure. With random sampling, the 95% confidence interval about this estimate is approximately

$$95\% \text{ confidence interval for estimated } p_F \cong p_F \pm 1.96\sqrt{\frac{p_F(1-p_F)}{N}} \tag{7.22}$$

In contrast to FORM, one of the advantages of Monte Carlo simulation is that the precision of the estimate for p_F can be assessed and can be controlled with N. A rough rule of thumb is that the number of realizations required for a practically precise estimate of p_F is $100/p_F$; this number provides for a 95% confidence interval that is approximately +/−20% about the estimated value. The precision can be evaluated in practice by plotting the estimated probability of failure versus the number of simulations and assessing its convergence.

The efficiency of Monte Carlo simulation in estimating small p_F values can be improved by sampling from $f_{X_2 \ldots X_n}(x_1, \ldots, x_n)$ with bias so that more of the realizations lead to the event of failure. This technique, known as importance sampling (e.g., Madsen et al. 1986; Melchers 1987), can improve the computational efficiency but requires greater expertise to implement and to ensure that the results are not biased.

Example 5. Monte Carlo Simulation for Suction Caisson Anchors in Spar Mooring System

The probability of failure estimated in Example 4 with FORM will also be estimated with Monte Carlo simulation. The FORM estimate for p_F is 0.0016, meaning that approximately 100/0.002 = 50,000 realizations will be required to achieve a practically precise estimate for p_F from Monte Carlo simulation. Ten sets of 5000 realizations each are utilized to assess the precision in the overall estimate of p_F (the sample mean of the ten estimates) from the variation in the estimates between the ten set (the sample standard deviation of the ten estimates).

The following steps describe the calculations for a single realization k:

1. A random number is generated for X_1, u_{1k}; the RAND() function generates random numbers in Excel®. The realization for x_{1k} is then obtained from its lognormal probability distribution (Figure 7.11): $x_{1k} = exp[\Phi^{-1}(u_{1k})\xi_{x_1} + \lambda_{x_1}]$.

2. A random number is generated for X_2, u_{2k}, and the realization for x_{2k} is obtained from its lognormal probability distribution (Figure 7.11): $x_{2k} = exp[\Phi^{-1}(u_{2k})\xi_{x_2} + \lambda_{x_2}]$.

3. From the realization for x_{2k}, the conditional mean and standard deviation for the natural logarithm of X_3 given x_{2k} are calculated from the bivariate lognormal distribution (Figure 7.11): $\lambda_{x_3|x_{2k}} = \lambda_{x_3} + \rho_{\ln x_2, \ln x_3}(\xi_{x_3}/\xi_{x_2})[\ln(x_{2k}) - \lambda_{x_2}]$ and $\xi_{x_3|x_{2k}} = \xi_{x_3}\sqrt{1 - \rho^2_{\ln x_2, \ln x_3}}$. Then a random number is generated for X_3, u_{3k}, and the realization for x_{3k} is obtained from its lognormal probability distribution: $x_{3k} = exp[\Phi^{-1}(u_{3k})\xi_{x_3|x_{2k}} + \lambda_{x_3|x_{2k}}]$.

4. A random number is generated for X_4, u_{4k}, and the realization for x_{4k} is obtained from its Type I maximum value probability distribution (Figure 7.11): $x_{4k} = -\ln[-\ln(u_{4k})]/\alpha_{x_4} + u_{x_4}$.

5. A random number is generated for X_5, u_{5k}, and the realization for x_{5k} is obtained from its normal probability distribution (Figure 7.11): $x_{5k} = \Phi^{-1}(u_{5k})\xi\sigma_{x_5} + \mu_{x_5}$.

6. A random number is generated for X_6, u_{6k}, and the realization for x_{6k} is obtained from its normal probability distribution (Figure 7.11): $x_{6k} = \Phi^{-1}(u_{6k})\xi\sigma_{x_6} + \mu_{x_6}$.

7. The set of six input variables, x_{1k} to x_{6k}, are then used to calculate the performance function in Equation 7.19.

8. The indicator function for realization k, $I[g(\vec{x}_k) \leq 0]$, is set to 0 if $g(\vec{x}_k) > 0$ and 1 if $g(\vec{x}_k) \leq 0$.

These calculations are repeated N times, and the probability of failure is estimated from Equation 7.21.

The results from the ten sets of 5000 realizations each are presented in Figure 7.16. The sample mean of these ten sets provides the estimate for p_F, 0.0013. The sample standard deviation from these ten sets $s_{pF,10}$ can then be used to estimate confidence bounds on the estimate $p_F = 0.0013 \pm 1.96 s_{pF,10}/\sqrt{10 \text{ sets}} = 0.0013 \pm 0.00042$ (Figure 7.16). This approach of sub-dividing the total number of realizations into sets provides

Set	p_F
1	0.0016
2	0.0012
3	0.0014
4	0.0008
5	0.0026
6	0.0006
7	0.0018
8	0.0002
9	0.0016
10	0.0012
estimate = sample mean	0.00130
sample standard deviation	0.000675
lower 95% bound	0.00088
upper 95% bound	0.00172

Figure 7.16   Results from Monte Carlo analysis.

a more general means to establish confidence bounds on the estimate than Equation 7.20 because it can be applied to any estimate from the Monte Carlo simulation (e.g., the standard deviation in the caisson capacity) and to any type of sampling, including importance sampling. Note that the FORM estimate for p_F, 0.0016, is similar to the Monte Carlo estimate and within its 95% confidence bounds.

7.3 Practical Implementation

The practical implementation of reliability-based methods for offshore foundation design is problem and project specific. The following discussion is intended to provide guidance on the various factors to consider in practice, including foundation capacities, foundation loads, systems versus components, and target reliability values.

7.3.1 Foundation Capacities

Foundation capacities are uncertain because of a variety of factors:

1. How do lateral and vertical variations in geotechnical properties affect the performance of the foundation?
2. How much variation is there in stratigraphy and properties between foundation elements and between available data (e.g., a soil boring) and the foundation elements?
3. How well do the available geotechnical data relate to the properties of interest for predicting the foundation capacity?
4. How will installation of the foundation affect its capacity?
5. How will the capacity vary with time after installation?
6. How will rate of loading and cyclic versus static loading affect the capacity?
7. How will the possible combinations of loads and moments applied to the foundation affect its capacity?
8. How much uncertainty is there in the model used to predict the foundation capacity?

In achieving a reliable foundation, it is important to consider any biases between the capacity used in design and the expected or most likely value for the capacity, as well as the range of possible values for the capacity when the load is applied.

Example 6. Consideration of Lower Bound in Axial Pile Capacity
A pile foundation for a tension leg platform will be subjected to a sustained tendon load with relatively small dynamic variations due to environmental conditions. The geologic setting is unusual, leading to a relatively large coefficient of variation in the capacity: δ_R = 0.5. Conversely, the possible variations in the maximum tendon load are small since the load is composed primarily of pretension, leading to a relatively small coefficient of variation in the load: δ_S = 0.15. The target probability of failure, 0.00001, reflects a large failure consequence.

The required median factor of safety to achieve this target level of reliability is obtained from Figure 7.2: $\delta_{total} = \sqrt{\delta_R^2 + \delta_S^2} = \sqrt{0.5^2 + 0.15^2} = 0.52$ and FS_{median} > 9. This required median factor of safety is very large and will require a very conservative design for the pile foundation. However, there is additional information about the capacity that has not been considered in this reliability analysis.

The lognormal distribution for pile capacity used to develop Figure 7.2 allows for the capacity to range down as low as zero. However, there is a physical lower bound on the uplift capacity that consists at a minimum of the submerged weight of the pile and plug.

In addition, the available side shear under undrained loading (the dynamic variations due to environmental conditions) will be at least as great as the remolded undrained shear strength. This lower bound can be incorporated into the probability distribution for the capacity using a mixed lognormal distribution, where the lognormal distribution for capacity is truncated at the lower bound value r_{LB}, and the probability that the capacity is equal to r_{LB} is the probability that the capacity is less than or equal to r_{LB} in the original lognormal distribution before truncation (Figure 7.17). This lower-bound form of the distribution for capacity matches well with pile load-test data (Najjar 2005). Equation 7.5 is then modified as follows to calculate the probability of failure with a lower bound on the capacity (Najjar and Gilbert 2009):

$$p_F = P(R \le S) = P(R \le S | r = r_{LB}) P(R = r_{LB}) + \int_{r_{LB}}^{\infty} P(R \le S | r) f_R(r) \, dr$$

$$= \left[1 - \Phi\left(\frac{\ln(r_{LB}) - \ln(s_{median})}{\sqrt{\ln(1 + \delta_S^2)}} \right) \right] \Phi\left(\frac{\ln(r_{LB}) - \ln(r_{median})}{\sqrt{\ln(1 + \delta_R^2)}} \right) \qquad (7.23)$$

$$+ \int_{r_{LB}}^{\infty} \left[1 - \Phi\left(\frac{\ln(r) - \ln(s_{median})}{\sqrt{\ln(1 + \delta_S^2)}} \right) \right] \frac{1}{\sqrt{2\pi}\sqrt{\ln(1 + \delta_R^2)}\, r} e^{-\frac{1}{2}\left(\frac{\ln(r) - \ln(r_{median})}{\sqrt{\ln(1 + \delta_R^2)}} \right)^2} dr$$

The required median factor of safety is shown as a function of the lower-bound capacity in Figure 7.18. Accounting for the submerged weight and the remolded shear strength, the lower-bound capacity will typically be within 0.3 to 0.6 times the median capacity. Therefore, this lower-bound capacity will allow for the required median factor of safety to be reduced by a factor of two to three times while still achieving the target level of reliability (Figure 7.18).

Example 7. Design of Suction Caisson without Site-specific Soil Boring
The soil boring drilled to design a suction caisson anchor for axial loading is located several kilometers away from the location of the anchor. If the anchor is at the location of the soil boring, then the coefficient of variation in the axial capacity is 0.3 to account for variations between the predicted and actual capacity. An analysis of data from numerous

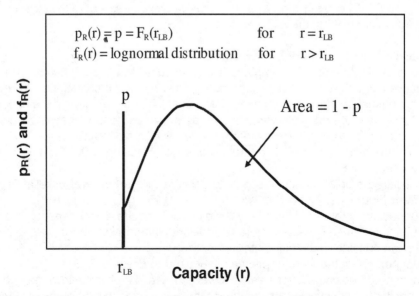

Figure 7.17   Mixed lognormal distribution for capacity with a lower bound.

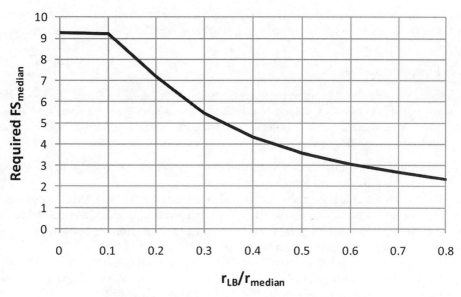

Figure 7.18   Required median factor of safety versus ratio of lower-bound to median capacity.

soil borings within this geographic region and in a similar geologic setting indicates that the coefficient of variation in design axial capacity for all boring locations is 0.15. Note that this magnitude of variation in capacity is smaller than that for the soil strength between measurements along the depth of a suction caisson anchor due to the effect of averaging in estimating the total side shear. An important project-specific consideration is how spatial variations in soil properties at different scales will affect the performance of the geotechnical system (e.g., Fenton and Griffiths 2008).

The total coefficient of variation in capacity at the anchor location is given by $\delta R = \sqrt{0.3^2 + 0.15^2} = 0.34$, assuming that the variations in capacity between predicted and actual and spatially between borings are independent. The coefficient of variation in the load is $\delta S = 0.2$, and the target probability of failure is $pF \leq 0.0001$. The required median factor of safety is then obtained from Figure 7.2: $\delta_{total} = \sqrt{\delta_R^2 + \delta_S^2} = \sqrt{0.34^2 + 0.2^2} = 0.37$ and $FS_{median} = 4.0$ (Figure 7.2). This median factor of safety would be applied to a design capacity obtained by taking the average of all of the borings in the field. A more detailed geostatistical analysis could also be performed to account for spatial correlations between locations (e.g., Keaveny et al. 1989). In accounting for spatial correlation, the capacity at the location of interest is represented by a weighted average of the values at all borings where the closer borings have more weight than the furthest borings (e.g., Gilbert et al. 2008). However, since the spatial variability in this field is relatively small (c.o.v. equals 0.15), this more sophisticated analysis may not be of significant practical value.

If drilling a site-specific boring is possible, then this analysis can be used to assess the value of drilling this boring. The benefit of a site-specific boring is that the additional uncertainty due to spatial variability will be eliminated, meaning that $\delta_R = 0.3$, $\delta_{total} = \sqrt{0.3^2 + 0.2^2} = 0.36$, and $FS_{median} = 3.8$ (Figure 7.2). Therefore, the expected benefit of drilling the site-specific boring is to reduce the required median factor of safety from 4.0 to 3.8, or by 5%. If the cost savings associated with a 5% reduction in the required design capacity is greater than the cost of drilling a site-specific boring, including the possible delay in installing the facility, then it will be worthwhile to drill it. Otherwise, it is preferable to account for the lack of a site-specific boring by increasing the required median factor of safety by 5%. If a site-specific boring is not drilled, then it is important to consider other possible consequences such as the impact on being able to install the caisson successfully.

7.3.2 Foundation Loads

Geotechnical engineers tend to focus more on the foundation capacity than on the loads. However, as illustrated in Figure 7.14, the most likely conditions leading to a failure may have as much to do with the load as the capacity. It is important that geotechnical engineers consider the range of possible loading conditions as carefully as the range of possible capacities:

1. How does the nominal design load used in the design check compare to the expected value or most likely value for the load?
2. How much uncertainty is there in the load and what contributes to that uncertainty, including variations with time, variations in construction and operational tolerances, uncertainties in final design parameters at the time the foundation is being designed, and uncertainties in the models used to estimate the loads?
3. Over what durations will different levels of loads be applied to the foundation?
4. What is the relative contribution of cyclic and static loading to the total loads?
5. How quickly will loads be applied after the foundation is installed?

Example 8. Characterization of Foundation Load for Production Riser

A foundation for a production riser will be loaded in tension. The maximum load will be applied when the riser is installed. The total tension load is dominated by the static buoyant force, with environmental loads due to waves and currents contributing only a small fraction of the total (less than 1%). The uncertainty in the total load is due to tolerance variations in the weights of components in the riser and to the possibility of making a ballasting error during installation. The nominal design load is established by assuming that a ballasting error occurs and adding a contingency for weight tolerances:

Design tension = Tension if ballasted properly
+ additional tension if ballasting error is made + contingency for weight tolerances

$$= T_{Expected} + 0.025T_{Expected} + 0.15T_{Expected} = 1.175T_{Expected} \qquad (7.24)$$

A probabilistic representation of the load is developed as follows:

$$P(T \leq t) =$$

$$P(T \leq t | \text{Proper Ballasting})[1 - P(\text{Error in Ballasting})] +$$

$$P(T \leq t | \text{Error in Ballasting})P(\text{Error in Ballasting}) = \Phi\left(\frac{t - T_{Expected}}{\left(\frac{0.023}{2}\right)T_{Expected}}\right) \qquad (7.25)$$

$$[1 - P(\text{Error in Ballasting})] + \Phi\left(\frac{t - 1.15T_{Expected}}{\left(\frac{0.025}{2}\right)1.15T_{Expected}}\right)[1 - P(\text{Error in Ballasting})]$$

where the weight tolerance variations are modeled as normal distributions with a standard deviation that is one-half of the contingency (i.e., the contingency is typically on the order of two standard deviations).

The cumulative probability distribution for the load is shown in Figure 7.19 for different probabilities of making a ballasting error. The distribution is shown as the probability that the tension will exceed a given value: $P(T > t) = 1 - P(T \leq t)$ where the tension is expressed as a proportion of the design tension. For example, if the probability of making

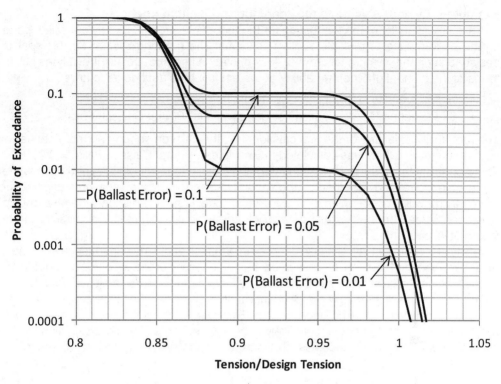

Figure 7.19 Probability distribution for foundation load.

a ballasting error is 0.05, then the probability the tension will exceed the design tension (i.e., Tension/Design Tension = 1) is equal to 0.002 (Figure 7.19). This analysis shows that there is considerable conservatism in the nominal design tension used in the design check. Even if there is a 10% probability of making a ballasting error, there is only a 0.4% probability that the tension will exceed the nominal design tension (Figure 7.19). Since the nominal design tension is factored up to establish the required foundation design capacity, the probability of a foundation failure in orders of magnitude smaller than 0.4%. This analysis also highlights the importance of human factors in the performance of the foundation. One way to improve the reliability of this foundation would be to implement operational procedures to reduce the probability of making a ballasting error.

7.3.3 System versus Components

While most design recipes are expressed in terms of components (such as a single pile), it is important to consider the reliability of systems as well as components. Generally, the consequences of a system failure are more significant than for a component failure, and the risk of system failure is ultimately what governs decision making.

The analytical and numerical methods described previously for assessing reliability can be applied generally to events of failure that involve individual components or systems. Monte Carlo simulation is particularly well suited for system analyses since it can become cumbersome to represent system performance analytically. Important considerations in predicting system performance are how system loads are distributed to individual foundation components and how system capacity is related to the capacities of individual foundation and other structural components.

Example 9. System Reliability for Spar Mooring System Anchors

The system reliability of the foundation for the mooring system in Example 4 will depend on the performance of multiple suction caissons. The mooring system for this spar is designed such that system failure (i.e., loss of station keeping) occurs if two lines fail. In terms of the foundation, the probability of system failure is defined as follows:

$$P(\text{System Failure}) = P(\text{Two Suction Caisson Failures}) =$$
$$P(\text{Second Caisson Fails}|\text{First Caisson Fails})\, P(\text{First Caisson Fails}) \qquad (7.26)$$

The probability of failure for the first caisson (i.e., the caisson supporting the most heavily loaded mooring line) was calculated in Example 4: $P(\text{First Caisson Fails}) = 0.0016$.

The probability that a second caisson fails if the first caisson fails will depend on how the two events are related. The load on the second caisson for a given sea state (significant wave height) will be greater since one line is missing. Figure 7.20 shows the relationship between the load on the next most heavily loaded line when one line is missing due to failure of its suction caisson.

The failure of the first caisson also provides information about the performance of the next most heavily loaded caisson. It is assumed that the capacity of the second caisson is independent of the first caisson since the majority of the uncertainty is due to spatial variations in soil properties and in seemingly independent variations between the actual capacity and the predicted capacity for each individual caisson. However, failure of the first caisson indicates that the loading environment is more likely to be severe. The probability distribution for the significant wave height is updated as follows based on the event that the first caisson failed:

$$f_{H_s|\text{First Caisson Fails}}(h_s\,|\,\text{First Caisson Fails}) = \frac{P(\text{First Caisson Fails}\,|\,h_s)\, f_{H_s}(h_s)}{P(\text{First Caisson Fails})} \qquad (7.27)$$

where $f_{H_s|\text{First Caisson Fails}}(h_s|\text{First Caisson Fails})$ is the updated or conditional probability density function for the significant wave height given that the first caisson failed, $P(\text{First Caisson Fails}|h_s)$ is the conditional probability that the first caisson will fail given a particular

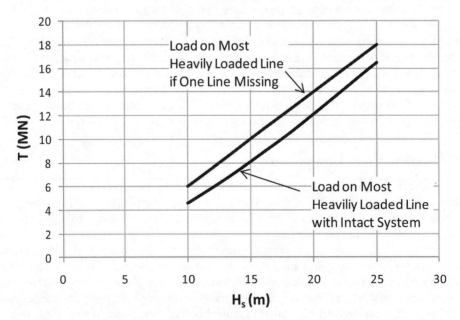

Figure 7.20 Expected maximum line tension versus significant wave height in 3-hour sea state for intact and damaged mooring system.

significant wave height, and $f_{Hs}(h_s)$ is the unconditional probability density function for the significant wave height (Figure 7.7). The conditional probability P(First Caisson Fails|h_s) is obtained by performing the FORM analysis presented in Example 4 with the significant wave height equal to a single value (i.e., setting the standard deviation to a very small number in the FORM accomplishes this condition). The results for these calculations are shown in Figure 7.21; the probability increases as the significant wave height increases. The unconditional (Figure 7.7) and updated (Equation 7.27) probability density functions for the significant wave are shown in Figure 7.22. Given that the first caisson has failed, it is more likely that the mooring system is being subjected to large waves (Figure 7.22).

The probability that the second caisson fails given that the first caisson fails is estimated using the same FORM analysis as in Example 4, replacing the probability distribution for the significant wave height with the conditional distribution in Figure 7.22 and using the relationship between the line load and significant wave height for the damaged system (Figure 7.20). The resulting probability of failure is P(Second Caisson Fails|First Caisson Fails) = 0.45. This system has some measure of redundancy because it is not certain that the mooring system will fail when the first anchor fails in a hurricane. The resulting probability of system failure in the 20-year lifetime of the mooring system is obtained from Equation 7.23:

$$P(\text{System Failure}) = 0.45 \times 0.0016 = 0.00073. \tag{7.23}$$

7.3.4 Target Reliability

Risk is the possibility of a loss. An appropriate risk is one that balances the cost of reducing the risk against that of accepting the risk. Appropriate risk is never zero risk, and excessive conservatism can be as troublesome as excessive risk.

Risk is quantified as the expected consequence of loss. In the simplest case, if the possibility of loss is a binary event that either does or does not occur, then the risk is equal to the probability

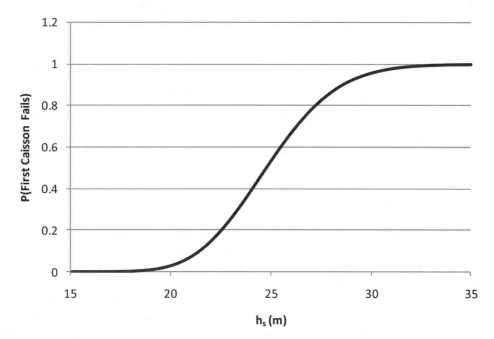

Figure 7.21 Probability that most heavily loaded caisson fails with intact mooring system versus significant wave height.

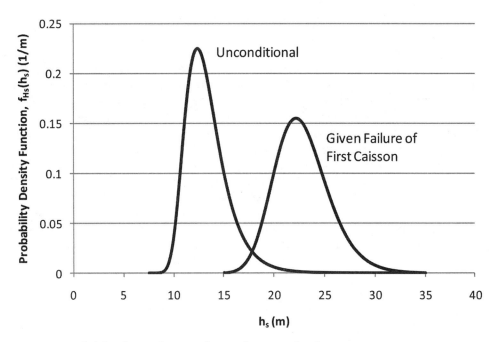

Figure 7.22   Probability density functions for significant wave height.

of the loss multiplied by its consequence. Therefore, risk is commonly represented by the probability and consequence of a loss. Consequences can include human safety and environmental and economic losses.

The balance between reducing versus accepting risk, and therefore the selection of a target reliability, needs to be considered on a project-specific basis. In the simplest sense, the balance is at the point where the incremental cost of reducing the risk is equal to the incremental reduction in the expected consequence of a loss.

General guidance for striking the balance between reducing versus accepting risk has been developed by different governments and industries. Examples include Whitman (1984) for a variety of civil engineering systems, USNRC (1975) for nuclear power plants, AIChE (1989) for chemical process facilities, ANCOLD (1998) and USBR (2003) for dams, and Bea (1991), Stahl et al. (1998), Goodwin et al., (2000) for offshore production facilities. An illustrative depiction of this type of guidance is shown in Figure 7.23 as risk tolerance thresholds, where the probability of a loss in the 20-year design life is considered to be tolerable if it is below the threshold for a given cost of failure.

7.4 Conclusion

Reliability is an important consideration in the design of offshore foundations. Both analytical and numerical methods for assessing reliability were described in this chapter. Examples were included to illustrate the application of these methods. In addition, guidance was provided for the implementation of reliability analyses in design practice.

One important conclusion in applying these methods to real problems is that each problem is unique. The information available, the geologic setting, the types of foundations, the composition of foundation and structural systems, the loads on foundations, the consequences of failure, and

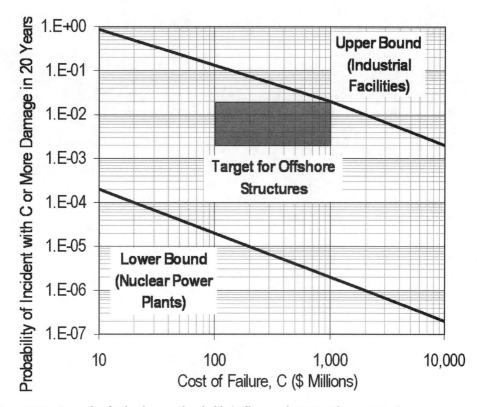

Figure 7.23 Example of risk tolerance thresholds (Gilbert et al. 2008, with permission).

the costs of reducing uncertainty or increasing reliability all vary from project to project. It is not possible nor is it constructive to provide generic values for input to a reliability analysis. Furthermore, while commercial software packages are helpful in facilitating calculations, they do not alleviate the substantive effort that is required in formulating a project-specific problem. Therefore, the emphasis in this chapter has been on fundamental principles and methods that can be readily tailored to suit individual problems.

Another important conclusion is that significant uncertainty arises from using necessarily simplified models to predict loads and capacities. There is always value in looking backward to evaluate how actual performance compares to predicted performance (e.g., Aggrawal et al. 1996; Bea et al. 1999; Gilbert et al. 2010). Looking backward guides the development of better understanding and predictive models in order to more effectively manage risk in going forward.

7.5 References

Aggarwal, R. K., Litton, R. W., Cornell, C. A., Tang, W. H., Chen, J. H., and Murff, J. D., 1996. Development of pile foundation bias factors using observed behavior of platforms during Hurricane Andrew, In *Proceedings of the Annual Offshore Technology Conference*, Houston, TX, OTC 8078:445-455.

AIChE, 1989. *Guidelines for chemical process quantitative risk analysis,* Center for Chemical Process Safety of the American Institute of Chemical Engineers, New York, New York.

ANCOLD, 1998. *Guidelines on risk assessment, working group on risk assessment,* Australian National Committee on Large Dams, Sydney, New South Wales, Australia.

Anderson, W. D., Silbert, M. N., and Lloyd, J. R., 1982. Reliability procedure for fixed offshore platforms, American Society of Civil Engineers, *Journal of Structural Engineering Division* 108(11): 2517-2538.

Ang, A. A-S. and Tang, W. H., 1975. *Probability concepts in engineering planning and design, Volume I—Basic principles*, John Wiley & Sons, New York.

———, 1984. *Probability concepts in engineering planning and design, Volume II—Decision, risk and reliability*, John Wiley & Sons, New York.

API, 1991. *Recommended Practice for Planning, Designing and Constructing Fixed Offshore Platforms—Load and Resistance Factor Design*, 1st Ed., American Petroleum Institute, RP-2A-LRFD.

———, 2000. *Recommended Practice for Planning, Designing and Constructing Fixed Offshore Platforms—Working Stress Design*, 21st ed., American Petroleum Institute, RP-2A-WSD.

Bea, R. G., 1991. Offshore platform reliability acceptance criteria, *Drilling Engineering, Society of Petroleum Engineers*, June, 131-136.

———, 1998. Reliability characteristics of a platform in the Mississippi River Delta, *Journal of Geotechnical and Geoenvironmental Engineering* 124(8): 729-738.

Bea, R. G., Jin, Z., Valle, C., and Ramos, R., 1999. Evaluation of reliability of platform pile foundations, *Journal of Geotechnical and Geoenvironmental Engineering* 125(8): 696-704.

Benjamin, J. R. and Cornell, C. A., 1970. *Probability, Statistics, and Decision for Civil Engineers*, McGraw-Hill, New York.

Brown, G., Brunner, M., Qi, X., and Stanley, I., 2006. Lateral buckling reliability calculation methodology accounting for buckle interaction, In *Proceedings of the Annual Offshore Technology Conference*, Houston, TX, OTC 17795.

Bruschi, R., Gudmestad, O. T., Blaker, F., and Nadim, F., 1996. Seismic assessment for offshore pipelines, *Journal of Infrastructure Systems* 2(3): 145-151.

Clukey, E. C. and Banon, H., 2000. Reliability assessment of deepwater suction caissons, In *Proceedings of Offshore Technology Conference*, OTC 12192:777-785.

Fenton, G. A. and Griffiths, D. V., 2008. *Risk Assessment in Geotechnical Engineering*, John Wiley & Sons, New York.

Gilbert, R. B., Chen, J. Y., Materek, B., Puskar, F., Verret, S., Carpenter, J., Young, A., and Murff, J. D., 2010. Comparison of observed and predicted performance for jacket pile foundations in hurricanes, In *Proceedings of the Annual Offshore Technology Conference*, Houston, TX, OTC 20861.

Gilbert, R. B., Choi, Y. J., Dangyach, S., and Najjar, S. S., 2005. Reliability-based design considerations for deepwater mooring system foundations, In *Proceedings of the International Conference on Frontiers in Offshore Geotechnics*, 317-324, Perth, Western Australia.

Gilbert, R. B., Najjar, S. S., Choi, Y. J. and Gambino, S. J., 2008. Practical application of reliability-based design in decision making, In *Reliability-Based Design in Geotechnical Engineering: Computations and Applications*, Phoon ed., Taylor & Francis Books, London.

Gilbert, R. B., Ward, E. G., and Wolford, A. J., 2001. A comparative risk analysis of FPSO's with other deepwater production systems in the Gulf of Mexico, In *Proceedings of the Annual Offshore Technology Conference*, Houston, TX, OTC 13173.

Goodwin, P., Ahilan, R. V., Kavanagh, K., and Connaire, A., 2000. Integrated mooring and riser design: Target reliabilities and safety factors, In *Proceedings of the Conference on Offshore Mechanics and Arctic Engineering*, 185-792.

Horsnell, M. R. and Toolan, F. E., 1996. Risk of foundation failure of offshore jacket piles, In *Proceedings of the Annual Offshore Technology Conference*, Houston, TX, OTC 7997:381-392.

Keaveny, J. M., Nadim, F., and Lacasse, S., 1989. Autocorrelation functions for offshore geotechnical data, In *Proceedings of the 5th International Conference on Structural Safety and Reliability*, 263-270.

Kraft, L. M. and Murff, J. D., 1975. A probabilistic investigation of foundation design for offshore gravity structures, In *Proceedings of the Annual Offshore Technology Conference*, vol. 3, Houston, TX, OTC 2370:351-366.

Madsen, H. O., Krenk, S., and Lind, N. C., 1986. *Methods of Structural Safety*, Prentice-Hall, New Jersey.

Marshall, P. W. and Bea, R. G., 1976. Failure modes of offshore platforms, In *Proceedings of the International Conference on Behavior of Offshore Structures* 2:579-635.

Melchers, R. E., 1987. *Structural Reliability Analysis and Prediction*, Ellis Horwood, Halsted Press, New York.

Muhlbauer, W. K., 2004. *Pipeline Risk Management Manual; Ideas, Techniques and Resources*, 3rd ed., Elsevier.

Nadim, F. and Lacasse, S., 1992. Probabilistic bearing capacity of jack-up structures, *Canadian Geotechnical Journal* 29(4): 580-588.

Nadim, F., Krunic, D., and Jeanjean, P., 2003. Probabilistic slope stability analyses of the Sigsbee Escarpment, In *Proceedings of the Annual Offshore Technology Conference*, Houston, TX, OTC 15203.

Najjar, S. S., 2005. The importance of lower-bound capacities in geotechnical reliability assessments, PhD diss., University of Texas at Austin.

Najjar, S. S., and Gilbert, R. B., 2009. Importance of lower-bound capacities in the design of deep foundations, *Journal of Geotechnical and Geoenvironmental Engineering* 135(7): 890-900.

Nodine, M. C., Gilbert, R. B., Cheon, J. Y., Wright, S. G., and Ward, E. G., 2009. Risk analysis for hurricane wave-induced submarine mudslides, In *Proceedings of 4th International Symposium on Submarine Mass Movements and Their Consequences*, 335-352, Springer.

OTRC, 2006. *Reliability of mooring systems for floating production systems,* Final report for Minerals Management Service, Offshore Technology Research Center, College Station, TX.

Ronold K. O., 1990. Reliability analysis of tension pile, *Journal of Geotechnical Engineering* 116(5): 760-773.

Stahl, B., 1986. Reliability engineering and risk analysis, In *Planning and Design of Fixed Offshore Platforms,* McClelland and Reifel eds., Van Nostrand Reinhold Publishing Company, New York.

Stahl, B., Aune, S., Gebara, J. M., and Cornell, C.A., 1998. Acceptance criteria for offshore platforms, In *Proceedings of the Conference on Offshore Mechanics and Arctic Engineering*, OMAE98-1463.

Sweeney, M., 2005. *Terrain and Geohazard Challenges Facing Onshore Oil and Gas Pipelines*, Thomas Telford Publishing, London.

Tang, W. H. and Gilbert, R. B., 1993. Case study of offshore pile system reliability, In *Proceedings of the Annual Offshore Technology Conference*, 677-686.

Tang, W. H., Woodford, D. L., and Pelletier, J. H., 1990. Performance reliability of offshore piles, In *Proceedings of the Annual Offshore Technology Conference,* Houston, TX, OTC 6379:299-307.

Thoft-Christensen, P. and Murotsu, Y., 1986. *Application of Structural Systems Reliability Theory*, Springer-Verlag, Berlin.

USBR, 2003. *Guidelines for Achieving Public Protection in Dam Safety Decision Making*, Dam Safety Office, United States Bureau of Reclamation, Denver, Colorado.

USNRC, 1975. *Reactor safety study: An assessment of accident risks in U.S. commercial nuclear power plants,* United States Nuclear Regulatory Commission, NUREG-75/014, Washington, D. C.

Whitman, R. V., 1984. Evaluating calculated risk in geotechnical engineering, *Journal of Geotechnical Engineering* 110(2): 144-188.

Wu, T. H., Tang, W. H., Sangrey, D. A., and Baecher, G. B., 1989. Reliability of offshore foundations—State of the art, *Journal of Geotechnical Engineering* 115(2): 157-178.

8

Soil-pipe Interaction for Subsea Flowlines

William McCarron, PhD, PE
ASGM Engineering

8.1 Introduction

The interaction of subsea flowlines with the seafloor has attracted much attention for many years, and that attention has increased as a result of costly failures and as more high pressure high temperature (HPHT) projects reach the development phase. Under constrained thermal expansion, the developing compressive force in a flowline laid on the seafloor may eventually initiate a lateral buckle. The amplitude of the lateral buckle grows in proportion to both the thermal strain and distance between lateral buckles. The flowline design must then account for the developing bending moments. At the ends of a flowline, the equipment interfacing with wellheads, manifolds, or foundations must be designed to accommodate the loads and displacements induced by thermal expansion.

In this chapter, the significant details to be addressed in the analysis of flowline thermal buckling are introduced, and illustrative examples are presented to provide some perspective. The discussion is limited to flowline interaction with cohesive soils.

8.2 Design Considerations

Among the first to develop detailed analysis procedures for lateral buckling of subsea flowlines were Hobbs (1984) and Hobbs and Liang (1989). Hobbs developed analytical procedures describing the buckling resistance of a flowline interacting with a seafloor. Hobbs' work is an extension of larger body of work investigating the buckling of railroad rails, specifically the work of Kerr (1978). Later, Tornes et al. (2000) described a flowline *walking* phenomenon where, under the influence of cyclic thermal operating events, a flowline may incrementally travel in the direction of the cooler end of the flowline. Palmer and King (2008) describe the full details of subsea pipeline engineering.

Lateral thermal buckles induce bending moments in the flowline that must be considered in strength and fatigue evaluations. HPHT flowlines may experience thermal expansion equivalent to 1 ft per 1000 ft of flowline. At pipeline end terminations (PLETs), the thermal expansion is either accommodated by equipment or restrained by a foundation anchor.

A detailed flowline lateral buckling investigation requires an empirical model describing the soil resistance to lateral movement in terms of a force-displacement relationship. The analytical methods presented by Hobbs assumed the lateral resistance is a frictional mechanism, but this is

247

a simplification. In reality, the soil resistance is generated by both a frictional mechanism at the base of the pipe section and by a passive soil resistance component. The passive soil component is related to either the initial pipe embedment as a result of the installation activity or a developing berm as the pipe moves laterally and sweeps up failed material.

Flowline walking is the result of the thermally induced motions exceeding that required to fully mobilize the shear resistance between the flowline and the soil in the axial direction. Hysteretic displacements result in the accumulation of finite axial displacement during asymmetric transient heating. Shorter flowlines are more susceptible to walking, as are flowlines with a smaller axial coefficient of friction.

PLET design must accommodate the effects of flowline expansion as well as the loads applied by jumpers between the flowline and subsea equipment. Jumpers will induce overturning moments on the PLET, which usually includes a mat foundation (Figure 8.1) to distribute the bearing forces. Occasionally, the PLET is interfaced with a foundation anchor to reduce its movement.

The safe design of hot on-bottom pipelines with lateral buckling (SAFEBUCK) proprietary joint industry project has developed design guidelines accounting for soil-pipe interaction as well as mechanical engineering aspects of HPHT pipelines. Bruton et al. (2005, 2006) described the scope of work for that project.

8.2.1 Lateral Buckling

Lateral flowline buckles are the result of the effects of internal (p_i) and external (p_e) fluid pressures, restrained thermal expansion, axial soil-pipe friction, and structural loads. The sum of these contributing loads is known as the *effective tension*. If the effective tension is sufficiently compressive, the flowline may buckle in the same manner as any beam-column structural member. The effective tension resulting in buckling is a function of the lateral soil resistance and the flowline bending stiffness. The flowline effective tension T_{eff} is evaluated from the expression (Sparks 1983, 1984):

$$T_{eff} = T_{mat} - p_i S_i + p_e S_e \tag{8.1}$$

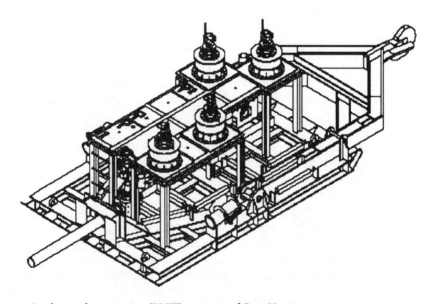

Figure 8.1   Pipeline end termination (PLET), courtesy of Gary Harrison.

where S_i and S_e are pipe section areas based on the inner diameter and exterior diameter, respectively, and T_{mat} is the true material tension, including the effects of restrained thermal expansion, applied loads, and accumulated soil-pipe friction loads. Flowline buckling will occur if T_{eff} is sufficiently compressive. Internal pressures create a compressive component and external pressures result in a stabilizing tensile contribution. Laterally unrestrained pipes may buckle under the effects of internal pressure even if the T_{mat} is tensile. Production tubing in producing oil and gas wells has been known to suffer fatigue failure as the result of repetitive buckling events.

Management of thermal buckling is a chief concern for subsea engineers. Mitigation efforts often include buckle initiators, such as the sleepers shown in Figure 8.2 that control the location of lateral buckles, thus reducing uncertainty in the flowline response.

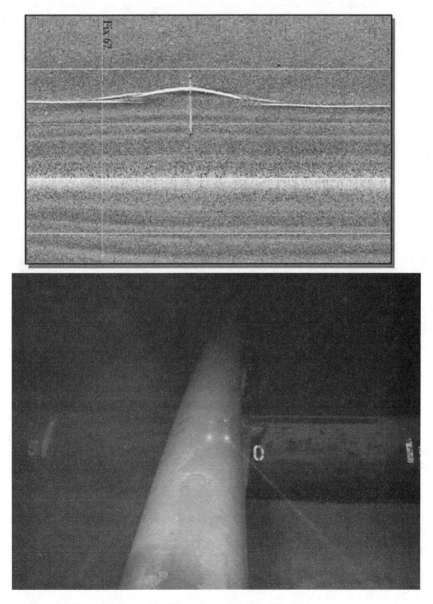

Figure 8.2   Sleeper initiation of lateral buckle, courtesy of Gary Harrison.

8.2.2 Flowline Walking

A detailed description of the phenomenon of walking is given by Tornes et al. (2000). Walking results in the accumulated translation of a flowline in the direction of its cool end. The discussion here briefly describes the source of the walking and some of its implications. Carr et al. (2006) presents a detailed discussion of flowline walking, including some analytical estimates of the walking displacements.

Flowline walking is a result of transient thermal heating of a flowline and the subsequent mobilization of axial soil-pipe friction due to thermally induced movement. Relatively small axial displacements (a few millimeters) of the flowline are required to completely mobilize the interface friction, and further displacement will result in some relative slip between the soil and pipe. On cooling, a residual relative interface displacement will be observed, the magnitude of which will vary along the flowline. Complementary residual interface friction will affect the distribution of axial forces (material and effective) along the flowline. If the ends of the flowline are anchored, residual foundation loads will develop, the magnitude of which can be extremely large.

The following conditions affect flowline walking:

- Longer flowlines are less susceptible to walking.
- Increasing axial friction coefficients reduce walking.
- Increasing mobilization displacement for the soil-pipe interface friction reduces walking.
- Walking is an outcome of the characteristics of asymmetrical thermal transients associated with flowline heating; steeper transients increase walking displacements.

As mentioned earlier, anchors for flowlines may experience increasing reactions as a result of the same mechanisms resulting in walking. Additional considerations are the effect of catenary riser loading at the end of flowlines and seabed slope (Brunner et al. 2006; Carr et al. 2006), which contribute to walking displacement.

Considering the importance of the soil-pipe frictional interface, there is relatively little experimental data in the normal stress regime of interest. Empirical equations relating the soil friction angle to its mineralogy are generally unreliable since they are predominately based on test data at very high normal stresses. Najjar et al. (2003, 2007) have used a tilt-table device to measure Gulf of Mexico cohesive soil internal friction angles, as well as soil-pipe interface friction angles, at low normal stresses. These tests are highly consistent and cost effective. Measurement of the interface friction characteristics on full-scale pipe is a lengthy and expensive process.

8.2.3 Upheaval Buckling

The need to maintain flowline product temperature above hydrate formation conditions has occasionally led to burial as a method of insulating it against heat loss. Burial has the added advantage of increasing the flowline downtime allowed before it is required to treat its contents with chemicals inhibiting hydrate formation (McCarron and Harrison 2008). The same mechanisms that generate lateral buckles also induce vertical buckles. However, the increased resistance associated with the backfill soil weight and strength results in the flowline buckling resistance increasing by a factor of two or more compared to a surface laid flowline. A recent DeepStar (2009) project has investigated the economic tradeoffs associated with conventional flowline insulation and the added installation costs of burial for insulation.

In the Gulf of Mexico, it is envisaged that the trench containing the flowline will be backfilled with native soil using a method involving undercutting the trench walls, which results in material falling onto the flowline. Available equipment can create a trench up to 6 ft deep. The resulting backfilling process provides as much as 3 ft clear cover above the flowline. There have been numerous laboratory investigations on the upheaval buckling resistance of buried flowlines and analysis methods (Schaminee et al. 1990; Palmer et al. 1990; White et al. 2001; Bransby, et al. 2002; Palmer et al. 2003).

Flow assurance investigations of buried flowlines require knowledge of the thermal properties of soils. Data for Gulf of Mexico cohesive soils are given by Young et al. (2001) and Garmon and Spikula (2003). Generic empirical correlations between thermal and soil properties are provided by Mitchell (1993).

8.3 Analysis Methodology

The numerical analysis of lateral and upheaval buckling follows standard procedures for any stability analysis. The success of the analysis depends on robust numerical techniques leading to a converged solution. Some general guidelines for the analyses are:

- Large displacement formulation is required to capture the so-called P-Δ effects leading to buckle initiation and amplified bending forces.
- The initial flowline alignment should include geometric imperfections necessary to initiate buckles.
- The lateral and axial soil-pipe interaction must be simulated with an elastic-plastic model that dissipates energy, otherwise unrealistically large stored strain energy may lead to numerical difficulties.
- When flowline plasticity is expected, the integration of the forces across the pipe section should include sufficient points to accurately capture the axial stress variation so that the bending resistance is accurately represented.
- A von Mises type material model is suitable for characterization of the flowline response, and the transition from elastic to elastic-plastic responses should be rounded/continuous and include a strain-hardening slope after the transition to preclude extraneous numerical results.

8.3.1 Lateral Buckling

Investigation of lateral buckling requires specific knowledge of the near surface soil strength profile. The soil strength profile and some estimate of the initial flowline penetration profile will allow the determination of the initial breakout soil resistance and whether the subsequent soil resistance will be strain softening or strain hardening as the lateral pipe displacement increases. The expected response in cohesive soils is determined by the normalized parameter $W/S_u D$ (Cheuk et al. 2007, 2008; McCarron 2008) where W is the submerged pipe weight/length, S_u is the shear strength at the base of the pipe, and D is the pipe outer diameter, including coating. Figure 8.3 shows schematically the soil resistance to lateral pipe displacement. Heavy pipes characterized by $W/S_u D > {\sim}2$ will dive to greater penetration while moving horizontally with a strain-hardening response, and light pipes, $W/S_u D < {\sim}2$, will rise and exhibit a strain-softening response. A neutral condition occurs at $W/S_u D = {\sim}2$, and the pipe will move horizontally.

For the special case of a monotonically increasing soil shear strength with depth, and accepting that a pipe will seek the neutral pipe elevation with $W/S_u D = {\sim}2$, then one can evaluate the final elevation a pipe will seek. Heavy pipes are by definition at an initially shallow penetration and will dive to the elevation with the appropriate S_u. Light pipes are by definition at an initially deep penetration and will rise to the elevation with the appropriate S_u.

The estimation of the initial pipe embedment is important to the evaluation of initiation of lateral buckling. The initial embedment is a function of the installation methodology, the pipe stiffness, and the soil strength profile. At the beginning of the design phase, when information is often

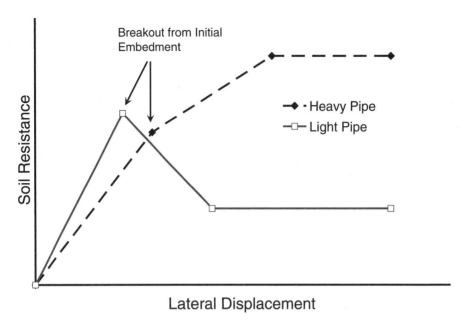

Figure 8.3   Lateral soil resistance to pipeline displacement.

lacking, the initial embedment is often assumed based on regional experience and observations. A number of bearing capacity estimates of penetration resistance for pipe shapes in cohesive soils are available (Murff et al. 1989; Randolph and White, 2008; Dingle et al. 2008) and have been used with varying success. A principal limitation of analytical limit analysis methods is the inability to adequately account for a depth-dependent strength (most solutions) and its strain-softening characteristics (soil sensitivities from 2 to 3), which are important for large vertical pipe penetrations. The need for convenient assessment of the penetration resistance leads to development of empirical correlations calibrated against laboratory tests and/or numerical data.

Calibration of the soil-pipe interaction model can be based on laboratory data or results from numeric simulations of the soil-pipe interaction. High quality laboratory investigations are time consuming. One particular setup at the Norwegian Geotechnical Institute is illustrated in Figure 8.4.

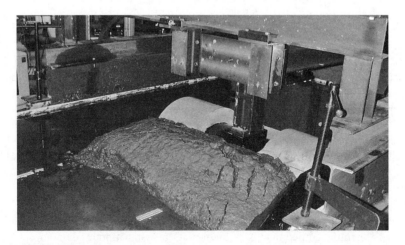

Figure 8.4   NGI pipe test facility.

Representative numerical simulations are increasingly possible and is illustrated in a later chapter. Laboratory tests should be designed to accommodate independent control of the vertical and lateral pipe motions. In the vertical direction, the ability to simulate a constant vertical force (weight of flowline) is mandatory.

Lambrakos (1985) performed field tests using towed pipe segments to measure the lateral soil resistance on 24-inch diameter pipes on clay and sand seafloors. The maximum pipe embedment was about 1 inch. The equivalent peak (as opposed to residual) lateral friction coefficients were typically in the range of 0.45 and 0.75 for clay and sand, respectively. Since the embedments were small ($\sim D/24$), these data may be construed as the resistances for a light pipe sliding across the seafloor.

Having performed numerical investigations of flowline buckling for various geometric imperfection amplitudes, it is useful for the designer to compare those results to the analytical solutions of Hobbs (1984), which are small displacement bifurcation investigations assuming the axial and lateral soil resistances are frictional. There are some unexpected results associated with the observed lateral soil-pipe interactions:

- Heavy pipes experience increasing lateral resistances under monotonic lateral displacements as a result of increasing pipe embedment.
- For a given flowline diameter and weight, larger ultimate lateral resistances are achieved in weaker soils as a result of the increasing pipe embedment.
- A stronger soil can result in increased bending moments in flowline buckles, due to a tendency for the buckle length to be shorter with increased curvature.

8.3.2 Flowline Walking

The mechanism of hysteretic translation (walking) of unrestrained flowlines as a result of thermal transients associated with flowline heating is critical in many situations. Walking results in the flowline moving in the direction from the hot to cold end of the flowline (Tornes et al. 2000). If unaccounted for, equipment interfaces at PLETs may be overstressed. If a flowline transitions to a catenary riser, the cumulative effect of the riser load and walking can pull curved flowlines out of their initial alignment.

The axial coefficient of friction is critical to walking evaluations. The most convenient, cost-effective, and robust determination of the pipe-soil interface frictional characteristics for cohesive soils is via a tilt-table test (Najjar et al. 2003, 2007). For several Gulf of Mexico clays, the soil-pipe residual friction coefficient has been found to be between 0.4 and 0.6 (friction angle 22° to 31°), with results skewed to the higher value. The tilt-table tests (Figure 8.5) are convenient for evaluating the drained internal soil friction angle, as well as the soil-pipe interface, including pipe coating.

It is important to correctly model the interface displacement required to mobilize the friction in walking analyses. Generally, the peak interface friction is mobilized after 2 to 5 mm displacement, and the residual friction is generally achieved after 25 to 50 mm of displacement (Najjar et al. 2003, 2007). Considering that axial flowline displacements are many times the mobilization displacement, it may not be too important to capture the peak friction value, and in any case, it is unknown if the peak friction is available after several operational cycles. Tilt-table results indicate that pipe coating can decrease the residual interface friction to 60 to 90% of the soil internal residual friction. Stable numerical evaluations of flowlines are achieved with mobilization displacements of 5 mm. Using a too small mobilization displacement may lead to convergence difficulties in numerical analyses.

Figure 8.5   Tilt-table test for soil-pipe interface friction, courtesy of Robert Gilbert.

The magnitude of the axial coefficient has different implications for walking and lateral buckle development. While high axial coefficients of friction reduce walking, they result in lateral buckles being generated at closer intervals.

8.3.3 Upheaval Buckling

The soil resistance to upheaval buckling includes two components: the weight of the backfill and its shear/material resistance. The flowline self-weight will also introduce some buckling resistance. There is some guidance in the literature on the evaluation of upheaval soil resistance (Palmer et al. 1990; Schaminee et al. 1990; Bransby et al. 2002), but the empirical equations describing the resistance vary considerably, and there is little guidance on the shape of the resistance-displacement curve.

It will generally be required to account for interaction of the buried flowline with the vertical walls of the trench. Sufficiently stiff and strong pipes will be able to push into the face of the trench wall and the intact native soil. Lateral motions between the trench walls will encounter a softer backfill material.

8.3.4 Modeling Scheme

Schematics of the soil-pipe interaction model described in this section are collected in Figure 8.6. For consistent and robust investigations of soil-pipe interaction, the implementation of the model

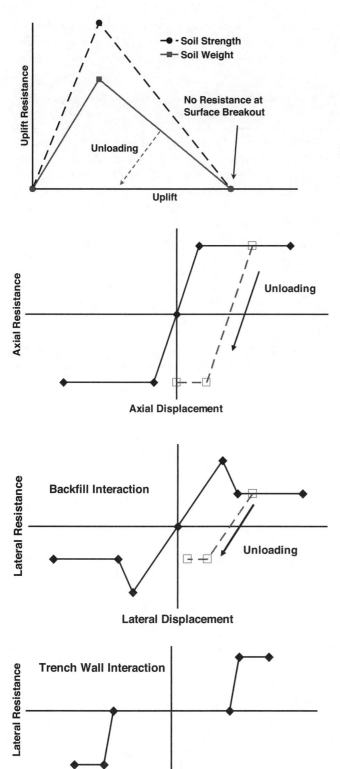

Figure 8.6 Model schematic for soil resistance to buried pipe displacement.

must be in the form of an elastic-plastic response. For buried flowlines, the lateral resistance is a combination of the backfill and intact materials. This is accomplished via an overlay modeling technique with one component based on the backfill resistance and the second being the incremental resistance of the intact material.

8.4 Illustrative Examples

The modeling of soil-pipe interaction is demonstrated here through the analysis of a flowline for walking, lateral buckling, and upheaval buckling. The analyses are performed with the finite element program Abaqus (2008). The flowline responses are driven by the effects of thermal expansion. The current industry design codes and design guidelines include a temperature dependent thermal expansion coefficient and strength de-rating (DNV 2010; SAFEBUCK 2004) for the flowline material. These important issues are not discussed here.

8.4.1 Lateral Soil-Pipe Resistance

The effects of the soil strength profile on the lateral soil-pipe interaction response are discussed for three soil strength profiles illustrated in Figure 8.7. We show here the very near surface strength characteristic, as this is the zone of influence for lateral pipe resistance. A hypothetical pipe diameter, including coating, of 12.625 inches is considered for two different submerged weights, 45 lbf/ft and 90 lbf/ft. Laboratory results (Cheuk et al. 2007, 2008; McCarron, 2008) indicate the initial breakout soil resistance is primarily determined by the pipe embedment and the soil

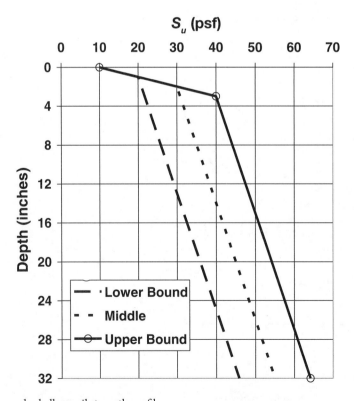

Figure 8.7   Example shallow soil strength profile.

strength at the pipe base. The lateral resistance for the two pipe weights and three strength profiles are shown in Figure 8.8 for an initial pipe embedment equal to $D/3$. The response for the 45 lbf/ft-weight pipe shows that for the two strongest soil profiles, the lateral response is strain softening, and for the lower-bound soil profile, the response is slightly strain hardening. The neutral shear strength in this case is 21.4 psf ($W/2D$). For the pipe weight of 90 lbf/ft, the lateral response is strain hardening for all three conditions, as the pipe dives to the neutral soil shear strength of 42.8 psf. For all but the strongest soil profile in Figure 8.7, this would result in a pipe penetration in excess of the diameter, which is outside the intended use of the empirical resistance models.

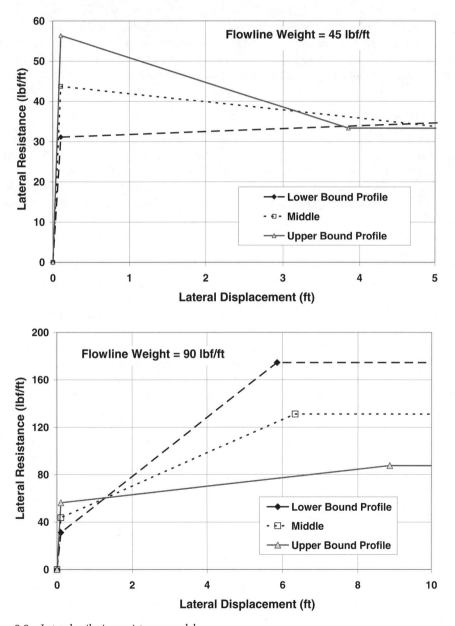

Figure 8.8   Lateral soil-pipe resistance model.

The breakout resistance is that associated with first movement of the pipe out of its initial embedment. In this case, it is assumed to occur at a pipe displacement equal to $D/10$ and the lateral resistance is $1.3S_u D$, as indicated by experimental data (McCarron 2008) for an initial embedment of $D/3$. For a so-called *light pipe* condition, this breakout resistance represents the peak resistance, while for a *heavy pipe* condition, the soil resistance increases as the pipe dives deeper. As indicated in Figure 8.8, for the pipe weight of 45 lbf/ft, the residual resistance for the three soil profiles is similar because the final pipe embedment is shallow. For a pipe weight of 90 lbf/ft, the final resistance is larger for the weakest soil profile, since the pipe dives to a greater penetration to reach the neutral shear strength. The shapes of the soil resistance load-displacement curves shown in Figure 8.8 are empirically based. In Chapter 9, we illustrate how to develop the resistance curves via empirical relationships calibrated against laboratory data and numerical analysis.

It is useful to point out that for light pipe conditions, the lateral soil resistance reduces with increasing displacement, which can lead to increased bending moments in the pipe during lateral buckling, as the pipe will tend to displace laterally more at locations of reducing resistance.

8.4.2 Lateral Buckling

A simplified lateral buckling investigation neglecting internal and external pressures is undertaken here. In this case, the effective tension and the material tension in the pipe are identical. The particular pipe investigated has an outer diameter (OD) of 6.625 inches and a wall thickness of 0.875 inches. The final insulated diameter is taken as 12.625 inches so that the lateral resistance curves above can be used. The analysis also investigates the effect of differing geometric imperfection magnitudes on the buckling resistance. The imperfection included is a continuous symmetrical shape of length 240 ft. The finite element model simulates a flowline having a total length of 5000 ft with fixed ends. The axial forces are generated only in response to thermally induced strains.

Table 8.1 and Figure 8.9 demonstrate the flowline buckling sensitivity to geometric imperfections. An alternative to geometric imperfections is the application of disturbing forces, but imperfections are generally easier to include and more closely simulate the reality of flowline fabrication and installation. Figure 8.10 shows the buckle force versus buckle amplitude history for the flowline-submerged weight of 45 lbf/ft and 6-inch imperfection amplitude. The post-peak softening response results in axial feed into the buckled region as the axial force is reduced and the buckled pipe segment moves laterally. The deformed shape illustrated in Figure 8.11 is similar to the Hobbs solutions (1984). Also shown in this figure are the bending moments. The displacements and bending moments indicate a slightly increased response in the center lobe for the upper-bound soil strength, due to the increased soil resistance following breakout (Figure 8.8) for this case. This effect can be more pronounced immediately after the buckle forms and before the pipe breakout in the negative lobe. The three-lobe buckle shape is a result of a local lateral equilibrium requirement; soil reactions in segments having positive displacements are balanced by those in segments with negative displacements.

Table 8.1 Buckling force (kips)

Imperfection (inches)	45 lbf/ft, lb	45 lbf/ft, middle	45 lbf/ft, ub	90 lbf/ft, lb	90 lbf/ft, middle	90 lbf/ft, ub
3	91	115	140			
6	66	84	100	69	85	100
12	45	56	66			
Hobbs	85	96	106			

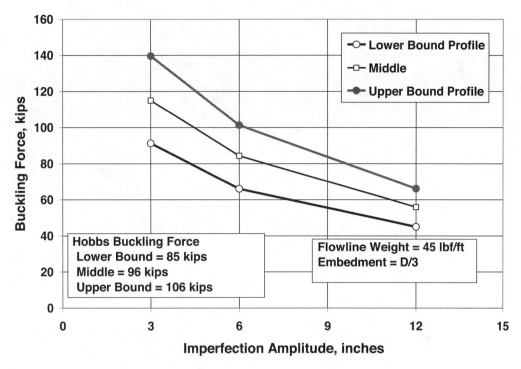

Figure 8.9 Buckle imperfection sensitivity.

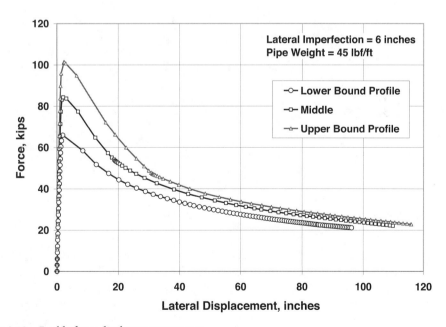

Figure 8.10 Buckle force-displacement response.

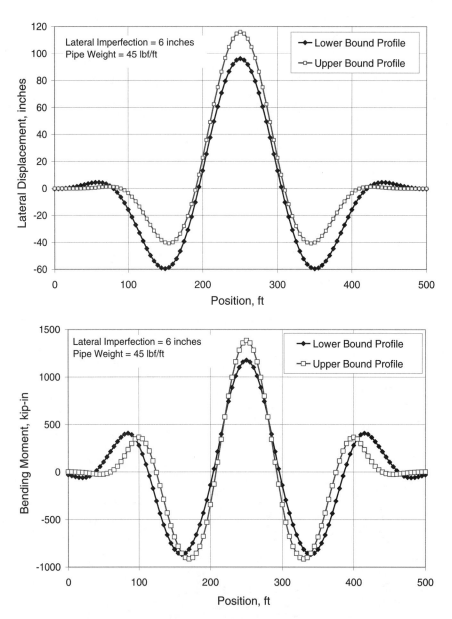

Figure 8.11   Buckle shape and bending moments, 45 lb/ft pipe weight.

The effects of pipe weight on the lateral buckle response are illustrated in Figure 8.12. The soil-pipe resistance for the two pipe weights is shown in Figure 8.8, where it is seen that the heavier pipe has a slight strain-hardening response in the upper-bound profile. This results in smaller lateral displacements in the center lobe of the buckle and slightly increased lateral displacements in the negative lobes. The negative lobe displacements increase to achieve lateral equilibrium in the buckled segment. The bending moments in the two conditions are not very different, as shown in Figure 8.12.

Table 8.2 summarizes the pipe embedment in terms of the normalized embedment z/D. In each case, the final embedment is determined as the elevation of the neutral shear strength for each soil strength profile.

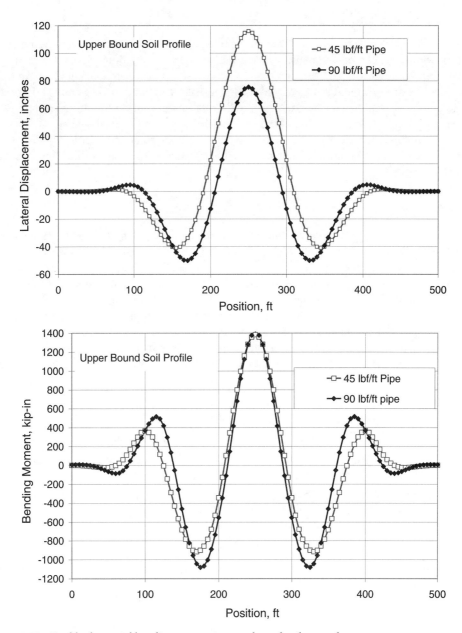

Figure 8.12 Buckle shape and bending moment, upper-bound soil strength.

Table 8.2 Initial and final flowline embedments (z/D)

Pipe weight (lbf/ft)	Neutral soil strength (psf)	Normalized initial embedment (z/D)	Lower bound	Middle	Upper bound
45	21.4	0.333	0.211	0.090	0.090
90	42.8	0.333	2.244	1.373	0.501

8.4.3 Upheaval Buckling

The upheaval buckling example uses the same pipe section as the lateral buckling example, but this time without insulation as it may not be required for buried conditions. The vertical soil resistance is shown in Figure 8.13. Development of the vertical resistance components is presented in Chapter 9. The finite element model of the pipe is 300-ft-long with a vertical geometric imperfection at the center to initiate the buckle. The force-displacement (buckle amplitude) responses are shown in Figure 8.14. Again, there is a softening response leading to axial feed-in to the buckle as the pipe unloads and displaces laterally. The buckle displacement pattern in Figure 8.15 for upheaval buckling is different (without negative lobes) than that for lateral buckling and has a shorter wavelength.

8.4.4 Walking

Examples of the effects of walking are illustrated with a flowline of length 34,500 ft. The pipe has an OD of 10.75 inches with a wall thickness of 0.67 inches. Applied thermal insulation with a thickness of 2.75 inches results in a submerged weight of 35.2 lbf/ft. The analysis here considers an axial coefficient of friction of 0.6. Due to its length, lateral buckles are introduced at five specific locations to control the bending moments. The buckles are introduced by the use of buoyancy, which reduces the lateral soil-pipe resistance and results in an asymmetric buckle shape (Figure 8.16). An alternative to the local application of buoyancy is the use of sleepers, which raise the pipe off the seafloor and completely eliminate lateral soil resistance. Brown et al. (2006) showed that the use of sleepers results in a marked increase in the reliability of the calculated flowline buckling load.

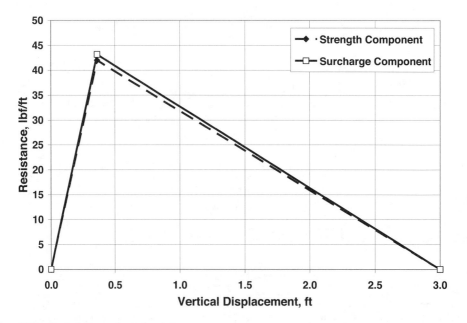

Figure 8.13   Buried pipe upheaval resistance.

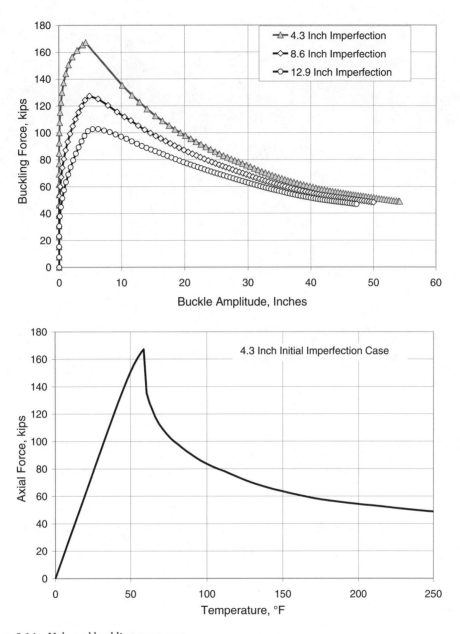

Figure 8.14 Upheaval buckling responses.

Walking was originally discussed in the literature with respect to flowlines having unrestrained ends. In practice, flowlines are physically attached to subsea equipment of some sort and, as a result, have some limitations on the displacements that can be accommodated. In this case, the analyses consider two situations with either an anchor at the hot end of the flowline or an anchor at both ends. The analysis considers 25 thermal cycles in the range of 39 to 250°F. The transient temperature history is illustrated in Figure 8.17.

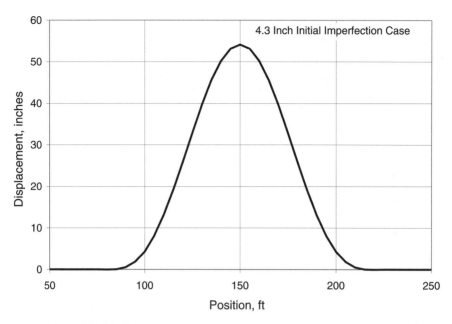

Figure 8.15 Upheaval buckle shape.

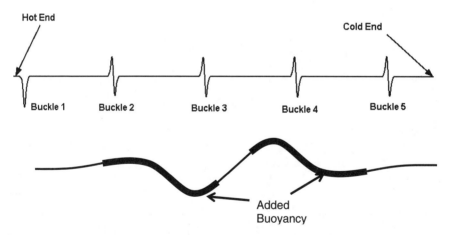

Figure 8.16 Flowline buckle intervals and buoyancy.

Figures 8.18 and 8.19 present the time histories of the end displacements for each analysis. The analyses with both ends piled shows the relatively constrained end displacements, while the analysis with the unrestrained cold end indicates rather large displacements increasing with each thermal cycle. Also shown are the effective tension diagrams for three specific cycles. In the hot condition, the effective tensions are relatively unchanged from cycle to cycle. Nonzero effective tensions at the end positions indicate the magnitude of the pile reactions. While the analysis with both ends restrained shows little cyclic effect on the cold condition effective tensions, the case with the hot end restrained indicates changes in the effective tension diagram and the cold end pile displacement with increasing cycles.

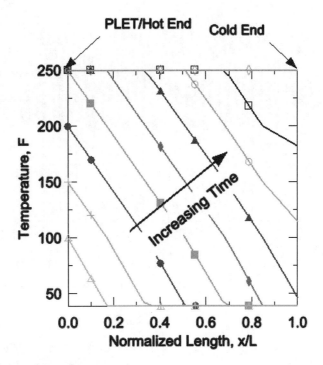

Figure 8.17 Walking analysis temperature transients.

The effective tensions at buckle locations in the hot condition are local minimums. Between the buckles, the effective tension increases (becomes more negative) as a result of soil friction. The design buckle spacing must consider the most compressive effective tension occurring between them, which is a function of the residual effective tension at the buckle locations and the soil-pipe interface friction. The compressive effective tension must be less than the buckle resistance of the flowline at any particular point. The slope of the effective tension diagrams between the buckles is due to the soil-pipe friction accumulated per unit length of the pipe.

8.5 Conclusions

The principal features controlling flowline buckling are the initial pipe embedment, the flowline submerged weight, and the shallow soil shear strength profile. These three features control the pipe trajectory as it moves laterally in a buckled segment. The lateral soil resistance is a function of the normalized pipeline weight W/DS_u. Regardless of the initial pipe embedment, its trajectory will result in a final neutral elevation value of W/DS_u = ~2 in the strength profile. Heavy pipes with initial normalized weights greater than 2 will dive to the neutral normalized weight of 2. Light pipes with initial normalized weights less than 2 will climb to the neutral weight elevation. Using this simple rule, the lateral pipe resistance is easily determined as either strain hardening (heavy pipe) or strain softening (light pipe).

The residual axial soil-pipe friction coefficient generally falls in a narrow range of 0.5 ± 0.1 for Gulf of Mexico cohesive soils tested on the tilt-table device. The tilt-table test operates in the regime of 1g soil stresses, so soil-pipe interface friction can be conveniently measured at the low stress values existing. The internal soil residual friction coefficient is higher, up to ~0.9, so the pipe coating effect is significant.

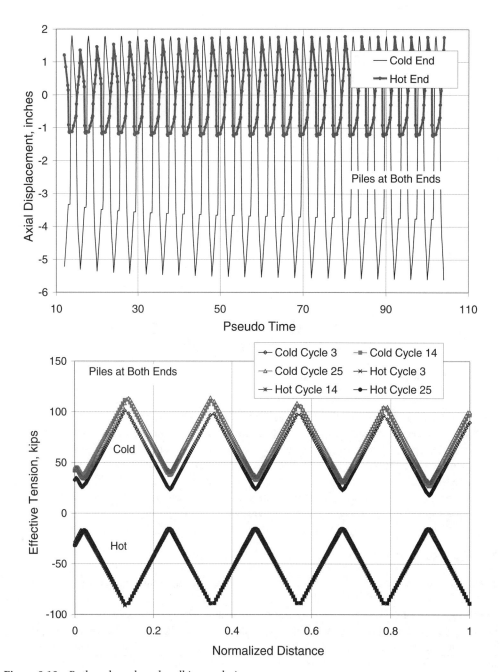

Figure 8.18 Both ends anchored, walking analysis.

The design of subsea flowlines requires consideration of the potential development of thermally induced buckles. For surface laid flowlines, the design process is typically to manage the occurrence of thermal buckles through the introduction of buckles at specific intervals. This is most easily accomplished through the use of sleepers or added flowline buoyancy that reduce the lateral soil resistance at the planned buckle locations. Without the use of these auxiliary features, the prediction of when and where lateral buckles form would be associated with a high degree of uncertainty and reduce the robustness of the design.

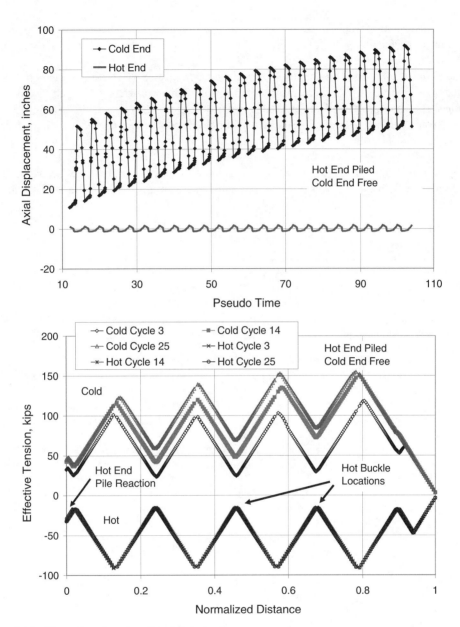

Figure 8.19　Hot end anchored, walking analysis.

Upheaval and lateral buckle shapes are different, with the lateral buckles of surface laid flow-lines developing some positive and negative lobes in response to local equilibrium requirements. Buckle development is accompanied by axial unloading and feed-in to the buckled region, increasing its lateral or vertical displacement. Buckles may develop in response to temperature increases as small as 50°F.

The initial flowline embedment after installation is a critical parameter controlling the buckling force. There are several methods for determining the initial embedment available to the designer. Care should be taken to ensure that the methodology adequately accounts for soil strength profile and its sensitivity (softening characteristics).

Flowline end displacements in response to thermal expansion may be substantial. To limit the accumulation of walking displacements and reduce demand on the PLET interface equipment, end anchors may be used.

Modeling of soil-pipe interaction is generally accomplished via the finite element method utilizing beam-column type elements in combination with empirically developed soil resistance models. The axial, lateral, and upheaval soil resistances are generally considered to be independent of each other. The soil resistance models should be elastic-plastic so that energy is dissipated; otherwise stored strain energy may result in numerical difficulties in models with several buckles.

While not discussed here, the designer should be aware that thermal effects may result in elongation of production casing, resulting in elevation changes of subsea wellheads. Such displacements will introduce loads in interface equipment spanning between the wellhead and the flowline.

8.5 References

Abaqus, 2008. Abaqus User's Manual, V6.8, SIMULIA, Providence, Rhode Island.

Bransby, M. F., Newson, T. A., and Brunning, P., 2002. The Upheaval capacity of pipelines in jetted clay back-fill, *International Journal of Offshore and Polar Engineering* 12(4): 280-287.

Brown, G., Brunner, M., Qi, X., and Stanley, I., 2006. Lateral buckling reliability calculation methodology accounting for buckle interaction, Paper presented at Offshore Technology Conference, OTC, Houston, Texas 17795.

Brunner, M. S., X. Qi, Zheng, J., and Chao, J. C., 2006. Combined effect of flowline walking and riser dynamic loads on HP/HT flowline design, Paper presented at Offshore Technology Conference, OTC 17806.

Bruton, D., Carr, M., Crawford, M., and Poiate, E., 2005. The safe design of hot on-bottom pipelines with lateral buckling using the design guideline by the SAFEBUCK joint industry project, Deep Offshore Technology Conference, Vitoria, Espirito Santo, Brazil.

Bruton, D. A., White, D. J., Carr, M., and Cheuk, C. Y., 2006. Pipe-soil interaction during lateral buckling and pipeline walking—The SAFEBUCK JIP, Paper presented at Offshore Technology Conference, OTC 19589.

Carr, M., Sinclair, F., and Bruton, D., 2006. Pipeline walking—Understanding the field layout challenges, and analytical solutions developed for the SAFEBUCK JIP, Paper presented at Offshore Technology Conference, OTC 17945.

Cheuk, C. Y., White, D. J., and Bolton, M. D., 2007. Large-scale modeling of soil-pipe interaction during large amplitude cyclic movements of partially embedded pipelines, *Canadian Geotechnical Journal* 44:977-996.

———, 2008. Large-Scale modeling of soil-pipe interaction during large amplitude cyclic movements of partially embedded pipelines, Reply to discussion by McCarron, *Canadian Geotechnical Journal* 45:744-749.

DeepStar, 2009. Flowline burial for insulation, CTR 9302, proprietary industry study.

DNV, 2010. Offshore standard OS-F101, Marine Pipeline Systems, Det Norske Veritas.

Dingle, H. R. C., White, D. J., and Gaudin, C., 2008. Mechanisms of pipe embedment and lateral breakout on soft clay, *Canadian Geotechnical Journal* 45:636-652.

Garmon, S. and Spikula, D., 2003. *Verification of the thermal performance of buried deepwater flowlines*, Global Petroleum Research Institute, Texas A&M University.

Harrison, G. E., Brunner M. S., and Bruton, D. A. S., 2003. King Flowlines—Thermal expansion design and implementation, Paper presented at Offshore Technology Conference, OTC 15310.

Hobbs, R. E., 1984. In-service buckling of heated pipelines, ASCE *Journal of Transportation Engineering* 110(2): 175-189.

Hobbs, R. E., and Liang, F., 1989. Thermal buckling of pipelines close to restraints, In *Proceedings of the 8th International Conference on Offshore Mechanics and Arctic Engineering*, 121-127.

Kerr, A. D., 1978. Analysis of thermal track buckling in the lateral plane, *Acta Mechanica* 30: 17-50.

Lambrakos, K. F., 1985. Marine pipeline soil friction coefficients from in-situ testing, *Ocean Engineering* 12(2): 131-150.

McCarron, W. O., 2008. Large-scale modeling of soil-pipe interaction during large amplitude cyclic movements of partially embedded pipelines, Discussion in *Canadian Geotechnical Journal* 44:977-996.

McCarron, W. O. and G. E. Harrison, 2008. High pressure-temperature flowline burial for insulation, In *Proceedings of the 3rd International Offshore Pipeline Forum*, IOPF2008-909.

Mitchell, J. K., 1993. *Fundamentals of soil behavior*, 2nd ed., Wiley-Interscience.

Murff, J. D., Wagner, D. A., and Randolph, M. F., 1989. Pipe penetration in cohesive soil, *Geotechnique* 39(2) 213-229.

Najjar, S. S., Gilbert, R., Liedtke, E. A., and McCarron, B. 2003. Tilt table test for interface shear resistance between flowlines and soils, Paper presented at International Conference on Offshore Mechanics and Arctic Engineering, OMAE2003-37499.

Najjar, S. S., Gilbert, R. B., Liedtke, E. A., McCarron, B., and Young, A. G., 2007. Residual shear strength for interfaces between pipelines and clays at low effective normal stresses, ASCE *Journal of Geotechnical and Geoenvironmental Engineering* 133(6): 695-706.

Palmer, A. C., Ellinas, C. P., Richards, D. M., and Guijt, J., 1990. Design of submarine pipelines against upheaval buckling, OTC Paper No. 6335: 551-560.

Palmer, A. C., White, D. J., Baumgard, A. J., Bolton, M. D., Barefoot, A. J., Finch, M., Powell, T., Faranski, A. S., and Baldry, J. A., 2003. Uplift resistance of buried submarine pipelines: Comparison between centrifuge modeling and full-scale tests, *Geotechnique* 53(10): 877-883.

Palmer, A. C. and King, R. A., 2008. *Subsea pipeline engineering*, 2nd ed., Pennwell Corporation, Tulsa, Oklahoma.

Randolph, M. F. and White, D. J., 2008. Upper-bound yield envelopes for pipelines at shallow embedment in clay, *Geotechnique* 58(4): 297-301.

SAFEBUCK, 2004. Safe design of pipelines and with lateral bucking, design guideline, Report BR02050, proprietary industry report.

Schaminee, P., Zorn, N. F., and Schotman, G., 1990. Soil response for pipeline upheaval buckling analyses: Full-scale laboratory tests and modeling, OTC Paper No. 6486:563-572.

Sparks, C. P., 1983. The influence of tension, pressure and weight on pipe and riser deformations and stresses, In *Proceedings of the 2nd International Offshore Mechanics and Arctic Engineering Symposium*, 443-452.

———, 1984. The influence of tension, pressure, and weight on pipe and riser deformations and stresses, Transactions of the ASME106:46-54.

Tornes, K., Ose, B. A., Jury J., and Thomson, P., 2000. Axial creeping of high temperature flowlines caused by soil ratcheting, In *Proceedings of ETCE/OMAE Joint Conference* 5055:1-11.

White, D. J., Barefoot, A. J., and Bolton, M. D., 2001. Centrifuge modeling of upheaval buckling in sand, *International Journal of Physical Modeling in Geotechniques* 2:19-28.

Young, A. G., Osborne, R. S., Frazer, I., 2001. Utilizing thermal properties of seabed soils as cost-effective insulation for subsea flowlines, Paper presented at Offshore Technology Conference, OTC 13137.

9

Modeling of Soil-pipe Interaction

William McCarron, PhD, PE
ASGM Engineering

9.1 Introduction

In Chapter 8, the significant details of subsea flowline lateral and upheaval buckling were described. In this chapter, methods to calibrate empirical lateral and upheaval soil-pipe resistance models with the aid of limit analysis procedures and numerical finite element (FE) analyses are described. Where available, results of physical tests are presented to augment and illustrate the validity of the empirical models.

9.2 Lateral Soil-pipe Resistance

The lateral soil-pipe resistance is dependent on its initial embedment, the soil strength at the pipe invert, and the flowline-submerged weight. In the Gulf of Mexico deepwater environment, the surface soils typically have undrained strengths in the range of 20 to 60 psf (Jeanjean et al. 1998). Occasionally, in regions of high seafloor currents or where landslides have occurred, the weaker material has been removed, and stronger materials are exposed, which results in a small pipe embedment and a reduced lateral resistance.

The available methods to determine pipe embedment are physical tests, limit analyses, or numerical analyses. All these methods may be confirmed by regional observations of installed flowline responses. The embedment is driven by both the flowline operating weight and the flowline installation method. The latter usually results in seabed contact pressures larger than the flowline operational weight (Wang et al. 2009).

Murff et al. (1989) presented a limit analysis bearing capacity solution for a pipe embedded in a homogenous Tresca material. This was later extended to account for a depth dependent strength and lateral motions by Randolph and White (2008). Merifield et al. (2008) augmented that work with numerical FE investigations of lateral soil-pipe interaction. Together, these significant analytical and numerical contributions result in a detailed phenomenological description of the response of flowlines under vertical and lateral loading.

9.2.1 Vertical Embedment Solution

Limit analysis solutions are available for normalized pipe embedments of $w = z/D \leq 0.5$. Figure 9.1 describes the vertical penetration resistance for a smooth pipe in foundations of uniform strength as well as depth dependent strength. The depth dependent strength is of the form $S_u = kz$. In all cases, the material is perfectly plastic without hardening or softening responses. For smooth and rough pipes in weightless homogenous strength cohesive soil, Merifield et al. (2009) produced the following analytical expressions for resistance to penetration with $0 \leq w \leq 0.5$:

$$N_{cV} = 5.66w^{0.32} \text{ smooth pipe} \tag{9.1}$$

$$N_{cV} = 7.40w^{0.4} \text{ rough pipe} \tag{9.2}$$

where $N_{cV} = V/DS_u$ and V is the vertical force.

Using the Abaqus Eulerian analysis capabilities (Abaqus 2008), vertical penetration analyses were conducted for a homogeneous shear strength (strain softening included), and the results are presented in Figure 9.2. The numerical results indicate a slightly higher resistance than does limit analyses at shallow embedment, likely due to the fact that the limit solutions neglect geometry changes associated with the surface soil heave and weight effects (although weight effects are small). The numerical analyses included a strain-softening material response with a two-thirds reduction in strength at a plastic strain of 500%.

9.2.2 Lateral Breakout Resistance

The lateral breakout resistance from its initial embedment controls the buckling capacity of the flowline. Early on, the breakout resistance was determined via empirical relationships developed from laboratory observations. Those laboratory tests provided valuable experience relating the effects of the pipe diameter, pipe weight, initial embedment, and lateral breakout resistance (see Chapter 8). Subsequently, numerical and limit analysis methods have been used to clarify aspects of the response (Dingle et al. 2008; Merifield et al. 2009; Wang et al. 2010). Today's numerical capabilities and previous laboratory programs have significantly reduced the need for new laboratory investigations. Figure 9.3 compares the breakout resistance determined via laboratory and numerical results. Merifield et al. (2008) described the lateral breakout resistance for smooth and rough pipes as:

$$N_{cH} = 2.7w^{0.64} \text{ smooth pipe} \tag{9.3}$$

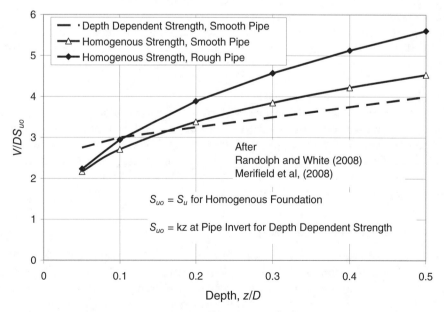

Figure 9.1 Pipe penetration resistance from limit analysis.

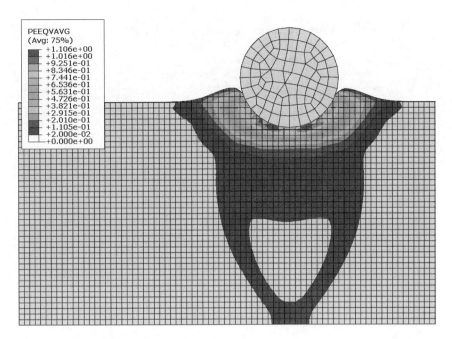

Figure 9.2 Pipe penetration resistance; Eulerian finite element analysis.

$$N_{cH} = 3.0w^{0.58} \text{ rough pipe} \qquad (9.4)$$

where $N_{cH} = H/DS_u$, w is the ratio of the embedment to diameter (z/D), and H is the lateral force. Also included in Figure 9.3 is experimental data from an offshore West African clay (Cheuk et al. 2007) and results of Eulerian FE analyses performed with the Abaqus program. Figure 9.4 illustrates the Eulerian FE model response. The numerical analyses included a strain-softening material response with a two-thirds reduction in strength at a plastic strain of 500%. The results of the empirical resistance equations have been extended beyond the limits of w for which they were developed so that they may be compared to the laboratory and FE results. We see in Figure 9.3 that the range in resistance results is in the vicinity of ±7%, which is encouraging considering the uncertainties and approximations involved.

Considering the results in Figure 9.3, a single empirical expression of the form of Equations 9.3 and 9.4 is adopted:

$$N_{cH} = 2.85w^{0.6} \qquad (9.5)$$

For small embedments, the lateral resistance should not be taken as less than ~0.5 times the weight of the pipe, which represents a minimal lateral friction resistance for light pipes. The implied coefficient of friction μ = ~0.5 is arrived at from the consideration of tilt-table tests (Najjar et al. 2007) and full-scale field tests (Lambrakos 1985).

9.2.3 Lateral Residual Resistance

The final component of the empirical lateral resistance model is the resistance expected at large displacements. Here we are concerned with the initial monotonic motion of the pipe during its lateral movement as a result of the formation of a lateral buckle. Associated with this resistance definition is the lateral displacement at which it occurs. In combination with the breakout

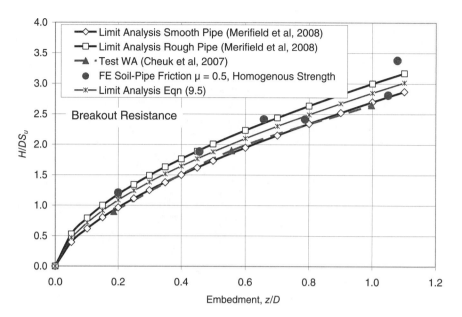

Figure 9.3   Breakout resistance as a function of pipe embedment.

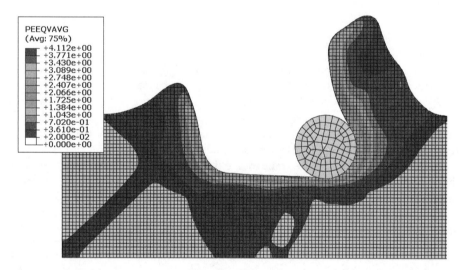

Figure 9.4   Lateral pipe resistance; Eulerian finite element model.

resistance, a force-displacement response is then developed for use in numerical calculations of the global flowline response. Based on numerical results of the same type described above for the breakout resistance, the residual lateral resistance response shown in Figure 9.5 is defined as

$$N_{cH} = 2.9w^{0.48} \tag{9.6}$$

The form of Equation 9.6 is modeled after the results presented above for the breakout resistance. The numerical analyses included a strain-softening material response with a two-third reduction in strength at a plastic strain of 500%. The parameters in Equation 9.6 differ slightly from those recommended by Wang et al. (2010), also based on numerical analyses. At shallow embedments, it is recommended that the resistance should be at least as large as a purely frictional response with $\mu = {\sim}0.5$.

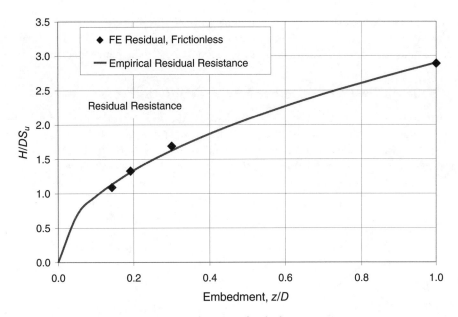

Figure 9.5   Radial lateral pipe resistance as a function of embedment.

Associated with the lateral movement of the pipe is a trajectory (including vertical and horizontal displacement components) dependent on pipe diameter, embedment, weight, and shear strength profile. As discussed in Chapter 8, a light pipe will rise in elevation until the W/DS_u ratio is in the range of 1.5 to 2, and a heavy pipe will dive in elevation to the same normalized pipe weight. The final elevation is referred to here as the neutral pipe elevation where the trajectory becomes horizontal. Figure 9.6 shows a recommended range of trajectory (vertical/horizontal displacement) as a function of the initial embedment, as well as some data extracted from the numerical work of Wang

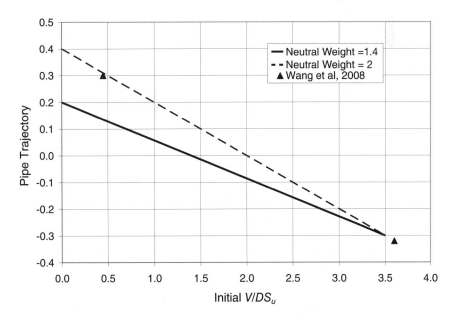

Figure 9.6   Pipe trajectory resistance during lateral loading.

et al. (2010). Trajectories extracted from this figure represent the average value between the initial embedment and the neutral pipe elevation. The two lines in the figure result in two different neutral pipe elevations with normalized pipe weights of 1.4 and 2 (V/DS_u where pipe trajectory is zero).

9.2.4 Example Lateral Resistance Model Calibration

We return to the example soil-pipe interaction first discussed in Chapter 8, which considered a pipe of diameter 12.625 inches with either a submerged weight of 45 or 90 lbf/ft and three different shear strength profiles. Using the empirical relationships developed in the previous sections, the lateral force-displacement relationships shown in Figure 9.7 have been developed. Tables 9.1 and 9.2 summarize characteristics of the problem.

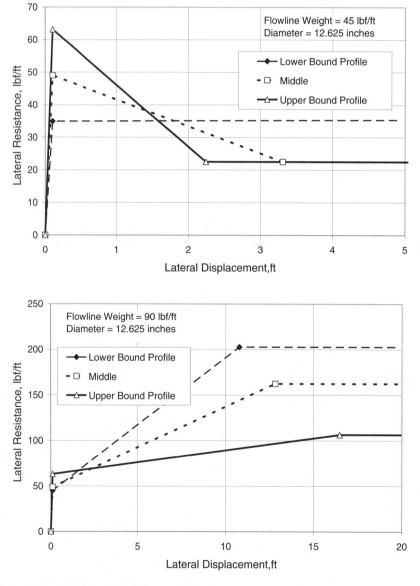

Figure 9.7   Example lateral resistance to pipe motion.

Table 9.1 Example lateral resistance; pipe weight = 45 lbf/ft

Soil profile	Initial condition		Final condition		Force-displacement	
	$w = z/D$	V/DS_u	$w = z/D$	V/DS_u (neutral)	Force, δ_{BO} lbf/ft, ft	Force, δ_{Res} lbf/ft, ft
LB	0.33	1.89	0.269	1.94	34.9, D/10	35.7, 9.1
Middle	0.33	1.34	0.095	1.94	49.0, D/10	22.5, 3.3
UB	0.33	1.04	0.095	1.94	63.1, D/10	22.5, 2.2

Table 9.2 Example lateral resistance; pipe weight = 90 lbf/ft

Soil profile	Initial condition		Final condition		Force-displacement	
	$w = z/D$	V/DS_u	$w = z/D$	V/DS_u (neutral)	Force, δ_{BO} lbf/ft, ft	Force, δ_{Res} lbf/ft, ft
LB	0.33	3.78	2.36	1.94	45.0, *D*/10	202, 10.8
Middle	0.33	2.69	1.49	1.94	49.0 , *D*/10	162, 12.8
UB	0.33	2.09	0.62	1.94	63.1, *D*/10	106, 16.5

Comparing the initial and final pipe embedments, the effects of the neutral pipe elevation are evident and control whether the pipe rises or dives from its initial condition. A trajectory slope of 0.129 has been used in these calculations. Light pipes with normalized weights V/DS_u less than the neutral value will rise in elevation when the soil shear strength is increasing with depth. In most cases, this leads to a reduced lateral soil resistance. The exceptions are due to the effects of berm formation or to a relatively small difference between the shear strengths at the initial and final pipe elevations.

9.3 Upheaval Soil-pipe Resistance

9.3.1 Upheaval Buckling Resistance

The soil resistance to upheaval buckling includes two components: the weight of the backfill and its shear/material resistance. The pipe self-weight will, of course, contribute additional resistance to uplift. The analyses here assume the disturbed backfill strength is defined by the soil sensitivity, strength, and submerged unit weight (22 pcf submerged weight was used here). The backfill material strength is reduced from the in situ value as a result of disturbance introduced as the trench sides are undercut and material falls into the trench.

While there are several methods of describing the upheaval buckling resistance (e.g., Schaminee et al. 1990), a common method is based on a simple bearing capacity type equation of the form

$$q = DS_u N_c \tag{9.7}$$

where D is the pipe diameter, S_u is the undrained strength of the backfill and N_c is a bearing capacity factor. For this application, S_u is the disturbed material shear strength, which is the result of the in situ shear strength of 33.5 psf and the sensitivity of 2 used here. The sensitivity of 2 is lower than typically encountered in the Gulf of Mexico (~3), in recognition that the surface material is already at a high natural water content and the backfill material will likely not be fully remolded.

Figure 9.8 shows results of an Eulerian FE (Abaqus 2008) analysis with ~26 inches of initial cover over the top of 9-inch diameter buried pipe. The FE determined reaction on the pipe at $t = 50$ in Figure 9.8 represents the effects of soil weight on the pipe. Based on a soil submerged weight

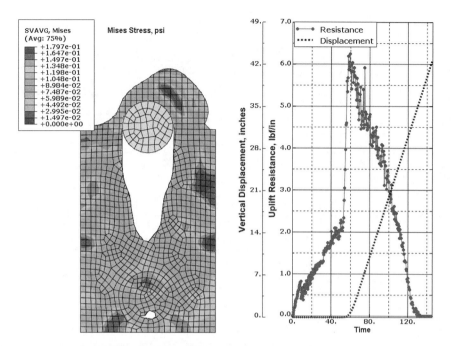

Figure 9.8   Determination of upheaval buckling resistance (t < 50 is simulation of self-weight application); Eulerian finite element analysis.

of 22 pcf, and a clear cover depth of 26 inches, a pipe reaction of 36 lbf/ft (resistance per length) or 3 lbf/in (corresponding to Figure 9.8 dimensions) is expected. The results indicate only 2 lbf/in, so the circular pipe shape evidently reduces the soil weight reaction force by one-third. The difference between the reaction force (load supported by pipe) at the end of the soil self-weight application and the peak resistance during uplift is ~4 lbf/in. Back calculation of the bearing factor for the soil strength of 16.7 psf (in situ value divided by sensitivity of 2) results in N_c = 3.86, below the factor for a circular pipe on a free surface (see Section 9.2.1). The numerical results also indicate that the pipe reaction goes to zero after ~30 inches of vertical displacement. Note that in this analysis, some further strain softening of the backfilled material is allowed (compared to the initial effects of disturbance that reduced the in situ strength by a factor of 2) to reduce the soil strength by 50%, compared to that at the beginning of the analysis.

In the FE analyses, the exterior vertical face boundaries are considered frictionless, which seems a reasonable assumption in light of uncertainties regarding the homogeneity of the back-filled soil mass.

The empirical model response is shown in Figure 9.9. The peak resistance is assumed to develop after the pipe displaces vertically by one-third of its diameter and softens to zero after displacing to the top of the backfill surface (36 inches).

The soil weight component of upheaval resistance to the pipe is smaller than the soil strength components for both the upper- and lower-bound soil strengths. The soil resistance is based on the combination of the material strength at a depth of 3 ft and the soil sensitivity as described above.

9.3.2 Buried Lateral Resistance

A FE analysis is conducted to determine the lateral soil resistance to buried pipe movement. Results are shown in Figure 9.10. The backfilled material is simulated with a shear strength one-

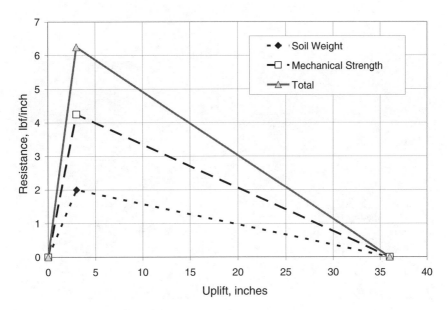

Figure 9.9 Empirical upheaval pipe resistance.

half the intact material shear strength of 33.5 psf. When the pipe intersects the intact material after ~6 inches of displacement, the reaction is 9 lbf/in, and the reaction at the end of the analysis when the pipe is fully penetrated into the intact material is 11 lbf/in. The corresponding bearing capacity factors for the backfill and intact material are 8.7 and 5.3, respectively. The larger than expected factor in the backfill material is likely associated with some interaction with the intact material due to the initial small distance between the pipe and simulated trench wall, which likely results in a constrained failure mechanism. The smaller than expected bearing factor in the intact material is due to the relatively close distance to the free top horizontal surface (26 inches), which experiences some upward displacement, indicating a lack of the restraint necessary to develop the maximum resistance available for deep conditions.

In the lateral pipe movement analysis, the backfill shear strength material was allowed to strain soften 33% while the intact material shear strength adjacent to the trench was allowed to strain soften from 50% (sensitivity of 2).

The empirical responses are described in Figure 9.11. In this case, a bearing capacity factor $N_c = 5.3$ for both the backfill and intact material is adopted.

The modeled resistance in the lateral direction is the sum of the backfill and intact material components. The backfill material has a softening response after the pipe moves past one-third of its diameter. The softening is fully developed when the pipe contacts the trench wall (used $N_c = 3.8$ at this point). The intact material reaches its peak resistance after the pipe penetrates it by one pipe diameter. In order to retain the proper intact material soil resistance, its peak value is reduced by an amount equal to the residual resistance of the backfill material. Since the empirical resistance model adds these two components together, the proper total value is maintained. This technique is adopted to include flowline resistance should it tend to unload (pipe movement toward initial unstrained location) during the analysis—for example, due to flowline cooling.

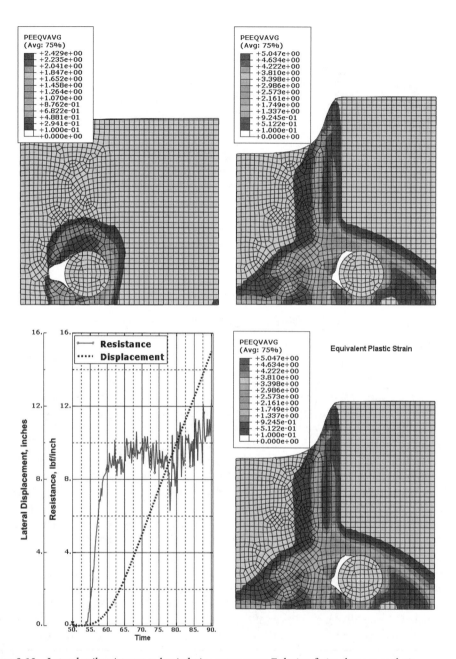

Figure 9.10   Lateral soil resistance to buried pipe movement; Eulerian finite element analysis.

9.4 Conclusion

This chapter has presented alternative methods to calibrate empirical soil-pipe interaction models used to analyze buried or unburied flowlines. The models are simple in nature, allowing them to be calibrated against limit solutions, test data, or responses observed in detailed numerical simulations of soil-pipe interaction. In the case of flowlines positioned on the seafloor surface,

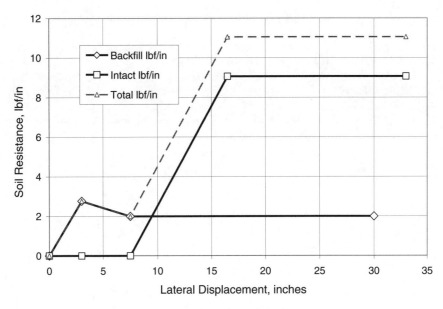

Figure 9.11 Empirical lateral soil resistance to buried pipe lateral motion.

a general consensus has developed over the last decade on the key parameters controlling the evolution of soil resistance to lateral pipe movement. Different organizations have developed empirical models to represent the responses, but no particular model has emerged as a favorite. The empirical model outlined here has been used successfully and has been proven robust. Omitted from this discussion is the important numerical implementation of the model for FE applications. While not challenging for this model, it is, of course, critical to robust numerical simulations.

9.5 References

Abaqus, 2008. *Abaqus analysis user's manual*, V6.8, Providence, Rhode Island.

Cheuk, C. Y., White, D. J., and Bolton, M. D., 2007. Large-scale modeling of soil-pipe interaction during large amplitude cyclic movements of partially embedded pipelines, *Canadian Geotechnical Journal* 44:977-996.

Dingle, H. R. C., White, D. J., and Gaudin, C., 2008. Mechanisms of pipe embedment and lateral breakout on soft clay, *Canadian Geotechnical Journal* 45:636-652.

Jeanjean, P., Andersen, K. H., and Kalsnes, B., 1998. Soil parameters for design of suction caissons for Gulf of Mexico deepwater clays, Paper presented at Offshore Technology Conference, OTC 8830.

Lambrakos, K. F., 1985. Marine Pipeline soil friction coefficients from in-situ testing, *Ocean Engineering* 12(2): 131-150.

Merifield, R., White, D. J., and Randolph, M. F., 2008. The ultimate undrained resistance of partially embedded pipelines, *Geotechnique* 58(6): 461-470.

———, 2009. Effect of surface heave on response of partially-embedded pipelines on clay, *Journal of Geotechnical and Geoenvironmental Engineering* 135 (6): 819-829.

Murff, J. D., Wagner, D. A., and Randolph, M. F., 1989, Pipe penetration in cohesive soil, *Geotechnique* 39(2): 213-229.

Najjar, S. S., Gilbert, R. B., Liedtke, E. A., McCarron, B., and Young, A. G., 2007. Residual shear strength for interfaces between pipelines and clays at low effective normal stresses, ASCE *Journal of Geotechnical and Geoenvironmental Engineering* 133(6): 695-706.

Randolph, M. F. and White, D. J., 2008. Upper-bound yield envelopes for pipelines at shallow embedment in clay, *Geotechnique* 58(4): 297-301.

Schaminee, P., Zorn, N. F., and Schotman, G., 1990. Soil response for pipeline upheaval buckling analyses: Full-scale laboratory tests and modeling, OTC 6486:563-572.

Wang, D., White, D. J., Randolph, M. F., 2009a. Numerical simulations of dynamic embedment during pipelines, In *Proceedings of the 28th International Conference on Ocean, Offshore, and Arctic Engineering*, OMAE2009-79199.

———, 2010. Large Deformation finite element analysis of pipe penetration and large-amplitude lateral displacement, Canadian Geotechnical Journal, 47(8): 842-856.

10

Constitutive Modeling for Geomaterials

William McCarron, PhD, PE
ASGM Engineering

10.1 Introduction

Modeling of soil, rock, and concrete—collectively referred to here as geomaterials—has an extensive history. This chapter presents some of the significant historical accomplishments related to the modeling of geomaterials, but at a relatively high level. Additional details are available in an abundant number of publications, several of which are referenced so the reader can refer to them as the need arises. The purpose of this presentation is to provide engineers with sufficient knowledge to evaluate the appropriate application of available geomaterial models. Chapter 11 presents several illustrative example finite element investigations with geomaterial models.

All geomaterials have pressure dependent strength characteristics. At low mean pressures, their response is generally dilative, and at high mean pressures, they experience plastic compaction. Rocks at high compressive geologic stresses may experience crushing of their grains and destruction of the cementing matrix. At high mean pressures, the response of geomaterials is generally strain hardening with homogeneous deformations. At low mean stresses, geomaterials experience strain-softening responses accompanied by post-peak localization of damage, and preferential shear planes are visible in laboratory tests.

The advanced material models in use today generally include differing strengths in triaxial compression (axial stress greater than confining stress in a triaxial test) and triaxial extension (axial stress less than confining stress). This is sometimes incorrectly referred to as strength anisotropy, but rather it is easily characterized by isotropic material models.

Plasticity-based models are the most popular and available geomaterial models. While there are many functional forms describing their strength characteristics, most realistic models fall within a *critical state* or capped-plasticity family. We confine the discussion to plasticity-based models.

There are numerous books discussing the implementation of plasticity models for geomaterials. The reader is referred to the books by Desia and his colleagues (Desai and Siriwardane 1984; Desai 2001) and Chen and his colleagues (Chen and Saleeb 1994; Chen 1994, 2007; Chen and Han 2007).

10.2 Critical State and Capped Plasticity Models

Plasticity models for geomaterials have their origins in the mathematical foundations developed for metal plasticity. The principal differences between metal and geomaterial plasticity modeling are the inclusion, for geomaterials, of a plastic compaction component at high mean stresses, plastic dilation at low mean stresses, a pressure dependent strength, and a noncircular trace of the yield surface in the deviatoric plane. The application of the so-called capped work-hardening plasticity models to soil mechanics was first suggested by Drucker et al. (1957) and then extended by the Cambridge University group including Schofield and Wroth (1968) with the development of the Cam-Clay family of models. These models, and many since, are commonly described as critical state models.

10.2.1 Basics of Plasticity Modeling

The three components of a classical plasticity model are: an elastic constitutive response, a yield surface, and a flow rule describing the evolution of plastic strains during plastic action. These are illustrated in Figure 10.1.

Within the yield surface, the response for classic plasticity models is elastic. When the stress path intersects the yield surface $F = 0$, plastic strains develop. Associated flow rule plasticity models define the direction of the plastic strain increments to be normal to the yield surface F. Geomechanics material models often use a nonassociated flow rule with the incremental plastic strains normal to a potential surface G. Nonassociated flow rules are adopted to reduce the excessive plastic dilation generally observed for associated flow rule models but result in nonsymmetric material stiffness matrices and increased numerical demand.

The yield surface F is defined in terms of the stress state and hardening parameters:

$$F = F(\alpha, \kappa, e, \varepsilon^p). \tag{10.1}$$

During plastic loading, the yield surface may evolve through a variety of hardening rules or remain fixed in the stress space, as is the case for perfectly plastic models. Isotropic hardening allows uniform expansion of the surface in all directions, kinematic hardening describes a translation of the yield surface, and mixed hardening combines the two effects. Figure 10.2 illustrates the concept for isotropic and mixed hardening surfaces. In practice, a number of hardening parameters

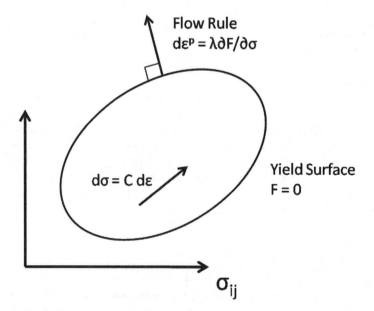

Figure 10.1 Plasticity modeling.

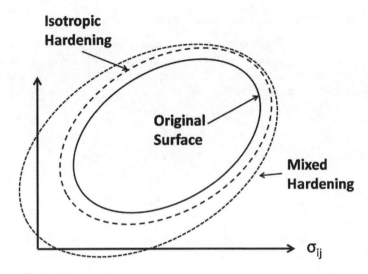

Figure 10.2 Yield surface hardening.

are possible, with the void ratio *e* and plastic strain most often used. Other potential hardening parameters κ are generally functions of the plastic strain.

The vast majority of geomaterial models are developed with the aid of isotropic elastic responses and isotropic yield surfaces. Stress-induced anisotropic behavior develops during plastic loading.

Plasticity models are developed in an incremental manner in the form:

$$d\sigma = C^e(d\varepsilon - d\varepsilon^p); \ d\varepsilon = d\varepsilon^e + d\varepsilon^p \tag{10.2}$$

The total strain *dε* is the sum of the elastic component *dε^e* and plastic component *dε^p*, and *C^e* is the elastic constitutive material stiffness. The incremental plastic strain component is defined as:

$$d\varepsilon^p = \lambda \frac{\partial G}{\partial \sigma}; \ \ G = F \text{ for associated plasticity.} \tag{10.3}$$

The scalar proportionality factor λ is determined via a consistency condition (Chen and Han 2007), resulting in the following for $F(\sigma, \varepsilon^p, \kappa)$:

$$\lambda = \frac{\dfrac{\partial F}{\partial \sigma} C^e d\varepsilon}{H + \dfrac{\partial F}{\partial \sigma} C^e \dfrac{\partial G}{\partial \sigma}}; \ \ \ H = -\frac{\partial F}{\partial \varepsilon^p} \frac{\partial G}{\partial \sigma} - \frac{\partial F}{\partial \kappa} \frac{\partial \kappa}{\partial \varepsilon^p} \frac{\partial G}{\partial \sigma} \tag{10.4}$$

The incremental elastic-plastic constitutive stiffness *C* can be expressed as:

$$d\sigma = Cd\varepsilon; \ \ C = C^e + C^p, \tag{10.5}$$

$$C^p = \frac{C^e \dfrac{\partial F}{\partial \sigma} \dfrac{\partial G}{\partial \sigma} C^e}{H + \dfrac{\partial F}{\partial \sigma} C^e \dfrac{\partial G}{\partial \sigma}}. \tag{10.6}$$

10.2.2 Geometric Representation of Stress

Understanding of plasticity models is facilitated by understanding the three-dimensional representation of stress. Plasticity models are most often described in terms of the stress invariants, resulting in an isotropic representation in the stress space. The six-dimensional stress state can be separated into hydrostatic and deviatoric components:

$$p = \frac{\sigma_1 + \sigma_2 + \sigma_3}{3}; \quad s_{ij} = \sigma_{ij} - p\delta_{ij}; \tag{10.7}$$

where σ_i are the three principal stresses, and δ_{ij} is the Kronecker delta.

$$\delta_{ij} = \begin{bmatrix} 1 & 0 & 0 \\ 0 & 1 & 0 \\ 0 & 0 & 1 \end{bmatrix} \tag{10.8}$$

The assumption that the yield surface is isotropic in the stress space allows it to be defined in terms of the stress invariants. Commonly used stress invariants are described as:

$$I_1 = \delta_{ij}\sigma_{ij} = \sigma_{11} + \sigma_{22} + \sigma_{33} = \sigma_1 + \sigma_2 + \sigma_3 = 3p = \sqrt{3}\xi \tag{10.9}$$

$$J_2 = \frac{1}{2}s_{ij}s_{ij}; \quad \rho = \sqrt{2J_2}; \quad \sigma_{eq} = \sqrt{3J_2} \tag{10.10}$$

$$J_3 = \frac{1}{3}s_{ij}s_{jk}s_{ki} \tag{10.11}$$

$$\cos(3\theta) = \frac{3\sqrt{3}}{2}\frac{J_3}{J_2^{3/2}} \tag{10.12}$$

The invariants σ_{eq} and θ are referred to as the Mises equivalent stress and the Lode angle. A useful procedure for calculating the principal stresses from the stress invariants is:

$$\begin{bmatrix} \sigma_1 \\ \sigma_2 \\ \sigma_3 \end{bmatrix} = \begin{bmatrix} p \\ p \\ p \end{bmatrix} + \frac{2\sqrt{J_2}}{\sqrt{3}}\begin{bmatrix} \cos(\theta) \\ \cos\left(\theta - \frac{2\pi}{3}\right) \\ \cos\left(\theta + \frac{2\pi}{3}\right) \end{bmatrix} \tag{10.13}$$

The stress state can be represented graphically in the three-dimensional principal stress space, as is shown in Figure 10.3. The deviatoric plane, sometimes referred to as the π plane and shown in Figure 10.3, is perpendicular to the hydrostatic axis $\sigma_1 = \sigma_2 = \sigma_3$.

The Lode angle varies in the range $0 \leq \theta \leq \pi/3$. Later we note that its introduction in strength surface descriptions leads to a three-fold symmetry in the deviatoric plane.

10.2.3 Cam-clay Model

Schofield and his colleagues understood that a geomaterial strength is a function of the stress state, stress history, and material density. They hypothesized a strength surface in the *e-p-q* space of the type shown in Figure 10.4, where *e* is the void ratio, *p* is the mean pressure, and *q* is the deviatoric stress. This model conveniently described the qualitative behavior of both overconsolidated and normally consolidated soils. While originally developed as a mechanism to describe the strength behavior of clays, similar models have been developed for sands and rocks and fall within the critical state category of models. The term critical state implies that the peak

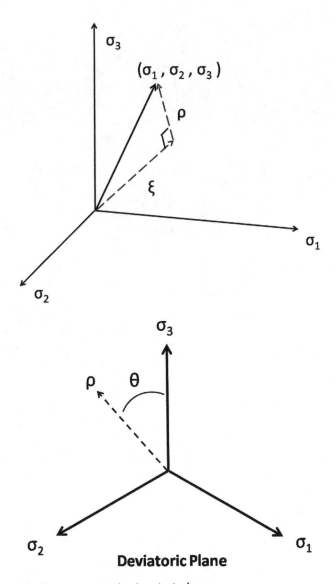

Figure 10.3   Geometric stress representation in principal stress space.

and residual strengths (for strain-softening responses) are a function of both the stress state and density (or voids ratio).

The Cam-clay model represented a framework to describe the combined effects of material density, stress path, and stress history on the effective stress response of soils. As an example, the drained and undrained stress paths for overconsolidated and normally consolidated soils are represented in Figure 10.5, which shows the state responses in terms of the voids ratio and mean pressure (*e-p* plane). An overconsolidated soil (1) and normally consolidated soil (2) begin at the same voids ratio and different mean pressures. During undrained triaxial compression, the voids ratio for both conditions remains the same, and the stress path ends at the same strength $q = \sigma_1 - \sigma_2$ (critical state A, Figure 10.5b). The difference between the mean pressures

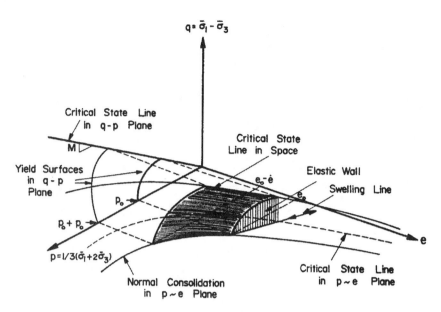

Figure 10.4 State boundary surface (Ko and Sture 1981, reprinted, with permission, from *ASTM STP 740-Laboratory Shear Strength of Soil*, copyright ASTM International).

on the total and effective stress paths is the excess pore pressure developed during the test. The normally consolidated soil test ends with a positive excess pore pressure u_2, and the overconsolidated soil ends with a negative excess pore pressure u_1. During drained loading, the normally consolidated specimen responds with a decreasing voids ratio ending at critical state B in Figure 10.5a-c, and the overconsolidated specimen first experiences a reduction in its voids ratio and then dilates (increasing voids ratio) before reaching the critical state D. All the stress paths end on the critical state line with a slope of M in the *p-q* plane.

10.2.4 Capped Plasticity Model

Drucker et al. (1957) presented the idea of a capped plasticity model as a method of limiting material dilation during plastic action. Yield or strength surfaces of the Mohr-Coulomb or Drucker-Prager type have a constant slope in the hydrostatic meridian, and as a result, plastic action with an associated flow rule can lead to unrealistically large dilation. The so-called cap models introduced a second surface, closing the yield surface. The cap surface intersection with the Drucker-Prager (or Mohr-Coulomb) surface occurred at a point where the cap surface tangent was horizontal, therefore resulting in zero dilation during plastic action (Figure 10.6).

The representation of the Cap model in Figure 10.6 assumes that compressive stresses and strains are positive. The Drucker-Prager surface with associated plasticity results in plastic dilation (expansive plastic volumetric strains), and the cap surface with associated flow results in plastic compaction. The plastic flow at the intersection of the two surfaces is usually defined perpendicular to the hydrostatic axis so that the plastic strains are completely deviatoric, thus providing some control on the dilation if the cap is allowed to contract as well as expand.

10.2.5 Single-surface Critical State Models

A main limitation of many of the plasticity models developed in the 1970s was that they included two surfaces, one to characterize the strength of a relatively dense state with a dilative response,

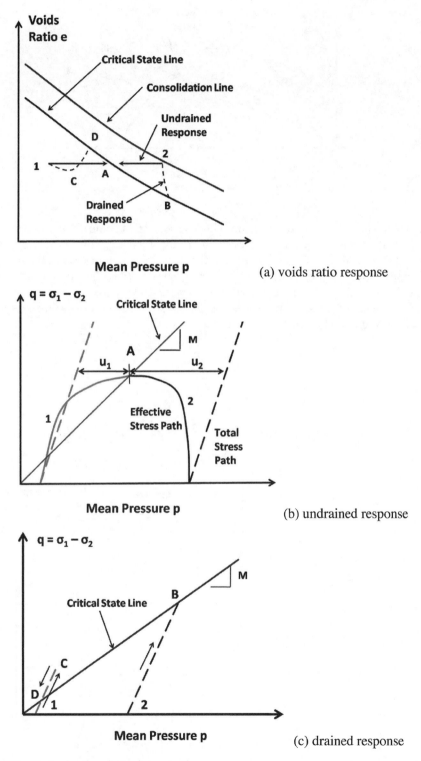

Figure 10.5 Drained and undrained stress paths.

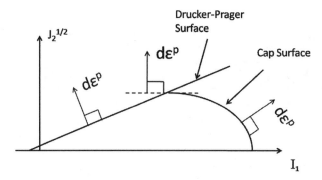

Figure 10.6   Drucker-Prager cap model.

and a second to characterize a normally consolidated state with a compactive response. The numerical implementation was complicated by the necessity of tracking stress paths relative to two surfaces and then managing the behavior when the stress loading point was at the intersection of both surfaces. Desai and his colleagues (Desai and Faruque 1984) introduced a closed single surface model, including a noncircular trace in the deviatoric plane, to capture the strength characteristics in triaxial compression, triaxial extension, and intermediate stress states. This model is shown in Figure 10.7 and has the functional form of:

$$F = \left(\frac{J_2}{p_a^2}\right) - F_b F_s = 0;$$

$$F_b = -\alpha\left(\frac{I_1}{p_a}\right)^n + \gamma\left(\frac{I_1}{p_a}\right)^2; \qquad (10.14)$$

$$F_s = (1 - \beta S_r)^n; \quad S_r = \frac{\sqrt{27}}{2}\frac{J_3}{J_2^{3/2}} = \cos(3\theta)$$

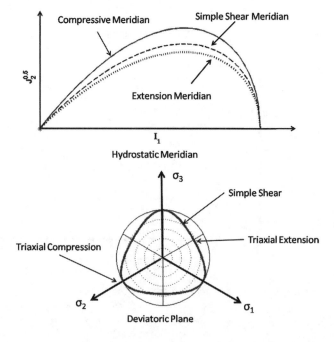

Figure 10.7   HISS yield surface
characteristics.

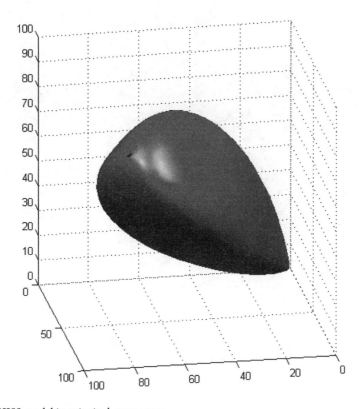

Figure 10.8 HISS model in principal stress space.

where p_a is the atmospheric pressure in the same units as the stresses. The stress invariant $S_r = \cos(3\theta)$ varies from 1 (triaxial compression) to −1 (triaxial extension).

The shapes of the yield surface in I_1-$J_2^{1/2}$ meridian and in the deviatoric plane (or π-plane) are shown in Figure 10.7. The surface is shown in the principal stress space in Figure 10.8. The cross-section in the deviatoric plane represents the strength in various types of loading—for example, triaxial compression or triaxial extension. The function F_s controls the shape of the surface in the deviatoric plane. The ratio of the triaxial compression strength to the simple shear or triaxial extension strengths increases with β (Figure 10.7). The introduction of the Lode angle through the parameter S_r results in a three-fold symmetry in the deviatoric plane. The value of m affects the angularity of the trace in the deviatoric plane. These parameters allow flexibility in modeling a wide range of materials (clay, sand, rock). Variants of this model were introduced with different hardening rules (isotropic, kinematic, mixed), associated and nonassociated flow rules, and were collectively termed hierarchical single surface (HISS) models to convey the possibility of including various attributes, as required, to capture the required material response characteristics.

10.3 Traditional Strength Models

Strength models such as the Mohr-Coulomb and Drucker-Prager surfaces have a long history of successful application to limit analysis (Chen 1975) and finite element investigations. Their primary application is for determining the capacity of foundations.

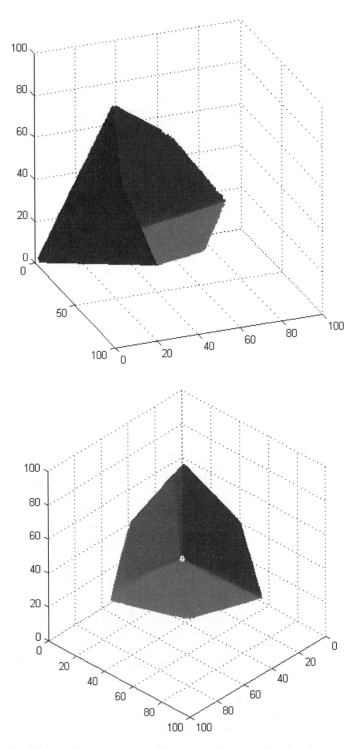

Figure 10.9   Mohr-Coulomb model in principal stress space.

The Mohr-Coulomb criterion may be expressed in terms of the principal stresses and stress invariants (Chen 1994) as:

$$F = \sigma_1 \frac{1 - \sin\varphi}{2c \cdot \cos\varphi} - \sigma_3 \frac{1 + \sin\varphi}{2c \cdot \cos\varphi} - 1 = 0 \tag{10.15}$$

$$F = I_1 \sin\theta - \frac{\sqrt{J_3}}{2}\left[3(1 + \sin\varphi)\sin\theta + \sqrt{3}(3 - \sin\varphi)\cos\theta\right] + 3c \cdot \cos\varphi = 0 \tag{10.16}$$

or

$$F = \sqrt{2\xi}\sin\varphi - \sqrt{3\rho} \cdot \sin\left(\theta + \frac{\pi}{3}\right) + \rho \cdot \cos\left(\theta + \frac{\pi}{3}\right)\sin\varphi + \sqrt{6} \cdot c \cdot \cos\varphi = 0 \tag{10.17}$$

The Mohr-Coulomb expression in terms of the principal stresses ($\sigma_1 > \sigma_3$) reveals its independence of the intermediate principal stress σ_2. For a frictionless material, the Mohr-Coulomb model reduces to the Tresca criterion.

The Drucker-Prager criterion expressed in terms of the stress invariants is:

$$F = \sqrt{J_2} - \alpha I_1 - \kappa = 0 \tag{10.18}$$

where α and κ are model strength parameters. The Mohr-Coulomb and Drucker-Prager surfaces are plotted in the principal stress space in Figures 10.9 and 10.10, respectively, and their shapes are compared in the deviatoric plane in Figure 10.11. Chen and Saleeb (1994) presented matching conditions for the two surfaces. These are reproduced here in Table 10.1. If the Drucker-Prager criterion is matched to the triaxial compression strength, then foundation strength calculations will exceed those for the Mohr-Coulomb model. The reverse is true when matching the triaxial extension strength. Plane strain matching conditions will result in equal capacity conditions for

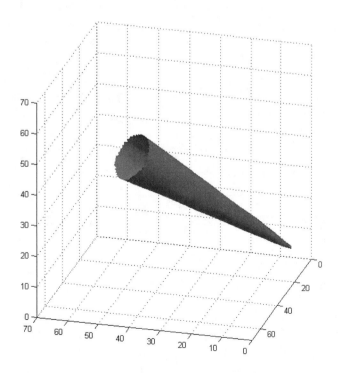

Figure 10.10 Drucker-Prager model in principal stress space.

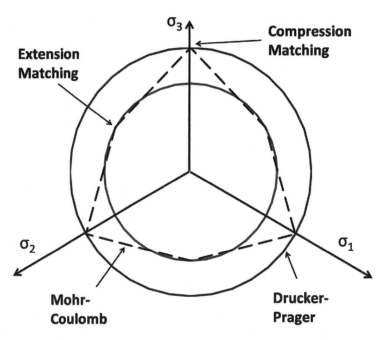

Figure 10.11   Drucker-Prager model compared with Mohr-Coulomb model in deviatoric plane.

two-dimensional plane strain simulations. Plane strain parameter matching is often used as a compromise for 3-D modeling.

From the matching conditions in Table 10.1, we see that the ratio of the triaxial compression to extension strength is:

$$\frac{\rho_{TC}}{\rho_{TE}} = \frac{3 + \sin \varphi}{3 - \sin \varphi} \qquad (10.19)$$

Table 10.1   Matching conditions for Drucker-Prager and Mohr-Coulomb criteria (Chen and Saleeb 1994).

Condition	α	κ
Triaxial compression	$\dfrac{2 \sin \varphi}{\sqrt{3}\,(3 - \sin \varphi)}$	$\dfrac{6c \cdot \cos \varphi}{\sqrt{3}\,(3 - \sin \varphi)}$
Triaxial extension	$\dfrac{2 \sin \varphi}{\sqrt{3}\,(3 + \sin \varphi)}$	$\dfrac{6c \cdot \cos \varphi}{\sqrt{3}\,(3 + \sin \varphi)}$
Plane strain	$\dfrac{\tan \varphi}{(9 + 12 \tan^2 \varphi)^{1/2}}$	$\dfrac{3c}{(9 + 12 \tan^2 \varphi)^{1/2}}$

10.4 Recent Developments

Geomaterials at low mean stresses display a strain-softening response that eventually manifests itself in the form of localized shear planes. In the context of standard stiffness based finite element computations, inclusion of such material responses can lead to mesh-dependent solutions (Bazant and Cedolin 1991; Bazant and Planas 1998). Several methods of alleviating this deficiency have been proposed. Regularization of the material strength based on element size was proposed (Bazant and Planas 1998). Nonlocal solution variables have been introduced (e.g., Bazant and Cedolin 1991), resulting in the so-called gradient plasticity methods. Desia et al. (1997) argued the introduction of a disturbance parameter in a plasticity-based model, the result being the disturbed state concept (DSC) could reduce or eliminate mesh dependencies. The DSC model behaves much as a damage mechanics formulation, except that the disturbed/damaged part continues to carry a stress, and at the material point level, the intact and disturbed portions interact. An advantage to this formulation is that the possibility of the evolution of material disturbance is clearly understood in the context of stress states relative to the yield/strength surface. Desai (2001) describes a number of implementations of this model in combination with the HISS formulation, collectively called HISS/DSC models (Desai et al. 1993; Desai and Toth 1996; Katti and Desai 1995; Wathugala and Desai 1993; Shao 1998; Shao and Desai 2000).

The DSC formulation was introduced to provide a mechanism to capture the disturbance, or damage, introduced to a geomaterial as a result of evolving micro cracks or material structure destruction during plastic action. In the HISS model, the hardening parameter α (Equation 10.14) is a function of the plastic deviatoric strains e^p and plastic volumetric strains ε^p_{vol}, as, in one possibility, defined by:

$$\alpha = \frac{h_1}{\left(\xi_v + h_3 \xi_d^{h_4}\right)^{h_2}}; \quad d\xi_v = \frac{d\varepsilon^p_{vol}}{\sqrt{3}}; \quad d\xi_d = (de^p \cdot de^p)^{1/2} \tag{10.20}$$

Larger values of α result in smaller sizes (lower strengths, smaller elastic domains) of the yield surface. The HISS/DSC model includes an auxiliary residual strength surface (alternatively called disturbance or damage surface), as shown in Figure 10.12. Damage evolves when the stress state is between the yield and residual strength surfaces. This zone has been characterized as the brittle zone in the rock mechanics literature (Menendez et al. 1996, Zhu and Wong 1997). As described by Desai (2001), the observed stress state σ^A is described in the form of damage mechanics relation:

$$\sigma^A_{ij} = (1 - D)\sigma^I_{ij} + D\sigma^D_{ij} \tag{10.21}$$

where σ^I is the stress state in the intact material (computed via the HISS model, in this case, assuming no disturbance), σ^D is the stress state in the disturbed material, and D is the disturbance parameter given by:

$$D = D_u\left[1 - \exp\left(-A\xi_d^Z\right)\right] \tag{10.22}$$

where D_u, A and Z are material parameters, and ξ_d is a measure of the deviatoric plastic strain given by the integration of $d\xi_d$ (10.20). As mentioned previously, the primary difference between the DSC and classical damage mechanics formulations is that in the DSC formulation, the disturbed/damaged material continues to carry a stress.

The bounding surface plasticity concept was introduced to reproduce a smooth transition between elastic and elastic-plastic behavior (Dafalias and Herrmann 1982). The amount of plasticity is a function of the distance between the stress state and an image stress on the yield surface, measured along a line passing through the stress state and a projection center. A primary advantage

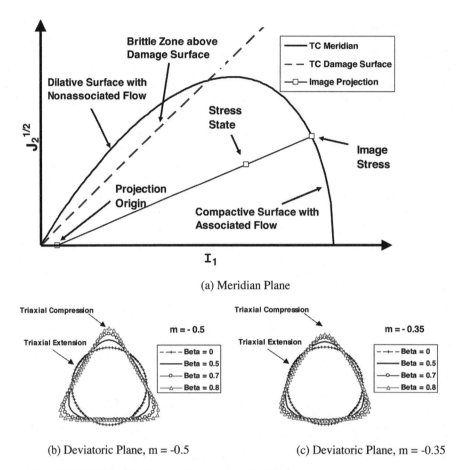

(a) Meridian Plane

(b) Deviatoric Plane, m = -0.5   (c) Deviatoric Plane, m = -0.35

Figure 10.12   HISS/DSC model with bounding surface plasticity.

of the bounding surface plasticity concept is that it allows the introduction of hysteretic plastic response during cyclic stress variations below the yield surface.

10.5 Conclusion

The key features required of a geomaterial model have been introduced. These include a pressure dependent strength and the ability to represent plastic dilation and/or compaction (as required by the specific problem to be addressed). There are many excellent papers with detailed discussions of modeling geotechnical material responses. The reader is referred to papers by Lade and Nelson (1987) and Lade (1988, 2005). While the pressure dependent behavior of sands and clays with respect to the observation of dilative or compactive response is well known, fewer engineers are aware of the same behavior in concrete and rocks. Both concrete and rock may experience plastic compaction, although the pressures required are generally greater than those encountered in buildings or foundations.

Several papers describing the pressure dependent response of Berea sandstones are available in the literature (Menendez et al. 1996, Zhu and Wong 1997). Berea sandstones are fairly uniform in properties, although a range of material strengths are available, and have been the subject of numerous laboratory experimental studies. In rock mechanics, the increasing practice of recording acoustic emissions during triaxial tests has improved understanding of the development of micro cracks in rocks (e.g., Baud et al. 2004). These observations indicate that micro crack development is predominantly a post-peak strength occurrence.

The capacity of foundations in normally consolidated clay materials, the dominate case for the Gulf of Mexico, is often determined using pressure-independent material models of the Mises or Tresca type in total stress simulations with a depth dependent strength definition. If accurate displacement determinations are necessary, then a critical state type model capable of replicating compactive or dilative response and with a closed yield surface should be employed in a coupled analysis, including pore pressure calculations. Situations including cyclic loading generally preclude the use of classical perfectly plastic or isotropic hardening material models. In this case, one should resort to models including kinematic hardening, mixed hardening, or subsurface plasticity. The choice between these options is not always straightforward, and the engineer is often hampered by availability of models including the necessary attributes. It is imperative that the engineer have a reasonable understanding of the stress path experienced in a particular situation and of the capabilities of material models to reproduce the expected significant characteristic responses of the material.

10.6 References

Abaqus, 2009. ABAQUS Analysis user's manual, V6.9, Providence, Rhode Island.

Baud, P., Klein, E., Wong, T-F., 2004. Compaction localization in porous sandstones; Spatial evolution of damage and acoustic emission activity, *Journal of Structural Geology* 26:603-624.

Bazant, Z. P. and Cedolin, L., 1991. *Stability of structures*, Oxford University Press.

Bazant, Z. P. and Planas, J., 1998. Fracture and size effect in concrete and other quasibrittle materials, Boca Raton, FL: CRC Press.

Chen, W. F., 1975. Limit analysis and soil plasticity, Elsevier, New York, NY.

———, 1994. *Plasticity and modeling*, vol. 2 of *Constitutive equations for engineering materials*, Elsevier, New York, NY.

———, 2007. *Plasticity in reinforced concrete*, Ft. Lauderdale, FL: J. Ross Publishing.

Chen, W. F. and Han, D. J., 2007. *Plasticity for structural engineers*, Ft. Lauderdale, FL: J. Ross Publishing.

Chen, W. F. and Saleeb, A. T., 1994. *Elasticity and modeling*, vol. 1 of *Constitutive equations for engineering materials*, Elsevier, New York.

Dafalias, Y. F. and Herrmann, L. R., 1982. Bounding surface formulation of soil plasticity, In *Soil Mechanics—Transient and Cyclic Loads*, eds. Pande and Zienkiewicz, Wiley and Sons, New York.

Desai, C. S., 2001. *Mechanics of materials and interfaces: The disturbed state concept*, Boca Raton, FL: CRC Press.

Desai, C. S., Basaran, C., and Zhang, W., 1997. Numerical algorithms and mesh dependence in the disturbed state concept, *International Journal for Numerical and Analytical Methods in Engineering* 40(40): 3059-3083.

Desai, C. S. and Faruque, M. O., 1984. Constitutive model for (geological) materials, *Journal of Engineering Mechanics* 110(9): 1391-1408.

Desai, C. S. and Siriwardane, H. J., 1984. *Constitutive laws for engineering materials with emphasis on geologic materials*, Prentice-Hall, Englewood Cliffs, NJ.

Desai, C. S. and Toth, J., 1996. Disturbed state constitutive modeling based on stress-strain and nondestructive behavior, *International Journal of Solids Structures* 11:1619-1650.

Desai, C. S., Wathugala, G. W., and Matlock, H., 1993. Constitutive model for cyclic behavior of clays, II: Applications, *Journal of Geotechnical Engineering* 119(4): 730-748.

Drucker, D. C., Gibson, R. E., and Henkel, D. J., 1957. Soil mechanics and work-hardening theories of plasticity, *ASCE Transactions* 122:338-436.

Katti, D. R. and Desai, C. S., 1995. Modeling and testing of cohesive soil using disturbed-state concept, *Journal of Engineering Mechanics* 121(5): 648-658.

Ko, H-Y. and Sture, S., 1981. State of the art: Data reduction and application for analytical modeling, *Laboratory Shear Strength of Soil*, ASTM STP 740, eds. R. Yong and F. Townsend, 329-386.

Lade, P. V., 1988. Effects of voids and volume changes on behavior of frictional materials, *International Journal for Numerical and Analytical Methods in Geomechanics* 12:351-370.

———, 2005. Overview of constitutive models for soils, *Soil Constitutive Models: Evaluation, Selection, and Calibration*, ASCE Geotechnical Special Publication 128, eds. J. Yamamuro and V. Kaliakin, 34.

Lade, P. V. and Nelson, R. B., 1987. Modeling of the elastic behavior of granular materials, *International Journal for Numerical and Analytical Methods in Geomechanics* 11:521-542.

Menendez, B., Zhu, W., and Fong, T-F., 1996. Micromechanics of brittle faulting and cataclastic flow in Berea sandstone, *Journal of Structural Geology* 18(1): 1-16.

Schofield, M. A. and Wroth, C. P., 1968. *Critical state soil mechanics*, McGraw-Hill Books, New York.

Shao, C., 1998. Implementation of DSC model for dynamic analysis of soil-structure interaction, PhD diss., University of Arizona.

Shao, C. and Desai, C. S., 2000. Implementation of DSC Model and application for analysis of field pile tests under cyclic loading, *International Journal for Numerical and Analytical Methods in Geomechanics* 24:601-624.

Wathugala, G. W. and Desai, C. S., 1993. Constitutive model for cyclic behavior of clays, I: Theory, *Journal of Geotechnical Engineering* 119(4): 714-729.

Zhu, W. and Wong, T-F., 1997. The Transition for brittle faulting to cataclastic flow: Permeability evolution, *Journal of Geophysical Research* 102(B2): 3027-3041.

11

Finite Element Applications

William O. McCarron, PhD, PE
ASGM Engineering

11.1 Introduction

In this chapter, the use of plasticity models to investigate and solve elementary as well as advanced engineering problems is demonstrated. First, we present fundamental bearing capacity analyses with the objective of validating the numerical techniques against problems for which classical solutions exist. We then branch into a few conventional and unconventional problems addressed in deepwater foundation geotechniques and other areas. In each case, the finite element (FE) solutions are compared to theoretical solutions, field observations, or laboratory observations to demonstrate the validity of the method.

11.2 Plane Strain Bearing Capacity Evaluation

Plane strain foundation bearing capacity investigations on soils obeying the classical Mohr-Coulomb strength criterion have both upper- and lower-bound limit analysis solutions. In some cases, the upper- and lower-bound solutions coincide, indicating an exact solution is available. Engineers new to plasticity analysis, or those experimenting with new analysis programs, are well rewarded with experience gained in investigating problems for which classical solutions are available. As an example, we present numerical results of bearing capacity investigations for c-φ (cohesion c and friction angle φ) materials, expanding the classical limit solutions to include the effects of nonassociated plasticity. Since not all engineers have access to FE programs with the Mohr-Coulomb model, a few comparisons between it and the Drucker-Prager model are presented.

11.2.1 Purely Cohesive Material

All the bearing capacity solutions presented in this section make use of the FE model shown in Figure 11.1. The analysis considers a strip footing 2-ft wide, but only half of it is included in the model, which accounts for symmetry. The foundation extends to a depth of 5 ft and 10 ft from the footing centerline. The capacities are determined by the finite investigation with the Abaqus (2009) program and compared with limit analyses presented by Chen (2007).

The Abaqus (2009) implementation of the Drucker-Prager model uses slightly different parameters than those introduced in Chapter 10 to describe the surface cohesion (d) and frictional property ($\tan \beta$), as a result of its formulation in terms of the mean pressure p and the Mises

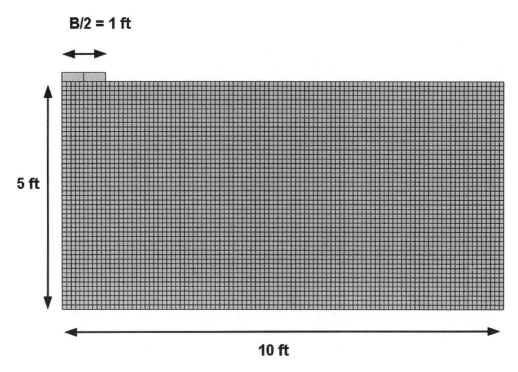

Figure 11.1   Finite element model for bearing capacity investigations.

equivalent stress σ_{eq} (Equation 10.10). For plane strain matching, these parameters are related to the Mohr-Coulomb parameters c and φ via:

$$\tan \beta = \frac{\sqrt{3} \cdot \sin \varphi}{\sqrt{1 + \frac{\sin^2 \varphi}{3}}}; \quad \frac{d}{c} = \frac{\sqrt{3} \cdot \cos \varphi}{\sqrt{1 + \frac{\sin^2 \varphi}{3}}}. \tag{11.1}$$

Figure 11.2 compares the limit solutions for a weightless purely cohesive material ($c = 1$ psf) modeled with the Mohr-Coulomb model (which simplifies to the Tresca criterion for $\varphi = 0$). Both analyses result in initially the same capacity and are within ~4% of the classical limit solution with the bearing capacity factor $N_c = 5.14$. The elastic response is simulated with a Young's modulus of ~500 psf and a Poisson's ratio of 0.3.

11.2.2 Material with Cohesion and Friction

The bearing capacity of a material with $c = 1$ psf and $\varphi = 30°$ is determined using both associated and nonassociated flow rules with the Drucker-Prager model. The nonassociated flow rule assumes zero plastic dilation. The results presented in Figure 11.3 show a bearing capacity factor $N_c = 31$ for the associated flow rule, which closely matches the upper-bound solution value of 30.2 (Chen 2007). The capacity for the nonassociated condition is a lower value of $N_c = 23.7$ due to the omission of energy dissipation associated with volumetric plastic strains. Included in Figure 11.3 are the velocity mechanisms for the analyses. The nonassociated failure velocity field displays a sharper boundary and encompasses a slightly smaller volume. Note that the failure mechanism in Figure 11.3 encompasses a larger volume than that for the purely cohesive material in Figure 11.2.

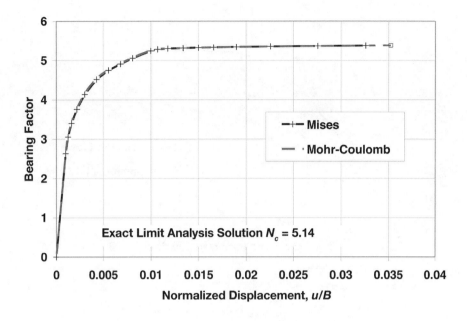

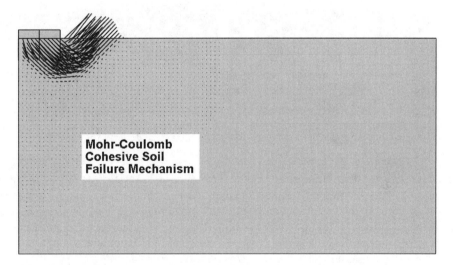

Figure 11.2 Bearing capacity determination for a weightless purely cohesive material.

11.3 Suction Pile Lateral Resistance

As the oil industry ventured into water depths greater than 4000 ft in the Gulf of Mexico, the choice of production structures shifted from vertically anchored tension leg platforms to SPARs (large diameter deep draft floating caissons), which rely on a flexible lateral mooring system. These systems employ suction piles to anchor the centenary shaped mooring lines with a predominate lateral load. Murff and his colleagues (Murff and Hamilton 1993; Aubeny et al. 2001, 2003a,b; Sharma 2004) developed a series of upper-bound limit solutions to calculate the resistance of suction piles under lateral and vertical loads. An industry project (Anderson and Murff

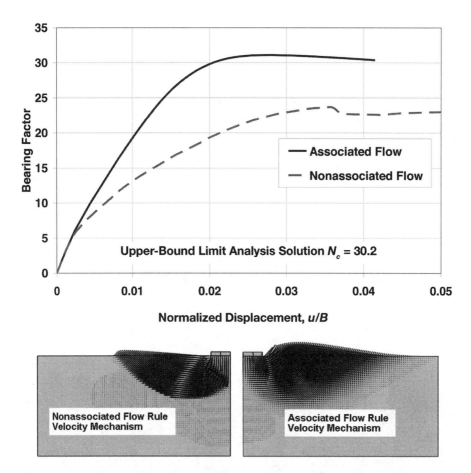

Figure 11.3   Bearing capacity determination with Drucker-Prager mode; c = 1 psf, φ = 30°.

2003; Andersen et al. 2003) summarized methods for calculating installation resistance and holding capacity for various deepwater foundation alternatives, including suction piles and drag embedment anchors. In the case of suction piles, various methodologies for calculating installation and ultimate foundation resistances were benchmarked against a common set of problems. In the following, we compare the FE and limit analysis solutions of a caisson with a length to diameter ratio of five in a foundation with a linear increasing strength profile.

The caisson has a diameter D of 5 meters and a length L of 25 meters. The foundation is a cohesive material with an undrained shear strength profile $S_u = 1.25z$ kPa, where z is the depth in meters, and the strength is determined via direct simple shear test (DSS) conditions. The FE analyses are undertaken with 2-D elements incorporating asymmetrical deformations. The mesh for the analyses is shown in Figure 11.4. The elements used are 8-node parabolic elements with reduced integration. The interface between the caisson and soil does not allow relative motion. The analyses were performed with the Abaqus von Mises pressure independent strength to approximate undrained conditions, the Mises equivalent uniaxial strength defined as $\sigma_{eq} = S_u\sqrt{3}$, where S_u is the DSS undrained strength.

Figure 11.5 compares the FE determined capacity for purely horizontal loading at different elevations with limit solutions using the method described by Aubeny et al. (2001). Also included are FE solutions for inclined loads at an angle of 30° to the horizontal. The limit analysis solution

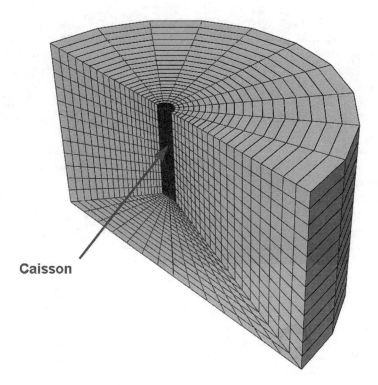

Figure 11.4 Finite element model for suction caisson analysis.

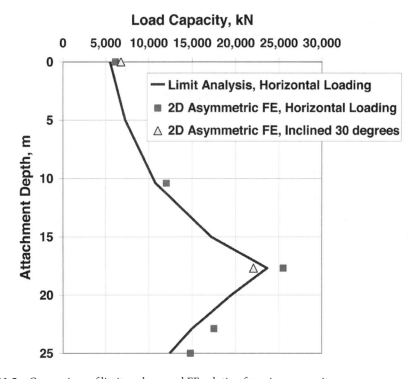

Figure 11.5 Comparison of limit analyses and FE solution for caisson capacity.

assumed the lateral resistance is computed using the N_c diagram in Figure 11.6. The two solutions correlate well, until the load application point is below the optimum depth of ~17 m. The FE solutions were computed with a small displacement solution. Large displacement solutions tended to provide larger capacities in the absence of a material softening component due to the effect of the inclining caisson, which resulted in a reduced overturning moment on the deformed geometry.

The caisson's holding capacity increases with the depth of load attachment until it reaches a maximum at ~70% of its length. The increased capacity is a result of less rotational displacement as the depth of loading increases.

Vertical load components affect the lateral capacity through two mechanisms. The most obvious is that the vertical component applied to the face of the caisson introduces a moment that will influence the rotation of the caisson. In addition, the shear stress mobilized on the face of the caisson reduces the capacity of the soil to resist lateral forces, eventually resulting in an axial pullout failure mechanism for sufficiently larger axial load components. In order to capture the caisson resistance associated with end bearing, the FE mesh should be suitably refined, perhaps to a greater degree than that required to capture the lateral resistance generated in this region.

It is common to specify a reduced soil resistance in the axial direction at the interface of the caisson to account for soil disturbance introduced during installation. The API industry project described by Anderson et al. (2003) reduced the interface soil strength by 35% to account for this effect. That project also noted that anchors with resultant loading at angles greater than 30° to 40° from the horizontal, depending on the L/D ratio, were controlled by the vertical load resistance. With respect to caissons controlled by the lateral load, Figure 11.5 indicates that the lateral holding capacity increases with the depth of application of the load. For soil strength profiles increasing linearly with depth, the optimum load attachment point for maximum resistance is at a ~70%

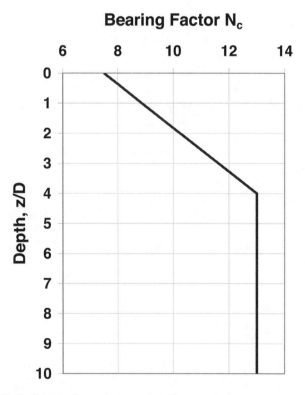

Figure 11.6 N_c profile for limit analysis solution to lateral capacity of suction caisson.

of the caisson depth. Aubeny et al. (2003a) summarizes results of FE analyses of a caisson of varying L/D ratios under combined vertical and horizontal loading.

The two-dimensional elements with asymmetrical deformations used here (Abaqus CAXA elements) have some limitations with respect to simulating relative deformations between the soil and caisson. As a consequence, the condition used here is that of a perfectly rough interface with no relative motion. More complicated conditions, including gapping or relative slip, are better treated with true three-dimensional elements and available contact modeling techniques for these elements in Abaqus.

The limit analysis solutions presented in Figure 11.5 were developed using the lateral resistance developed with the N_c profile shown in Figure 11.6. The upper-bound solution minimizes the energy associated with a rigid pile motion (displacement and rotation). The incremental lateral soil force is computed as $D{\cdot}S_u{\cdot}N_c{\cdot}\Delta L$ at discrete intervals. The N_c factors are larger than those used for conventional lateral pile design in the offshore industry (API 2000). Recent numerical (Murff and Hamilton 1993; Randolph and Houlsby 1984; Sukumaran and McCarron 1999; Aubeny et al. 2003a) and physical test data (Jeanjean 2009) indicate the current industry recommended practice likely underestimates the lateral resistance.

Aubeny et al. (2003b) investigated the effects of soil gapping on the backside of the caisson and anisotropic material strength. Gapping can reduce the capacity by up to 50%, depending on specific caisson characteristics, while the effect of anisotropic strength was minimal for the specific conditions investigated.

11.4 Torpedo Pile Penetration

The installation of anchors via dynamic penetration—dropping them from some elevation above the seafloor—has been undertaken in offshore locations. The installation penetration analysis is a combination of empirical and numerical methods. Here we compare results obtained by two FE procedures: the first involving ALE (arbitrary Lagrangian-Eulerian) remeshing and the second an Eulerian analysis. Both analyses are performed with the Abaqus/Explicit program. The results are compared to small-scale laboratory test data.

The physical tests involved dropping a 20-inch long, 1.25-inch diameter pile weighing 4.3 lbm with a 60° tip cone into a cohesive soil with an undrained shear strength of ~51 psf. The pile struck the target soil mass at a velocity of 31 ft/s. Figure 11.7 compares some aspects of the two FE models used. The one-eighth symmetry ALE model on the left includes contact between the modeled pile and soil. The model on the right automatically accounts for interaction between the torpedo, modeled with conventional Lagrangian elements, and the soil modeled in an Eulerian mesh.

The laboratory program included three physical tests, with resultant pile penetrations between 26 and 29 inches. The ALE and Eulerian analyses resulted in pile penetrations of 27 inches. Figure 11.7 includes some details of the deformed mesh, and Figure 11.8 compares acceleration records for the physical tests and numerical models. The numerical results bound the test results. The numerical analyses were ended when the pile velocity was arrested.

The FE analyses included interface friction coefficient of 0.9 on the wall of the pile as well as strain rate and softening effects for the soil. Thus, the initial kinetic energy of the pile is dissipated through soil plastic action, frictional dissipation, and elastic strain energy absorption in the soil mass. Figure 11.9 shows the modeled soil strain-rate response in terms of the Mises equivalent stress. Based on a review of the peak stresses in the ALE analyses, the maximum strain rate experienced is believed to be ~10^4/s, occurring near the cone tip over a relatively thin layer of soil. In both analyses, the interaction of fins on the exterior of the pile is included. The ALE FE model indicates that frictional effects on the face of the pile accounts for ~55% of the energy dissipation (Figure 11.10). The fact that the soil strain energy is relatively small implies it can be ignored when using empirical energy balance methodologies to determine the pile penetration.

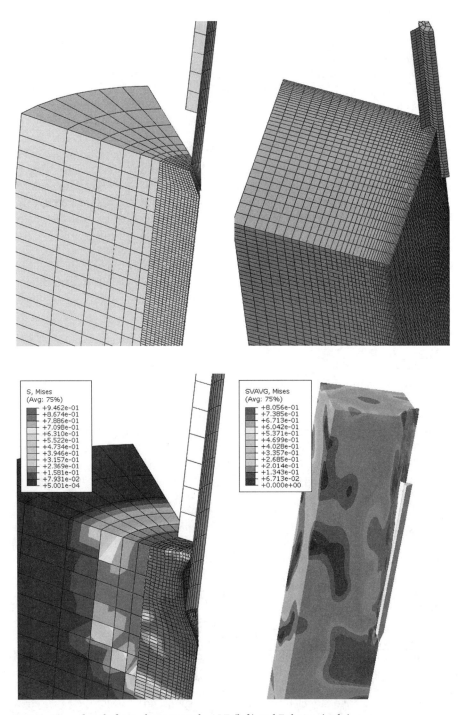

Figure 11.7   Torpedo pile finite element mesh; ALE (left) and Eulerian (right).

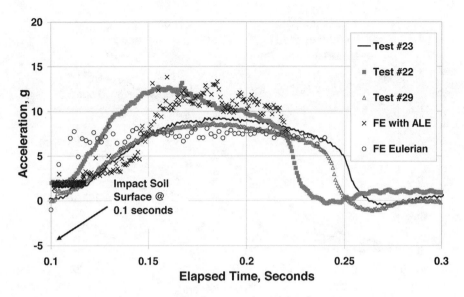

Figure 11.8 Comparison of experimental and numerical torpedo pile penetration response.

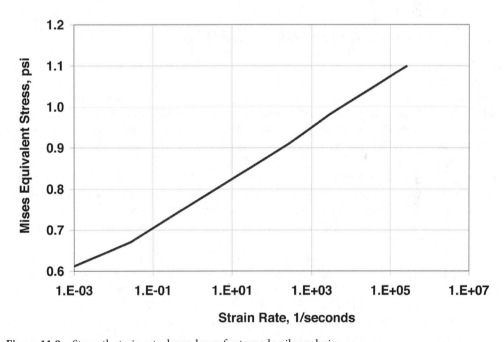

Figure 11.9 Strength strain rate dependency for torpedo pile analysis.

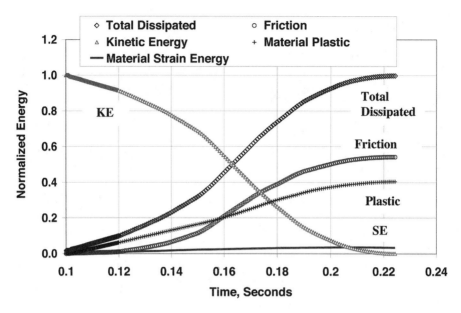

Figure 11.10   Torpedo pile energy histories; ALE analyses.

11.5 Cyclic Response of Axially Loaded Piles

Axially loaded piles subjected to tension loads have received much attention, particularly in the early period of deepwater hydrocarbon exploration, when tension leg platform technology was under development. A number of joint industry projects examining the response of tension piles were undertaken. Summary reports for several of these are available at the Mineral Management Service web site http://www.mms.gov/tarprojects/048.htm. One of those projects included axial capacity tests on small (model) diameter piles with diameters up to 3 inches at Sabine Pass, Texas. In the following, we present data from reports on those tests (Audibert and Bogard 1986) and compare the observed responses with those predicted under element states of stress and for a continuum FE model using the hierarchical single surface (HISS)/disturbed state concept (DSC) model. Additional investigations of the material response have been presented by Desai et al. (1993), Wathugala and Desai (1993), Katti and Desai (1995), Desai (2001), Shao (1998), and Shao and Desai (2000).

The Sabine Pass program included cyclic one- and two-way axial loading of several 3-inch piles extending to a depth of more than 50 ft in a normally consolidated marine clay with a plasticity index of 72 and a liquid limit of 100. The instrumented pile section recorded the response at a depth of ~52 ft, where the undrained strength is in the range of 500 to 750 psf. The instrumentation included axial strain gauges to determine the load transfer and sensors to record pore pressures and radial (horizontal) total stresses on the face of the pile.

Over a two-day period, the pile discussed here was first installed and then axially displaced with a series of one- and two-way tests. Results of the tests (Audibert and Bogard 1986) are included in Figures 11.11-11.13. Figure 11.11 illustrates the effect of pile setup on its axial capacity. Pile setup is the process of dissipation of installation induced excess pore pressures and the resultant consolidation of the soil mass. The increasing radial effective stress results in an increased load transfer at the soil-pile interface. The response of the soil-pipe interface shear stress transfer is illustrated in Figure 11.12 for two-way cyclic conditions. Here it is seen that the interface experiences some strength loss after the first cycle, and subsequent cycles result in a stabilized

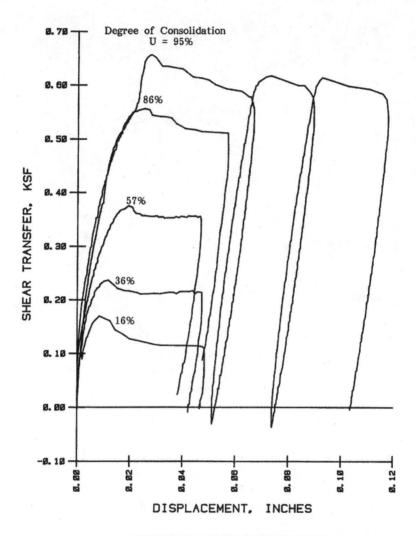

Figure 11.11   Sabine pass axial capacity tests (Audibert and Bogard 1986), reprinted with permission from AECOM.

response. Different soils and pile geometries may result in different responses. The loss of strength after the first cycle is likely associated with three main effects: 1) reduction in the mean effective stress at the soil-pile interface, 2) damage to the soil structure at the soil-pile interface, and 3) relative slip at the soil-pile interface. We should note that the Sabine Pass pile tests were evidently manually controlled, so the displacement cycles imposed at the pile top were not necessarily of the same magnitude, rate, or duration in each cycle.

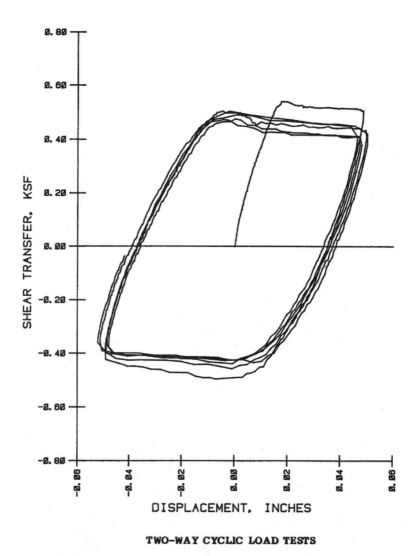

Figure 11.12   Sabine Pass interface stress-displacement transfer results (Audibert and Bogard 1986), reprinted with permission from AECOM.

The stress responses recorded at the soil-pile interface are shown in Figure 11.13. There is a gradual reduction in the radial effective stress during cyclic loading. The shear stress history also shows a strain-softening response when the strength of the interface is fully mobilized and, at the same time, a corresponding sharp drop of the pore pressure.

The HISS/DSC model developed by Desai (2001) and his colleagues is implemented here, with some modifications and enhancements, to demonstrate that it can replicate the observed pile test responses in element state of stresses. The present implementation includes subsurface plasticity

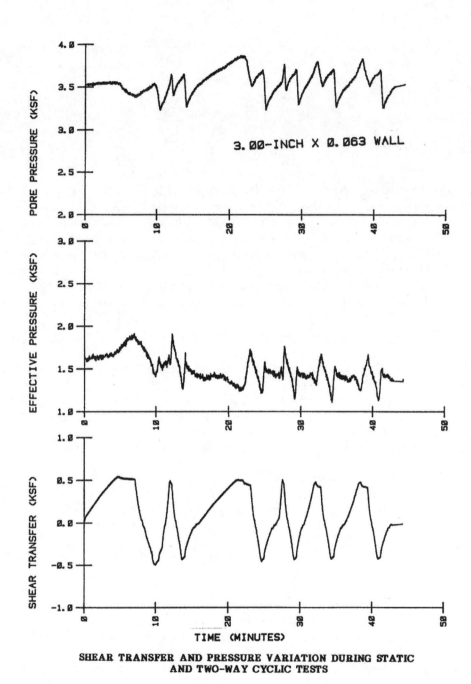

Figure 11.13 Sabine Pass interface response histories (Audibert and Bogard 1986), reprinted with permission from AECOM.

through a bounding surface formulation (Dafalias and Herrmann 1982) and the evolution of a nonassociated flow rule that reduces dilation. The hypo-elastic response is characterized by the following relationships for the bulk modulus and Poisson's ratio:

$$B_{\mathrm{mod}} = B_1 \left(\frac{l_1}{3p_a} \right)^{B_2} p_a + B_3 p_a \le B_4 p_a \tag{11.2}$$

$$\upsilon^L = \nu_1, \upsilon^{UL} = \upsilon_1 + (\upsilon_2 - \upsilon_1) \left(\frac{J_2}{J_{2\mathrm{max}}} \right)^{0.5}, \upsilon_1 \ge \upsilon_2.$$

The loading (υ^L) and unloading (υ^{UL}) definitions of Poisson's ratio allow a reasonably simple characterization of hysteresis during cyclic loading. The unloading variable Poisson's ratio allows for a stiff response with gradual stiffness reduction that is often observed in the unloading segments of cyclic tests. For loading conditions, Poisson's ratio is based on an average value. Lower Poisson's ratios result in a stiffer shear modulus. The parameter $J_{2\mathrm{max}}$ corresponds to the peak of the yield surface in the compression meridian.

The HISS/DSC model hardening parameter α (see Chapter 10, Equation 10.14) is a function of the plastic deviatoric strains e^p and plastic volumetric strains $\varepsilon^p_{\mathrm{vol}}$, as defined:

$$\alpha = \frac{h_1}{\left(\xi_v + h_3 \xi_d^{h_4} \right)^{h_2}}, \, d\xi_v = \frac{d\varepsilon^p_{\mathrm{vol}}}{\sqrt{3}}, \, d\xi_d = \left(de^p_{ij} de^p_{ij} \right)^{1/2}. \tag{11.3}$$

Larger values of α result in smaller sizes (lower strengths, smaller elastic domains) of the yield surface.

The subsurface plasticity is implemented as a function of the ratio of the distance l between the stress and image point to distance L between the image point and center of projection $XR = l/L$. To accomplish this, the plastic hardening modulus H for the surface includes a term H_{BS}, as shown in Equation 11.4 (Chapter 8 of Chen 1994; Dafalias and Herrmann 1982):

$$H = H_{BS} + H_p; \, H_{BS} = \left[1 - XR^{\mathrm{Sub2}} \right] \cdot XM1 \cdot p_a \left[1 + \left| \frac{m}{\eta} \right|^{XM2} \right] \tag{11.4}$$

where H_p is the classical plasticity hardening term in the plastic stiffness:

$$C^p = -\frac{C^e \dfrac{\partial F}{\partial \sigma} \dfrac{\partial G}{\partial \sigma} C^e}{H + \dfrac{\partial F}{\partial \sigma} C^e \dfrac{\partial G}{\partial \sigma}}; \, H_p = -\frac{\partial F}{\partial \varepsilon^p} \frac{\partial G}{\partial \sigma} - \frac{\partial F}{\partial \kappa} \frac{\partial \kappa}{\partial \varepsilon^p} \frac{\partial G}{\partial \sigma} \tag{11.5}$$

and *Sub2*, *XM1*, and *XM2* are modeling parameters. *M* is the ratio of $J_2^{1/2}/I_1$ at the peak of the yield surface, and η is the corresponding actual stress ratio. The image point on the yield surface is determined by the intersection of a line with the yield surface that is projected through a point on the hydrostatic axis and the stress state (Figure 11.14). The origin of the projection is defined as a fraction (model material parameter *Sub1*) of the point of intersection of the yield surface with the hydrostatic axis.

The model implementation includes the gradual evolution of a nonassociated flow response on the dry (overconsolidated or dilative) side of the yield surface. The flow is associated in the deviatoric plane; only the hydrostatic component is nonassociated. This modeling procedure was suggested by Desia et al. (1991). The gradual reduction in dilation is included by multiplying the hydrostatic component of the plastic strain gradient by a term of the form:

$$D1 = 1 - \frac{\xi_d}{\xi_{d\mathrm{max}}} \tag{11.6}$$

When *D*1 reaches zero, the dilation is reduced to zero.

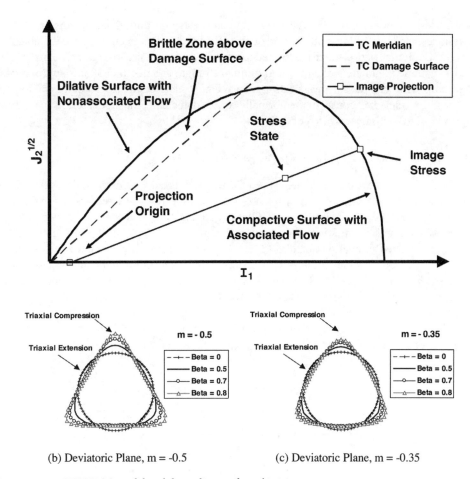

(b) Deviatoric Plane, m = -0.5   (c) Deviatoric Plane, m = -0.35

Figure 11.14   HISS/DSC model with bounding surface plasticity.

Table 11.1 Sabine Pass clay HISS/DSC model parameters

Surface parameters		Hardening parameters		Elastic parameters	
β	0.5	$h1$	0.0035	$B1$	3
γ	0.123	$h2$	0.78	$B2$	1
n	2.8	$h3$	0	$B3$	113
m	−0.5	$h4$	1.1	$B4$	200
initial α	0.065	ξ_{dmax}	0.05	v_1	0.35
				v_2	0.05
Disturbance parameters				Subsurface yield parameters	
XM	0.21	Du	0.75	$Sub1$	0.01
eo	0.9	A	1.73	$Sub2$	4
λ	0.1692	Z	0.3092	$XM1$	0.2
				$XM2$	0.02

The material model calibration relies on odometer tests and undrained-unconsolidated tri-axial tests reported by Audibert and Bogard (1986). The model parameters are listed in Table 11.1. The Mohr-Coulomb friction angle is assumed to be 29°, which is typical of Gulf of Mexico marine clays. Using these parameters, the in situ undrained strength at a depth of 50 ft is approximately 600 psf, consistent with the site investigation report. Figure 11.15 compares laboratory oedometer tests with the modeled response. In the normally consolidated case, the model results in a ratio of the horizontal to vertical stress (K_o) of 0.53.

11.5.1 Modeling of Element States of Stress

To mimic the cyclic pile tests, an element state of stress is analyzed with initial radial, vertical, and circumferential effective stresses of 1.6, 1.4, and 1.1 ksf, respectively. These stresses are based on evaluation of the test data (Figure 11.13) and investigations of the effects of soil disturbance during pile installation using the strain path method (Baligh et al. 1987; Whittle and Satubutr 2005). The strain-controlled simulation includes 13 two-way shear strain cycles, each with an amplitude of 0.025. The element volume change is zero, resulting in a response similar to a DSS test. The results are shown in Figures 11.16-11.18. The peak shear resistance is ~600 psf, reducing to 420 psf in the third cycle, after which the resistance is stable (Figure 11.16). Also shown in Figure 11.16 is the accumulation of damage with cyclic loading.

The first cycle results in generation of positive excess pore pressure (reduction in effective mean stress as a result of plastic compaction, Figures 11.17 and 11.18). The second cycle shows

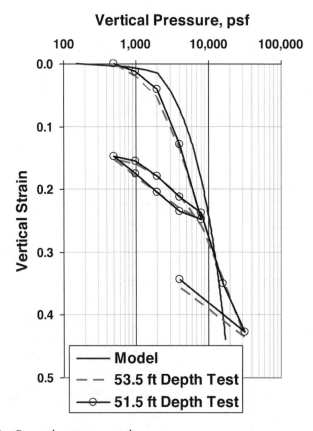

Figure 11.15   Sabine Pass oedometer test result.

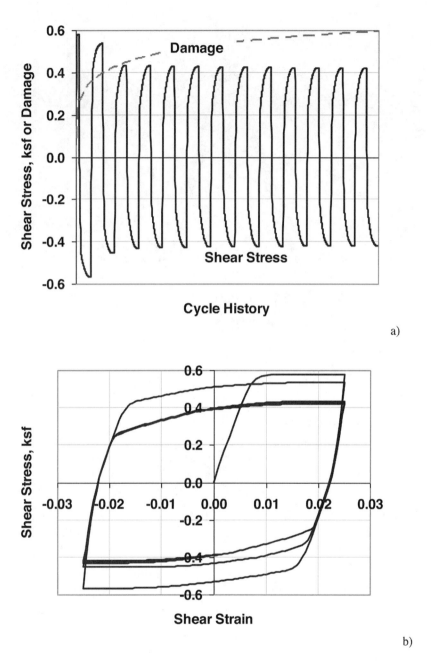

Figure 11.16 Cyclic DSS response; a) damage and shear stress history, b) shear stress-strain response.

some further subsurface plastic compaction with dilation at the peak conditions as a result of the soil becoming overconsolidated. After approximately five cycles, further subsurface plastic compaction ceases. At the peak shear stress in each cycle, the stress path approaches the triaxial compression state, as indicated by the Lode angle of 10° (triaxial compression Lode angle is 0°) in Figure 11.17. The reduction in effective radial stress (Figure 11.18) is larger than that observed in the field tests (Figure 11.13), perhaps due to the element state of stress investigated, which neglects the global equilibrium requirements.

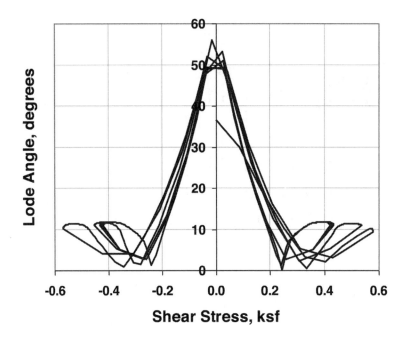

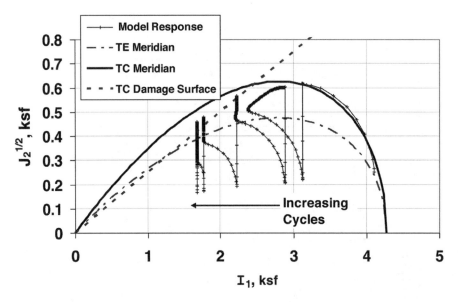

Figure 11.17 Cyclic stress path in invariant stress space.

From these results, we can conclude that the HISS/DSC model is capable of reproducing the significant characteristics observed in the Sabine Pass pile tests, namely, reduction of mean effective stress, reduction in strength associated with reduction in mean effective stress and the accumulation of damage, and gradual stabilization of material response. The inclusion of subsurface plasticity via a bounding surface formulation results in a smooth transition from elastic to elastic-plastic response and captures adequately the cyclic response for this problem.

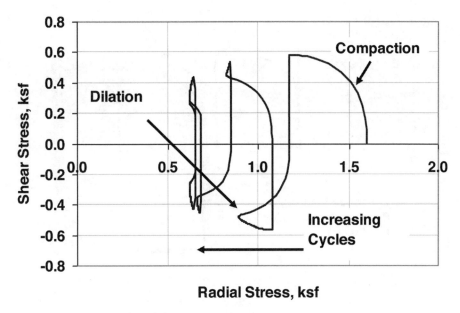

Figure 11.18 Cyclic DSS radial and shear stress path.

11.5.2 Modeling with Continuum Model

We continue the investigation of the Sabine Pass pile tests by modeling the responses under consecutive one-way and two-way cyclic loading segments. The effects of relative slip at the soil-pipe interface are more apparent, for illustration purposes, during one-way loading. The axisymmetric FE model used is shown in Figure 11.19. It includes the ability to accommodate relative soil-pipe

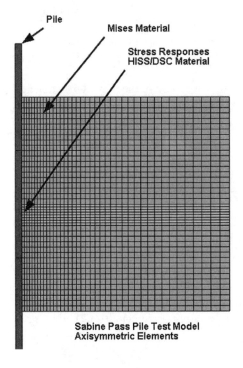

Figure 11.19 Continuum model for cyclic pile tests.

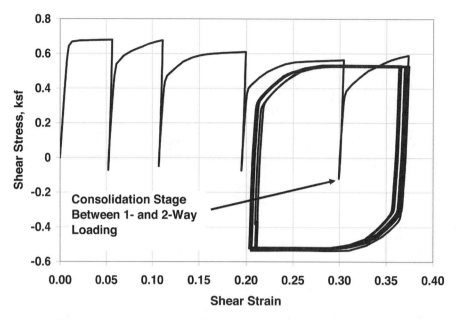

Figure 11.20 Shear stress transfer for one- and two-way loading segments.

slip through the use of a Coulomb frictional interface with a 0.5 coefficient of friction. The results in Figure 11.20 show gradual loss of the interface shear stress transfer capacity during the initial loading segment. The peak one-way loading resistance is approximately that observed during the field test (Figure 11.11). Following a consolidation period, the subsequent two-way cyclic loading shows a stable response at a capacity less than the initial one-way loading but approximately that observed in the field two-way loading test (Figure 11.12).

The computed response history data shown in Figure 11.21 illustrate the response of soil pore pressures and shear stress transfer during interface slip. At initiation of slip, a drop in soil pore pressure is observed, which allows a corresponding increase in the stress transferred across the Coulomb frictional interface. The similarity to the field data shown in Figure 11.13 is readily apparent.

The combined presentation of the element state of stress response in the previous section and the continuum model discussed here illustrates the capability to capture the significant response characteristics of a complex problem, including the effects of pile setup, plasticity, cyclic loading, and soil-pipe interface relative slip.

11.6 Rock Mechanics Application

The HISS/DSC model allows a realistic simulation of the post-peak development of micro cracks in sandstone and concrete. Widespread laboratory observations (acoustic emissions and visual) indicate that micro cracks begin to develop near the peak specimen load-carrying capacity and coalesce into visual cracks during post-peak softening. Figure 11.22 shows acoustic emissions (AE) data reported by Baud et al. (2004) for Berea sandstone indicating a spike of events at the peak resistance.

The accumulation of damage in the HISS/DSC model is consistent with laboratory observations, as illustrated in Figure 11.23, where results of a plane strain Berea sandstone compression

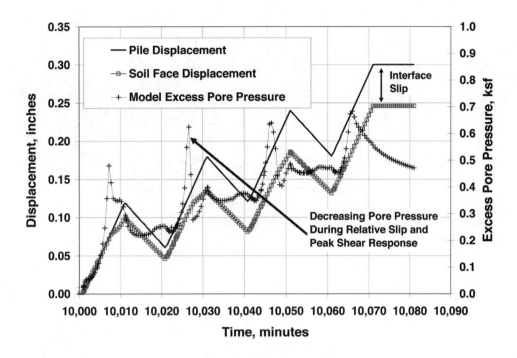

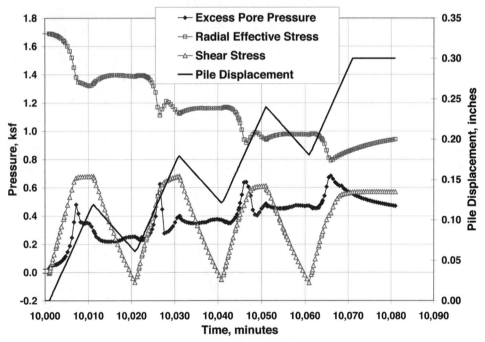

Figure 11.21   Soil-pile interface responses for one-way loading.

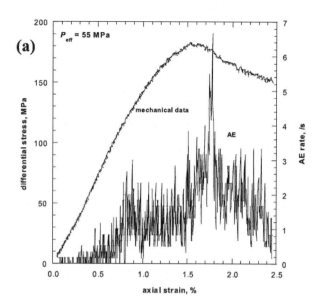

Figure 11.22 Berea sandstone AE data (Baud et al. 2004), reprinted with permission from Elsevier.

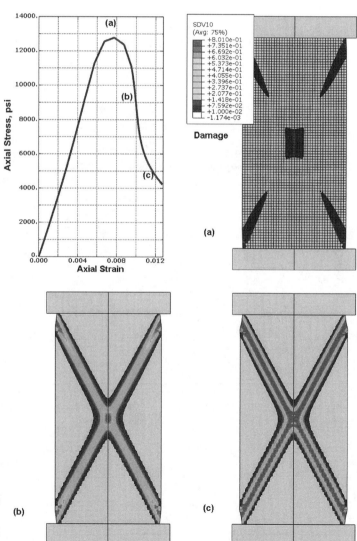

Figure 11.23 Accumulation of damage for plane strain simulation, without confining stress.

analysis without lateral confining stress are shown. There is negligible damage at the peak resistance, but it increases rapidly in the post-peak regime. Van Geel (1998) reported results of several plane strain compression tests on concrete specimens, including illustrations of the development of visible fracture surfaces (Figure 11.24).

Figure 11.25 compares the modeled responses for monotonic and load-unload stress paths in the post-peak regime. The HISS/DSC model captures the behavior observed for concrete, as reported by Spooner and Dougill (1975). Those authors presented the results shown in

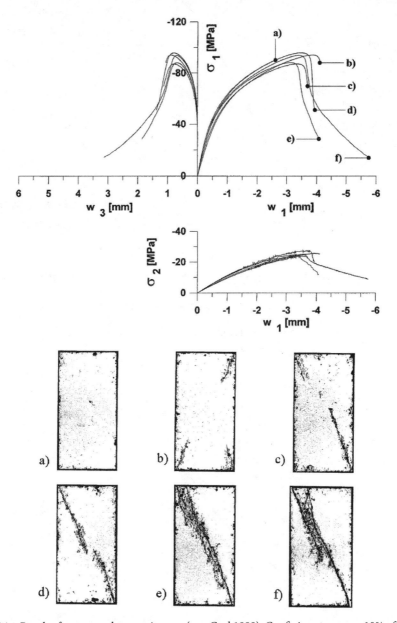

Figure 11.24 Result of concrete plane strain tests (van Geel 1998); Confining stress σ_3 = 10% of axial stress, reprinted with permission from Structural Design and Construction Technology Department, Eindhoven University of Technology.

Figure 11.26, indicating reloading in the post-peak regime generally reaches the residual response observed during monotonic loading. The same response has been observed by others for concrete (van Mier 1984).

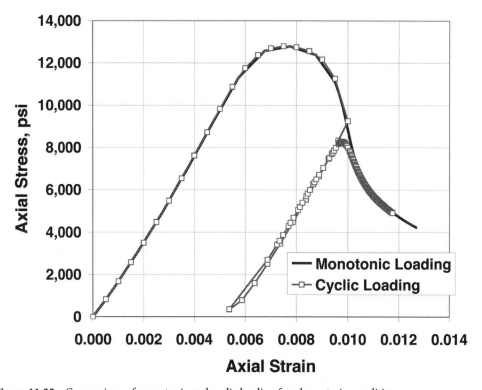

Figure 11.25   Comparison of monotonic and cyclic loading for plane strain condition.

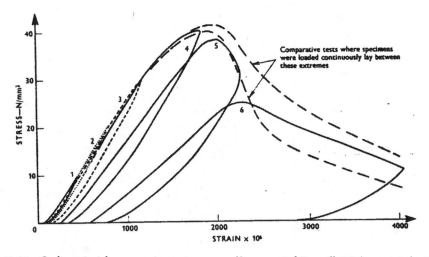

Figure 11.26   Cyclic uniaxial compression test response (Spooner and Dougill 1975), reprinted with permission from ICE Publishing.

11.6.1 HISS/DSC Model Calibration for Berea Sandstone

There is a large quantity of data from triaxial compression data in the literature for Berea sandstone. For the purpose of this study, the HISS/DSC yield surface has been calibrated against a subset of those data. The results are shown in Figure 11.27, where the presence of a compactive yield surface for rocks is illustrated.

The material calibration parameters used in Berea analyses are listed in Table 11.2.

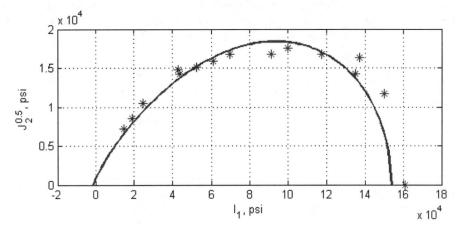

Figure 11.27   Calibrated yield surface compared to Berea triaxial compression data; (*) test data.

Table 11.2 Berea sandstone HISS/DSC model parameters

Surface parameters		Hardening parameters		Elastic parameters	
β	0.5	$h1$	0.2907	$B1$	350
γ	0.61	$h2$	0.001	$B2$	0.79
n	2.08	$h3$	0	$B3$	45000
m	−0.57	$h4$	1.1	$B4$	1e5
initial α	0.2907	ξ_{dmax}	0.4	v_1	0.15
R3	1500 psi			v_2	0.05
Disturbance parameters				Subsurface yield parameters	
XM	0.25	Du	0.8	$Sub1$	0.05
eo	0.1	A	200	$Sub2$	4
λ	0.2	Z	2	$XM1$	1.8e8
				$XM2$	0.02

11.7 Conclusion

This chapter demonstrated the use of geomaterial constitutive models with simple and complex boundary value problems under drained, undrained, and coupled consolidation conditions. The ability of classical plasticity models implemented in FE programs to replicate the results of limit analysis solutions for bearing capacity problems was illustrated. The analysis of high strain-rate projectile penetration problems with an explicit FE program was undertaken with Lagrangian and Eulerian solution techniques. The HISS/DSC model response under cyclic conditions was examined for an element state of stress and FE continuum domains, including relative slip at the soil-pile interface. Finally, the HISS/DSC model was used to investigate the response of sandstone at low mean stresses in the brittle response regime. The validity of the analysis procedures in each case was demonstrated by comparison with limit analysis solutions or physical test observations. Engineers with sufficient experience can be expected to analyze practical problems of significance. The success of such analyses is dependent on the use of a constitutive model including the significant material response features, good quality data with which to calibrate the material model, and the availability of a suitable FE program.

11.8 References

Abaqus, 2009. *Abaqus analysis user's manual*, V6.9, Providence, Rhode Island.

Andersen, K. H. and Murff, J. D., 2003. Study on deepwater anchor design practice, First year report to API.

Andersen, K. H., Murff, J. D., and Randolph, M., 2003. Deepwater anchor design practice, Phase II report to API/Deepstar.

API, 2000. *Recommended Practice for Planning, Designing and Constructing Fixed Offshore Platforms—Working Stress Design*, API RP2A, 21st ed., American Petroleum Institute.

Aubeny, C. P., Murff, J. D., and Moon, S. K., 2001. Lateral undrained resistance of suction caisson anchors, *International Journal of Offshore and Polar Engineering* 11(3): 211-219.

Aubeny, C. P., Han, S., and Murff, J. D., 2003a. Inclined load capacity of suction caissons, *International Journal for Numerical and Analytical Methods in Geomechanics* 27(14): 1235-1254.

Aubeny, C. P., Han, S., and Murff, J. D., 2003b. Suction caisson capacity in anisotropic, purely cohesove soil, *International Journal of Geomechanics* 3(2): 225-235.

Audibert, J. M. and Bogard, D. 1986. Some aspects of the fundamental behavior of axially loaded piles in clay, Final report, Earth Technology Corporation.

Baligh, M. M., Azzouz, A. S., and Chin, C. T., 1987. Disturbance due to ideal tube sampling, *Journal of Geotechnical Engineering* 113(7): 739-757.

Baud, P., Klein, E., and Wong, T. F., 2004. Compaction localization in porous sandstones: Spatial evolution of damage and acoustic emission activity, *Journal of Structural Geology* 26:603-624.

Chen, W. F., 1994. *Plasticity and modeling*, vol. 2 of *Constitutive Equations for Engineering Materials*, Elsevier, New York.

———, 2007. *Limit analysis and soil plasticity*, Ft. Lauderdale, FL: J. Ross Publishing.

Dafalias, Y. F. and Herrmann, L. R., 1982. Bounding surface formulation of soil plasticity, In *Soil Mechanics—Transient and Cyclic Loads*, eds. Pande and Zienkiewicz, Wiley and Sons, New York.

Desai, C. S., 2001. *Mechanics of materials and interfaces: The disturbed state concept*, CRC Press, Boca Raton, Florida.

Desai, C. S., Sharma, K. G., Wathugala, G. W., and Rigby, D. B., 1991. Implementation of hierarchical single surface δ_0 and δ_1 models in finite element procedure, *International Journal for Numerical and Analytical Methods in Geomechanics* 15:649-680.

Desai, C. S., Wathugala, G. W., and Matlock, H., 1993. Constitutive model for cyclic behavior of clays, II: Applications. *Journal of Geotechnical Engineering* 119(4): 730-748.

Jeanjean, P., 2009. Re-assessment of p-y curves for soft clays from centrifuge testing and finite element modeling, Paper presented at Offshore Technology Conference, OTC 20158.

Katti, D. R. and Desai, C. S., 1995. Modeling and Testing of cohesive soil using disturbed-state concept, *Journal of Engineering Mechanics* 121(5): 648-658.

Murff, J. D. and Hamilton, J. M., 1993. P-ultimate for undrained analysis of laterally loaded piles, *Journal of Geotechnical Engineering* 119(1): 91-107.

Randolph, M. F. and Houslby, G. T., 1984. The limiting pressure on a circular pile loaded laterally in cohesive soil, *Geotechnique* 34(4): 613-623.

Shao, C., 1998. Implementation of DSC model for dynamic analysis of soil-structure interaction, PhD diss., University of Arizona.

Shao, C. and Desai, C. S., 2000. Implementation of DSC model and application for analysis of field pile tests under cyclic loading, *International Journal for Numerical and Analytical Methods in Geomechanics* 24:601-624.

Sharma, R. R., 2004. Ultimate capacity of suction caisson in normally and lightly overconsolidated clays, MS thesis, Texas A&M University.

Spooner, D. C. and Dougill, J. W., 1975. A quantitative assessment of damage sustained in concrete during compressive loading, *Magazine of Concrete Research* 27(92): 151-160.

Sukumaran, B. and McCarron, B., 1999. Total and effective stress analysis of suction caissons for Gulf of Mexico conditions, *Geotechnical Special Publication* 88, Analysis Design, Construction and Testing of Deep Foundations, ed. J. Roesset, 247-260.

Van Geel, E., 1998. Behavior of concrete in multiaxial compression, PhD thesis, Eindhoven University.

Van Mier, J. G., 1984. *Strain-softening of concrete under multiaxial loading conditions*, Deft University.

Wathugala, G. W. and Desai, C. S., 1993. Constitutive model for cyclic behavior of clays, I: Theory, *Journal of Geotechnical Engineering* 119(4): 714-729.

Whittle, A. J. and Sutabutr, T., 2005. Parameters for average gulf clay and prediction of pile setup in the Gulf of Mexico, in *Soil Constitutive Models: Evaluation, Selection, and Calibration*, ASCE, *Geotechnical Special Publication* 128, eds. J. Yamamuro and V. Kaliakin.

Index

centrifuge testing and, 189, 190
installation of, 143–144, 202, 203
velocity profiles, 203, 204
Torpedo pile penetration, 305–308
Torvane shear test, 165
Tow fishes, 22
Tresca material model, 271, 293, 297, 300

U

Ultimate bearing capacity equation, 150
Ultimate pullout capacity (UPC), 199
Ultra Short Base Line (USBL) transducers, 30
Uncertainty in data interpretation, case study, 72, 73–75, 76
Unconformities, 27
Unconsolidated undrained triaxial (UU) shear test, 165
Underwater hammers, 120, 163. *See also* Hammer
    breakdown
Undrained shear strength, 46–48
    reference strengths, 47
    VST data and, 53–54
Unified Soil Classification System, 67
United States geological survey, 174
University of Colorado, 197
University of Western Australia, 196
UPC. *See* Ultimate pullout capacity
Upheaval buckling
    analysis methodology, 254
    design considerations, 250–251
    illustrative example, 262
    offshore pipelines and, 1
Upheaval soil-pipe resistance, 277–280
    buried lateral resistance, 278, 279–280, 281
    upheaval buckling resistance, 277–278, 279
USBL. *See* Ultra Short Base Line (USBL) transducers

V

Vane shear testing (VST), in situ, 10, 35, 50–53. *See also*
    VST data, use of
    down-hole VST tool, 51, 52–53
    seabed VST, 50, 51, 52
Vertically loaded anchors (VLA), 138–142
    anchor installation plan/performance, 141–142
    design factors, key, 140
    drag embedment anchors and, 138
    installation of, 139, 140–141
    VLA geometry, 140
VLA. *See* Vertically loaded anchors
Volumetric creep, 191
von Mises type material model, 251, 286, 297, 299–300, 302, 305
Vryhof Anchor's Stevmanta, 139, 140
VST. *See* Vane shear testing (VST), in situ
VST data, use of, 53–54

W

Walking, subsea flowlines. *See* Flowline walking
Wellheads, 4
West Nile Delta, 14
West Nile Project, 80, 81
Williamson and Associates, 22
Wison system, 33–34, 39
Wison XP system, 54

Y

Young's modulus, 300

Z

Zinc Development, 12